NEUROBIONICS

NEUROBIONICS

The Biomedical Engineering of Neural Prostheses

Edited by

ROBERT K. SHEPHERD
Bionics Institute & The University of Melbourne, Australia

WILEY Blackwell

Published by John Wiley & Sons, Inc., Hoboken, New Jersey
Published simultaneously in Canada

Library of Congress Cataloging-in-Publication Data:

Names: Shepherd, Robert K., editor.
Title: Neurobionics : the biomedical engineering of neural prostheses /
edited by Robert K. Shepherd.
Other titles: Neurobionics (Shepherd)
Description: Hoboken, New Jersey : John Wiley & Sons, Inc., [2016] | Includes
bibliographical references and index.
Identifiers: LCCN 2016002807 (print) | LCCN 2016004110 (ebook) | ISBN
9781118814871 (cloth) | ISBN 9781118816141 (pdf) | ISBN 9781118816035
(epub)
Subjects: | MESH: Nervous System Physiological Phenomena | Bionics | Neural
Prostheses | Brain-Computer Interfaces
Classification: LCC R857.M3 (print) | LCC R857.M3 (ebook) | NLM WL 102 | DDC
610.28–dc23
LC record available at http://lccn.loc.gov/2016002807

Cover image: Getty/Hemera Technologies

Inset image: Inset Photo used with permission of the Bionics Institute, East Melbourne, Australia

Typeset in 10/12pt TimesLTStd by SPi Global, Chennai, India
Printed and bound in Singapore by Markono Print Media Pte Ltd

10 9 8 7 6 5 4 3 2 1

This book is dedicated to my wife, Ursula, for her wonderful support, encouragement and counsel over the last 40 years; to our children Damon and Anna; their partners Jo and Junior; and our grandchildren Harley, Michaela, Jordan and Heidi who enrich our lives daily.

CONTENTS

CONTRIBUTORS

David Borton Department of Engineering and Physics, Brown University, Providence, RI, USA

David M. Brandman Department of Neuroscience, Brown University, Providence, RI, USA

Giles S. Brindley (Retired) Implanted Devices Group, Department of Medical Physics & Bioengineering, University College London, London, UK

Paul M. Carter Cochlear Ltd, Macquarie Park, NSW, Australia

Stuart F. Cogan Department of Bioengineering, University of Texas at Dallas, Richardson, TX, USA

Nick Donaldson Implanted Devices Group, Department of Medical Physics and Bioengineering, University College London, London, UK

James B. Fallon Bionics Institute & Medical Bionics Department, University of Melbourne, East Melbourne, Victoria, Australia

David J. Garrett Department of Physics, The University of Melbourne, Parkville and The Bionics Institute, East Melbourne, Victoria, Australia

Rylie A. Green Graduate School of Biomedical Engineering, UNSW, Sydney, NSW, Australia

Leigh R. Hochberg Department of Neuroscience, Brown University, Providence, RI, USA

Frank J. Lane Illinois Institute of Technology Rehabilitation Psychology, Chicago, IL, USA

Kristian P. Nitsch Department of Clinical and Rehabilitation Psychology Lewis College of Human Sciences Illinois Institute of Technology Rehabilitation Psychology Chicago IL, USA

Arto Nurmikko Department of Engineering and Physics, Brown University, Providence, RI, USA

Douglas McCreery Neural Engineering Program, Huntington Medical Research Institutes, Pasadena, CA, USA

Hugh McDermott Bionics Institute & Medical Bionics Department, University of Melbourne, East Melbourne, Victoria, Australia

John L. Parker Saluda Medical Pty Ltd, Artarmon, NSW, Australia

Marcia Scherer University of Rochester Medical Center, Rochester, NY, USA

Jens Schouenborg Neuronano Research Center, Experimental Medical Science and Nanometerconsortium, Lund University, Lund, Sweden

Peter M. Seligman Bionics Institute & Medical Bionics Department, University of Melbourne, East Melbourne, Victoria, Australia

Robert K. Shepherd Bionics Institute & Medical Bionics Department, University of Melbourne, East Melbourne, Victoria, Australia

Mohit N. Shivdasani Bionics Institute & Medical Bionics Department, University of Melbourne, East Melbourne, Victoria, Australia

Ming Yin Blackrock Microsystems, Salt Lake City, UT, USA

PREFACE

Neural prostheses are active implantable devices designed to: (i) provide therapeutic intervention, sensory feedback or motor function via electrical stimulation of nerves or muscles following trauma or disease; and/or (ii) record the electrical activity from nerve or muscle to detect disease states, enable the voluntary control of external devices such as prosthetic limbs, or to provide closed-loop feedback to modulate neural prostheses.

Since the introduction of the first commercial heart pacemakers in the late 1950s, there have been many devices approved for clinical use, resulting in a dramatic impact on the quality of life of millions of people around the world. Implantable heart pacemakers and defibrillators are a multi-billion dollar per annum industry. While the neural prosthesis industry is much younger, with an early wave of commercial devices appearing in the late 1970s, this is now a flourishing industry with impressive annual growth rates (Cavuoto *et al.* 2016). Four devices dominate this field: spinal cord stimulation for treatment of chronic pain; cochlear implants for stimulation of the auditory nerve in deafness; vagal nerve stimulation to treat epilepsy; and deep brain stimulation (DBS) to control motor disorders associated with Parkinson's disease and essential tremor.

Significantly, the development of neural prostheses is currently undergoing unprecedented expansion. There are a large number of devices in development or an early stage of commercialisation. These include visual prostheses for stimulation of the retina or visual cortex in blind patients; functional electrical stimulation to provide coordinated activation of nerve and muscle to assist with movement of the hand, arm and gait in stroke and spinal cord injury; DBS to treat pain, epilepsy or severe depression and related psychiatric disorders; vestibular prostheses to assist patients with balance disorders; and neural interfaces that record from the central or peripheral nervous system to monitor for the onset of seizures or to control external devices for amputees and severe spinal cord injured patients.

Recently neural prostheses have experienced an exciting new phase of innovation generated by the Obama Brain Initiative that encompasses the National Institutes of Health and the Defense Advanced Research Projects Agency, as well as GlaxoSmithKline's entry into the field to develop "electroceutical" techniques (Birmingham *et al.* 2014). These initiatives call for greater multidisciplinary collaboration, including the development of detailed anatomical and physiological maps of neural circuits associated with disease and treatment combined with neural modelling to optimise the development of therapeutic stimulation strategies. While outside the scope of this book, we will watch with great interest as outcomes from these initiatives are delivered to the clinic over the next decade.

Given the multidisciplinary nature of neural prostheses, the field has adopted multiple terminologies that are reflected across the 11 chapters. "Bionics", "medical bionics" or "neuroprosthesis" are used synonymously here with "neural prostheses". We have used additional application-specific terms: "neuromodulation" refers to the stimulus-induced modulation of neural activity for therapeutic purposes – DBS for the control of motor symptoms associated with Parkinson's disease, or spinal cord stimulation to alleviate back pain are examples; "functional electrical stimulation" refers to stimulation of peripheral nerve and muscle to assist in the movement of limbs following paralysis; "sensory neural prostheses" refers to devices that operate under sensory control such as cochlear (auditory) and retinal (vision) implants; "neurobionics" refers to neural stimulation treatments for disorders of the central nervous system (e.g. DBS for the treatment of movement disorders, epilepsy and pain); and "closed-loop" describes a feedback mechanism, typically based on electrophysiological recordings, used to modify the electrical stimulation parameters delivered via a neural prosthesis for improved efficacy.

New developments in neural prostheses are built on advances in electronics, materials science, electrochemistry, battery technology, neuroscience, clinical and surgical practice, and rehabilitation techniques. This book provides a comprehensive historical overview of the field (Chapter 1); it covers the key sciences underpinning the technology including the electrode-tissue interface (Chapter 2); electrochemical principles of safe electrical stimulation (Chapter 3); principles of recording from and stimulating neural tissue (Chapter 4); wireless technology (Chapter 5); and preclinical device testing (Chapter 6). Subsequent chapters describe specific clinical applications, citing devices that are both commercially available and in development, including cochlear implants and vision prostheses (Chapter 7); neurobionics in the treatment of Parkinson's disease, severe depression, obsessive compulsive disorder, pain and epilepsy (Chapter 8); and brain machine interfaces for the control of external devices such as prosthetic limbs (Chapter 9). The final two chapters provide important insight into the process of regulatory approval and commercialisation – issues critical to the successful translation of research to the clinic (Chapter 10); and the key ethical considerations associated with the development of these devices (Chapter 11). Finally, the Appendix provides a list of companies and research organisations currently developing and/or manufacturing neural prostheses.

There are many individuals who have been instrumental in ensuring the successful completion of this book. I gratefully acknowledge the authors of all the chapters – it has been a privilege to work with such a professional and knowledgeable group of individuals without whose efforts and attention to detail this publication would not have existed. In acknowledging our authors I would like to highlight Professor Giles Brindley's contribution to the chapter on the historical foundations of bionics (Chapter 1). Professor Brindley is a pioneer of the field – developing the first visual prosthesis in the 1960's (Brindley and Lewin 1968) – it is to his great credit that almost 50 years after this seminal work – and now in his 90th year – he continues to make important contributions to the advancement of neural prostheses. I am very grateful to Berenice Hale, Lyndal Borrell and Lauren Hill from the Bionics Institute for providing important administrative assistance; Justin Jeffryes, Stephanie Dollan and Allison McGinniss from Wiley for their endless advice and support for the project; and finally I acknowledge the staff and students of the Bionics Institute for providing such a stimulating environment in which to work.

Robert K. Shepherd
Melbourne, Australia.

REFERENCES

Birmingham, K., Gradinaru, V., Anikeeva, P., Grill, W.M., Pikov, V. et al. (2014) Bioelectronic medicines: a research roadmap. *Nat. Rev. Drug. Discov.*, **13**: 399–400.

Brindley, G.S. and Lewin, W.S. (1968) The sensations produced by electrical stimulation of the visual cortex. *J.Physiol.*, **196**: 479–493.

Cavuoto, J. (2016) The market for neurotechnology: 2016–2020, *Neurotech Reports*, 1–350.

PART I

FUNDAMENTALS OF NEURAL PROSTHESES

1

THE HISTORICAL FOUNDATIONS OF BIONICS

Nick Donaldson and Giles S. Brindley

Implanted Devices Group, Department of Medical Physics & Bioengineering, University College London, London, UK

1.1 BIONICS PAST AND FUTURE

In 1973, Donaldson and Davis published a paper called "Microelectronic devices for surgical implantation" in which they listed neuroprostheses in use and under development: pacemakers for the heart (fixed-rate, atrial-triggered and demand), incontinence devices, visual prostheses, dorsal column stimulators and electromyogram (EMG)) telemeters[1]. The field of bionics was then very young, the idea of surgically implanting an electronic device was new and very few people had worked on the technical difficulties entailed. Only pacemakers were then commercial products and there were no regulations in force. Now, 40 years later, there are many more types of device, both in clinical use and under development. A number of these devices will be described in Chapters 7–9 and include implants for addressing sensory loss (e.g. hearing, sight, balance), disorders of the brain and the mind (e.g. epilepsy, migraine, chronic pain, depression), as well as brain-machine interfaces. Manufacturing these devices and going through the process of regulation is now a multi-billion dollar industry.

The year 2013 may be remembered as the year in which GlaxoSmithKline (GSK) announced that they were to invest in the development of neurobionic devices, which

[1]The phrenic nerve stimulator ("Diaphram Pacer") of Glenn *et al.* (1973) was described in the same year.

Neurobionics: The Biomedical Engineering of Neural Prostheses, First Edition.
Edited by Robert K. Shepherd.

they call *Electroceuticals* or *Bioelectronic Medicines*[2] (Famm *et al.* 2013; Birmingham *et al.* 2014). The notion is that these will interact with the visceral nerves that innervate the internal organs to treat specific diseases. These diseases are not normally thought of as neurological (e.g. inflammation), but nevertheless there is some neural control. The announcement by GSK shows that the company thinks that implanted devices may become an alternative to some drug treatments. The motivations for their development no doubt include the rising costs of new drugs, better targeting of the causes of disease, and the realisation that implants might treat some of the increasingly prevalent diseases that threaten to overwhelm healthcare budgets (obesity, diabetes). They cite an example as the recent trial of a treatment for rheumatoid arthritis by stimulation of the vagus nerve (Koopman 2012). Some of the new implants will require surgical techniques new to human surgery, for example the splitting of spinal nerve roots in continuity into many fine strands. Only time will tell whether this vision is realistic, but it shows the huge rise in confidence that implanted bionic devices may be practicable and important in future healthcare.

The first electrical device implanted into a patient was the cardiac pacemaker of Elmqvist (1958), so the field is now nearly 60 years old (Figure 1.1). While Chapters 7–9 will review some of the types of implant with respect to their clinical

Figure 1.1 Elmqvist-Senning pacemaker of 1958. It is powered by two nickel-cadmium cells (arrowhead) which can be recharged by induction. The two transistors are on the right (arrows). The encapsulant is epoxy resin. An external valve oscillator was used for recharging at a frequency of 150 kHz. Scale bar = 1 inch.

[2] It will be interesting to see whether one of these names sticks, because both seem a misleading description of surgically-implanted devices.

function, Chapters 2–6 will review the field on which implant engineering is based, much of which has been built in this 60-year period. If we consider that the construction work in that period is the history of neurobionics, the purpose of this chapter is to look back to the pre-history, the foundation of the field, from the time before work began and probably before it was even conceived.

We have worked in London during the historical period (see Box 1.6: MRC Neurological Prostheses Unit) and the story is slanted toward our view of the significant technology.

1.2 HISTORY IN 1973

Donaldson and Davies (1973) suggested that neurological prostheses were the confluence of four streams of development: biomaterials (known from literature dating as far back as 1000 BC), electrical stimulation of nerves (Galvani 1791), electrophysiological recording (Matteucci 1842) and transistors (1948).

1.2.1 Biomaterials

A textbook by Susrata from 1000 BC describes the use of catgut for sutures. In Europe, from the 16th to the mid-19th century, linen and silk were the normal materials for sutures and ligatures; for sutures, horse hair, catgut and cotton were tried occasionally, and for ligatures, strips of leather. But these seem to have been passing fashions, and most surgeons continued to use silk or linen. Whatever the material, it was not a biomaterial in the modern sense; it was not expected to remain in the body for years, but either to be removed by the surgeon within a week or two, or to be extruded through the skin as part of the healing process within a few months.

The first internal fixation of a fracture with a metal plate and screws was performed by Lane in 1895, but Lane's plate and screws were of ordinary steel, and would certainly corrode. Stainless steel (*18-8* 18% chromium, 8% nickel) was patented in 1912, but the original stainless steel corroded badly in sea-water. It was not until about 1926 that a modified stainless steel, *18-8-SMo*, which had an additional 2–4% of molybdenum was developed, which resisted corrosion in sea-water and so could reasonably be expected to remain uncorroded in the body. This stainless steel was widely used in the internal fixation of fractures in the 1930s, and sometimes remained uncorroded for years (Haase 1937).

The variability remained mysterious, but it was made unimportant by the invention (1932) and introduction into bone surgery (1937) of Vitallium, an alloy of cobalt, chromium and molybdenum, which has never been reported as corroding in the body (Venable and Stuck (1938). The first widely successful artificial hip (though not absolutely the first artificial hip) was the cup arthroplasty (Smith-Peterson 1939). It used a *Vitallium* cup which was not bonded either to the head of the femur or to the acetabulum. Modern artificial hips have a ball bonded to the femur and a cup bonded to the pelvis. Problems of fixing the ball and cup to the bones and of wear at the articulating surfaces have been largely overcome. For artificial finger joints, it has been possible to avoid articulating surfaces by using adequately flexible silicones (Williams and Roaf 1973). Silicones were first used in medicine as coatings for syringe needles for reduced blood clotting (1946). In the same year, silicone rubbers were first used for surgical repairs and, in 1956, for the first hydrocephalus shunts (Colas and Curtis 2004). Thus by 1973 the field of biomaterials was established as a collaboration between surgeons, biologists and materials scientists, who had made progress by innovation with new materials, better designs and improved surgical techniques.

Less was known about implantable electrical materials: the first electrical implant in an animal was described by Louks (1933) and that was simply a coil, insulated with Collodion varnish, connected directly to electrodes; the experiments continued for 12 days. Clearly the idea that artificial materials can be implanted into the body was well established by 1973, but the specific difficulties of electrical devices were new.

1.2.2 Nerve stimulation and recording

It was established by Galvani in 1791 that nerves could be stimulated. The idea that nerves carried sensory messages to the brain and commands back to the muscles was stated in the 1st century AD by Galen, who argued for it against contrary opinions of some classical Greek authorities; he thought that the nerve signal was transmitted by fluid flow. However, when Leeuwenhoek looked at nerves in cross-section using his new microscope (1674), he was not convinced that there was any tubular structure to carry the fluid.

Newton wrote in 1678 about "a certain most subtle spirit which pervades and lies hid in all gross bodies, by the force and action of which … all sensation is excited and the members of animal bodies move at the command of the will, namely by the vibrations of this spirit, mutually propagated along the solid filaments of the nerves, from the outward organs of sense to the brain, and from the brain into the muscles." For the optic nerve, Newton repeated this opinion in his "Opticks" (Newton 1730): "Do not the rays of light in falling upon the bottom of the eye excite vibrations in the tunica retina? Which vibrations, being propagated along the solid fibres of the optic nerve, cause the sense of seeing?"

Since 1745, when the Leyden jar was invented, it was well known that electricity passing through human skin causes strong and often painful sensations. At least since 1738 (Swammerdam) it was known that if, in a preparation consisting of a frog's gastrocnemius muscle and sciatic nerve and little else, the nerve was pinched, contraction of the muscle followed immediately. Galvani (1791), using just such a preparation, showed that passing electricity from a frictional machine through the nerve had the same effect. He also did experiments using dissimilar metals, which he misinterpreted. Volta confirmed and extended Galvani's experiments, interpreted them correctly, and used them as the basis of his invention of the battery (1800), which quickly led to the discovery of the relation between electricity and magnetism, the work of Oersted, Ampere, Ohm and Faraday, and the great advances in electro-technology from which we all benefit today.

The action potential of the nerve was first detected by Matteucci (1842). The speed of conduction of the nerve message was measured by Helmholtz (1850) by comparing, in frog nerve-muscle preparations, the difference in timing of the muscle contraction according to whether the near or the far end of the nerve was stimulated electrically. He found it to be about 20 m/sec. In 1856, Herrmann measured the speed of movement of the action potential directly, and found that it was the same as that of the message as measured by Helmholtz, thus making it almost certain that the action potential was a true sign of the message.

The time course of the action potential at any one point on the nerve was known only very roughly until the development of valve amplifiers during the First World War. Gasser and Newcomer (1921) were the first to apply such amplifiers to nerve action potentials, and to display them on a cathode-ray oscilloscope. During 1921–1930, Gasser and Erlanger, in a long series of papers in the *American Journal of Physiology*, described these techniques and others to elucidate the form of the action potential and the influence of fibre diameter and myelination on it and on the speed of conduction. It was already known, from theory and from observations made with older equipment, that if both recording electrodes were placed on an intact nerve, a biphasic action potential was found, the potential difference reversing as the active region moved from one electrode to the other. However, if the end of the nerve was crushed and one electrode placed on it, a nearly-monophasic response was found. Gasser and Erlanger, with amplification, cathode-ray oscilloscope, a limb nerve (ulnar) and one recording electrode on an intact nerve at least 20 cm from the stimulating electrodes and the other on the crushed end of the nerve, found a monophasic response when they used weak stimuli, but with strong stimuli it became polyphasic, the additional peaks coming later than the one that was already present with weak stimuli. By good arguments from the results of further exploration, taking into account what was already known about the anatomy of limb nerves, they concluded that their nerve contained fibres of many different diameters. The largest conducted fastest and were most electrically sensitive. Smaller fibres were slower and less sensitive. The speeds of conduction did not follow a Gaussian distribution; they were strongly grouped into five classes, called Aα, Aβ, Aγ, B and C, by Erlanger and Gasser (1930). It soon became clear that the C fibres were unmyelinated and the A and B fibres were myelinated.

From about 1910–1930, there was much interest in how the amplitude of a rectangular pulse just sufficient to stimulate a nerve, nerve fibre, muscle or muscle fibre, varied with the duration of that pulse. Such measurements could be (and were) made with great accuracy, and easily showed that long pulses favoured unmyelinated nerve fibres and skeletal and cardiac muscle fibres, and that short pulses favoured myelinated nerve fibres, which were the most sensitive even to long pulses (say 10–20 milliseconds), but immensely so to short pulses (<0.5 msec). These experiments added little to our understanding of how the nervous system works, but are useful to the designers of bionic devices.

In 1939, A.L. Hodgkin made two steps towards understanding the nature of the nerve impulse. First he proved what had been suspected before but never proved: that the fraction of the action current of one node of Ranvier that is conducted along the axoplasm to the next node of Ranvier in a vertebrate myelinated nerve fibre is sufficient to stimulate this (next) node. Then, in the same year, Hodgkin succeeded in recording the action potential of the giant nerve fibre of the squid from an electrode inserted into the fibre. Further research was interrupted by the war, but in 1952 Hodgkin and A.F. Huxley used intracellular recording from squid giant fibres to establish a thorough understanding of the electrical and ionic basis of the nerve impulse.

In contrast to the purely electrical transmission within a nerve cell and its processes, transmission from one neurone to another, sometimes excitatory but sometimes inhibitory, is almost always carried out by means of chemical transmitters. There are at least 20 of these. A few were discovered in the 1930s, many more in the 1950s and 1960s, and there may still be a few unidentified. One transmitter may have different actions on different postsynaptic neurones. Often (perhaps always) these different actions depend on different receptor molecules.

Much of our knowledge of the function of structures in the brain comes from observations of the effects of lesions, occurring in disease or (less often) produced experimentally. Observations of the effects of disease have led to new neurophysiological knowledge almost only when followed by good postmortem examination of the brain.

It was widely (though not universally) believed throughout the first two-thirds of the 19th century that all parts of the cerebral cortex were alike in function, with the reservation (going back to Hippocrates) that the left hemisphere was more concerned with the right half of the body and the right hemisphere with the left half. Such "equipotentiality" within each hemisphere was not disproved until 1863, when Broca observed that lesions of one small area of the left hemisphere caused inability to speak, and in 1871, when Fritsch and Hitzig showed that electrical stimulation of different parts of the cerebral cortex caused movements of different parts of the contralateral half of the body.

The effects of electrical stimulation *within* the brain became known only when Horsley and Clarke (1908) designed their apparatus for stereotaxic surgery, which allowed the end of a probe to be accurately placed almost anywhere within the brain. The tip of the probe carried an electrode, so the brain structure in which it lay could be stimulated electrically, or electrical activity recorded from it, or a lesion of controlled size made in it by diathermy. The Horsley-Clarke apparatus, originally for the human brain, was soon adapted for use in experimental animals.

1.2.3 Transistors

The transistor was essential for pacemakers and in fact the first human pacemaker was made just after silicon transistors became available with their lower leakage current. However, inductively-powered stimulators with tuned coils and solid-state rectifiers, not requiring implanted transistors, could have been made earlier; such devices have been very valuable in the development of neuroprostheses because of their simplicity and reliability. For example, the first visual prosthesis did not use implanted transistors, and the inductively-powered sacral anterior root stimulator uses them only in external equipment, including the oscillators that provide the radio-frequency magnetic fields. However, the arrival of transistors in the 1950s clearly showed the possibility for future small low-powered electronic devices, small enough to implant. This sense of anticipation was increased by the development of the integrated circuit (patented in 1959).

1.2.4 Conclusion

Before a new type of bionic device is implanted into a patient, an ethical committee must be convinced that there is a reasonable chance that it will be effective and the risks of implantation are not too great. In the case of nerve stimulators and neural signal amplifiers, understanding the mechanisms is bound to be helpful, so scientific knowledge from biophysics and neurophysiology, that line of scientific endeavour that included Galvani, is valuable. The transistor allowed the necessary miniaturisation. Regarding risk, it is essential to know that the materials implanted in the body are harmless and provoke no more than a mild response that does not jeopardize the device or the patient's health.

It was true that in 1973 medical bionics depended on progress in all these fields, biomaterials, nerve recording and stimulation, and transistors, but surely there were several other antecedents and even if some were not apparent at the time that Donaldson and Davies were writing, we should now acknowledge them as having been essential to the success of the field. These other historical antecedents are the subject of the following sections.

1.3 ANAESTHESIA

Two hundred years ago, the idea of implanting an artificial device into the body would surely have been regarded as at best fanciful and at worst a horror. Surgeons had a desperate job to do while the patient tried to endure the extreme pain and good surgeons were those who were quick; amputations might be completed in seconds but remained agonising. Pain-killing drugs had long been used, notably opium and mandrake, but these (especially mandrake) had undesirable side effects.

Given that surgery caused so much suffering, it is perhaps surprising that the use of anaesthesia by nitrous oxide, ether and chloroform in surgery came as late as it did. Humphrey Davy, the chemist, discovered the pain-killing effect of nitrous oxide – laughing gas – and used it on himself while having a tooth removed. This was part of his earliest work, published in 1800 (Routledge 1881). Subsequently, not only did he demonstrate the effects of this gas in lectures but other chemists did too. Faraday pointed out that ether had similar effects to nitrous oxide in 1818 (Routledge 1881). However, the method was not immediately tested by clinicians and it was not until the 1840s that two small-town Americans, Crawford Long (doctor) and Horace Wells (dentist), anaesthetised patients with ether and nitrous oxide respectively. Neither gained recognition for their achievement. In Britain, nothing was done until J.Y. Simpson, Professor of Midwifery at Edinburgh, started experimenting with chloroform, first on his mother's dog, then on his friends as an evening amusement, before starting to use it on his patients during childbirth.

Chloroform had been discovered in 1831. Despite the apparent alleviation of pain, there was considerable resistance from traditionalists to use of chloroform during childbirth. This was largely overcome by the intervention of Queen Victoria who was expecting her seventh child. She commanded Simpson to act as midwife and, after delivery, gave royal approval for chloroform. In America, the breakthrough in surgery occurred when the dentist Dr William Morton, who had started experimenting with Wells, anaesthetised a patient with ether who was about to have a tumour removed by Professor Warren, Chief Surgeon at the Massachusetts Hospital in Boston in 1846: the operation in front of many witnesses was completely convincing. The first use during surgery in Britain was an amputation done under chloroform by Mr Robert Liston in 1846 at University College Hospital. Since a process for synthesizing ether was discovered in 1540 (Routledge 1881), there appears to be no reason why this huge advance in surgery could not have been discovered 300 years earlier.

1.4 ASEPTIC SURGERY

Until past the middle of the 19th century, death rates from infection following major surgical operations were very high, and such operations were done only for very strong reasons. The death rates fell greatly with *antiseptic surgery* (carbolic acid spray), introduced by Joseph Lister in Glasgow in 1867, and much further still in the 1880s with the development of *aseptic surgery* in which everything that touched or might touch the patient was sterilized in advance by heat.

1.5 CLINICAL OBSERVATION AND EXPERIMENTS

It would be completely wrong to think that the medical application of electricity was waiting for neuroscientific theory before attempting the treatment of patients. During the 18th century, science and particularly electrical science was of great popular interest and the fact that muscle could be stimulated through the skin by electric shocks from the friction generators of the time, caused widespread interest in its possible curative effects. Therapy was offered by conscientious practitioners such as John Wesley, as well as charlatans (Fara 2002). For example, paralysed patients travelled long distances to be treated by Benjamin Franklin who treated them with strong shocks by discharging large Leyden jars, but generally this was regarded as only a temporary cure[3]. Treatment of this sort continued right through to the early 20th century. McNeal (1977) reported that almost every American doctor's consulting room in the late 19th century had at least one electrical machine. In 1919, St Bartholomew's

[3]With our modern understanding, we would expect that these people would have upper motor neuron damage and the shocks only demonstrated that the paralysed nerve and muscle was still capable of contraction.

Hospital in London had an Electrical Department with a Medical Officer in Charge who wrote a book in which he divided the medical applications of electricity into the categories of: electrochemical cauterisation (destruction by caustic solution at the cathode); iontophoresis for introducing drugs into the body; diathermy; galvanic acupuncture (pain relief); and treatment for paralysis (Cumberbatch 1929). The list of conditions that he claimed could be treated is very long, from acne and angina, via moles and sciatica, to warts and writer's cramp; the list includes many infectious diseases but no evidence of efficacy was presented. Lumping the treatment of such diverse conditions together under the electrical umbrella seems to have been abandoned after the Great War.

Bionic devices are based on a foundation of neuroscience, but neuroscience is far more than the results of animal experiments, such as nerve-muscle preparations; a very large part of it is accumulated clinical observation. For most parts of the brain, clinical observations provide more than half of what is known about their function; for example, without clinical observation we should have no knowledge whatsoever of what the cerebellum does, even if we knew all that we now know about its connections and its chemical transmitters. Insight into the development of some bionic devices has followed such clinical observation, or experiments done with the patients' consent during surgical procedures. Two examples are mentioned in Box 1.5. One is Gordon Holmes's mapping of the visual cortex during the First World War; clinical observations on brain-damaged soldiers, which came near to being also an experiment, in that Holmes knew all that was already known about anatomical investigations of the geniculo-striate tract, and almost certainly adjusted the details of his examination of each patient so as to make each patient yield the greatest possible amount of information about the projection of the retina on the striate cortex. The second, also mentioned in Box 1.5, is the stimulation of the visual cortex during surgical removal of an epileptic focus from one occipital lobe; this was done by Foerster (1929) and by Krause and Schum (1931) a few years later. Stimulating electrically was not normal practice, and probably did not influence how much brain was removed. Both surgeons were experimenting, doing something that was very unlikely to cause harm and from which they were likely to learn something new. In fact, all these observations led to the idea of stimulating the visual cortex to give sight to the blind. Another example is the treatment of Parkinsonian patients by Deep Brain Stimulation. This derived from a chance clinical observation that a person, who had administered himself an illicit drug, methyl phenyl tetrahydropyridine (MPTP), developed symptoms like Parkinsonism (1983). The poison was administered to animals, which also developed the symptoms, and could be used as models. It was found that lesions in the sub-thalamic nucleus could reverse the symptoms (1990), and in 1993 treatment by stimulation of the nucleus had been demonstrated in patients (Limousin *et al.* 1995; see also Box 1.1).

Box 1.1 The treatment of pain

Suffering pain was always the lot of man and pain treatment is a topic in some of the earliest texts, such as the Egyptian papyri and clay tablets of Babylon. It was known from these early times that electric fish could provide relief by numbing the area affected: a Nile Catfish is shown in a tomb picture from the Egyptian 5th Dynasty (Kellaway 1946). Aristotle and others refer to the numbness produced by the shocks from these fish and the Roman writer Scribonius Largus (46 AD) described a treatment for headache. "Headache ... is taken away ... by a live black torpedo placed on the spot which is in pain, until the pain ceases. As soon as the numbness is felt, the remedy should be removed lest the ability to feel is taken from that part." (Rawlings *et al.* 1992).

The therapeutic use of electric fish continued and perhaps still continues, but after the invention of the friction electrostatic generator and the Leyden jar in 1745, the similarity between the two types of shock was clear. However, it was difficult to understand how this shock could be delivered under water, so Henry Cavendish (the man who discovered hydrogen and measured the weight of the Earth with a torsion balance) made an underwater model of the Torpedo fish that, while connected to a friction electrostatic generator, was able to give powerful shocks to peripheral nerves and induce numbness in those who came to see the demonstration (Fara 2002). In the 19th century, Duchenne treated neuralgia, sciatica and rheumatism by electricity; and after 1858, electro-anaesthesia was used in dentistry, the current being passed through the region of the affected tooth. By about 1870, a body of literature had been published (Rawlings *et al.* 1992), but the method then went into decline until it was rediscovered in about 1930 and after that a more scientific approach was taken with studies of dermatomes and physiological pathways, in particular by stimulation of the spinothalamic tract, the brainstem and the thalamus for pain relief. In 1965, Melzack and Wall published their *gate theory* of pain, which provided rationales for peripheral nerve and brain stimulation treatment (Melzack 1973). The first implants for treating chronic pain (dorsal column stimulators) were described by Sweet and Wepsic (1968).

There are now many different implant treatments for pain (Sakas *et al.* 2007), including: trigeminal nerve stimulation for craniofacial pain; occipital nerve stimulation for migraine; epidural stimulation of motor cortex for deafferentation pain; spinal cord stimulation for pains of the back; phantom limb and others types; and deep brain stimulation for many types of pain including spinal cord injury and peripheral neuropathies. Sakas *et al.* (2007) comment that management of chronic pain has been the greatest success of the neuromodulation treatments. Neuromodulation is an important addition to the treatment of pain by tissue ablation, neurotomy or drug delivery treatments, which are available to neurosurgeons. It is interesting that a significant step toward deep brain stimulation was an observation by Pool (1954) of an analgesic effect of stimulating the forniceal columns while carrying out psychosurgery, an effect that Pool and Heath found they could repeat in non-psychiatric patients (Raslan *et al.* 2007).

Box 1.2 The conventional implanted device since 1970

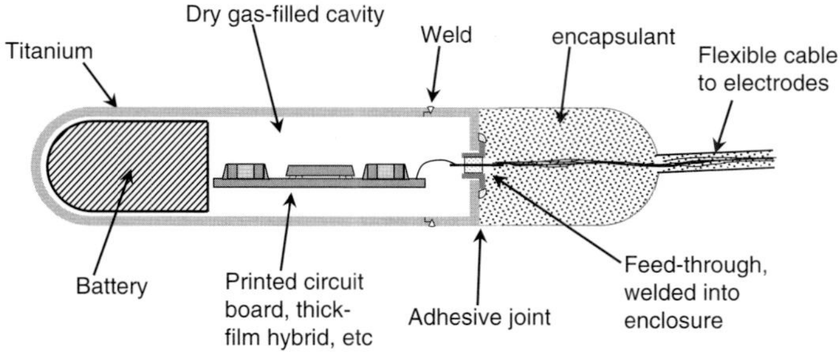

Figure Box 2 A non-scale drawing showing the main features of the most common type of implant during the last 40 years. The electronic components, and sometimes a battery are inside a metal enclosure which is "hermetic", meaning that leak rate of moisture is low enough that the inside will remain dry for the required lifetime of the device. Conductors are brought out through annular "feed-throughs", which comprise a metal pin, a glass or ceramic bush and a metal ring. On the outside of the enclosure, wires are joined to the pins of the feed-thoughs which may be part of the output cable (as shown) or may go to a surgical connector. The cables are usually either multi-strand wires (as shown) or helical single-strand wires, and there may be one or more wires in each cable. The polymer encapsulant is essential to insulate the exposed wires where they are joined to the pins of the feed-throughs: to be an effective insulator, the encapsulant must remain bonded to the enclosure and the feed-through.

1.6 HERMETIC PACKAGES

This section and the following describe technology that was developed prior to the first modern implants (Figure 1.2). The important features of almost all commercial implants since 1970 are shown in Box 1.2. This design, with a weld-sealed titanium enclosure and feed-throughs, soon became the norm despite the fact that even in 1970 there were clearly other possible methods using ceramics, glasses and polymers. In the following sections, we describe the wide range of technology which was then available.

The hermetic package is an impermeable enclosure which, by acting as a barrier to water vapour, maintains the electronic components inside in a dry environment. The origins of this technology are in vacuum science: the feasibility of sealing electrical connections into evacuated glass vessels was demonstrated throughout the 20th century by incandescent lights bulbs and later electronic vacuum valves (tubes).

Figure 1.2 The first modern pacemaker was made by Telectronics in 1971. The engineer David Cowdrey, who had been charged with developing a hermetic enclosure, selected Ti because of its light weight, strength, corrosion resistance and weldability. He used deep-drawn Ti half-cases that were welded together using TIG welding to keep the inside cool. He also developed ceramic/Ti alloy feed-throughs for the package. Interestingly, the technology was never patented because the company's patent attorney advised them that it "was obvious"! This picture shows a Telectronics 'Slimline' device from 1977.

1.6.1 Vacuum methods

The earliest known apparatus for creating a vacuum was made by Berti in 1641: an 11 metre long-vertical tube was filled with water before taps at the top and bottom were shut and opened respectively, allowing the water to descend, evacuating the top of the tube. von Guericke started working on reciprocating pumps in the same decade, which enabled him to evacuate a barrel and later pairs of hemisphere to demonstrate the existence of air pressure. Boyle and Hooke improved the pump and added a manometer in 1658/9, so we know that they achieved a pressure of 6 Torr. Little progress was made in the next two centuries: the first prize for vacuum pumps at the Great Exhibition in London in 1851 went to Newman, whose pump only reached 0.5 Torr. However, in the remainder of the 19th century, progress was rapid, mainly due to pumps with liquid pistons, the lowest pressure reaching less than 10^{-5} Torr. McLeod invented his vacuum gauge in 1874, allowing pressures down to 10^{-4} Torr to be measured. The diffusion pump which was independently invented by Gaede in Germany and by Langmuir in the United States during the First World War, produced even lower pressures, the ultimate being 10^{-8} Torr until the 1950s. The need to mass produce evacuated light bulbs from the 1870s meant that vacuum pumps had to be made for industrial use, with much higher pumping rates as well as low ultimate vacua. The Edison Company used manual pumps at first, but by the end of the 19th century mechanical pumps were in use. Methods for measuring low pressures were the subject of much work and by 1920 the minimum measurable pressure had reached 10^{-8} Torr[4].

1.6.2 Welding[5]

Before the late 19th century, blacksmiths had joined metals by hammering the hot parts, forming what might now be called a solid-state weld. The development of modern industrial processes started with the development of the oxy-hydrogen torch and then the hotter oxy-acetylene torch by Fouch and Picard, which was being used for commercial welding in 1903. Electric arc welding soon followed and joints were improved following the invention of the coated electrode by Kjellberg (patented in 1908), which kept atmospheric oxygen and nitrogen away from the molten metal, preventing brittleness and porosity in steel. Many other types of welding soon followed. In 1914, a 34-mile-long oxyacetylene-welded pipeline was fabricated in Oklahoma for the oil industry. Radiographic inspection of welded joints was first described in a paper from the US Naval Research Laboratory in 1926. In the same year, the first thick-walled pressure vessel was fabricated by welding. British shipbuilders made the first all-welded warship in 1923 and the first all-welded submarine in 1943, as welds were shown to be superior to rivets. Thus welding was well established by 1970 and not only could many alloys be joined in this way but evidently gas-tight or waterproof joints were possible. By that date, there was already literature on titanium welding and how to avoid porous joints. Titanium, and similar metals like tantalum, have the advantage that they are protected from corrosion by stable surface passivating films but at elevated temperatures they absorb atmospheric gasses and alter their properties, making them more brittle and possibly leading to porosity. However, if welding is done under inert gas or in a vacuum, strong gas-tight joints are possible. A thorough description of the metallurgy and welding methods for titanium and other metals can be found in Kearns (1984).

1.6.3 Glass

Glass technology is very ancient. Glass vessels and glazed pottery appeared early in human history. The material has been of great importance in science due to its useful properties, particularly transparency, impermeability and the fact that it can easily be shaped at high temperatures to form glassware, including chemical glassware. Macfarlane and Martin (2004) suggested that over half the most important experiments in the history of science would have been impossible without glass.

Glasses are supercooled liquids that lack long-range order; they are metastable and crystallisation rates may be negligible on a human time scale. Because there is

[4]This paragraph is based on (Redhead, undated).

[5]This paragraph is based on *(The History of Welding, undated)*

no change of state when they are cooled from a temperature at which they are liquid to room temperature, they are characterised by their progressive change in viscosity with temperature. Most glasses are mixtures of oxides and are categorised first by the major oxide constituent, the three most significant being silicate, phosphate or borate. Silicate glasses are by far the most important and only silicate has any technological value without admixed oxides, in the form of vitreous silica or "Fused Quartz", the others being too soluble in water.

Forming a joint between glass and metallic electrical conductors has been an important technological problem, at least since the mid-19th century experiments on ionisation of gasses. If a metallic wire is fused into the glass while it is softened by heating, and then allowed to cool, the glass is likely to crack unless the thermal expansion curves of the two materials are well-matched. Figure 1.3, taken from a book published for experimentalists in 1938, shows expansion curves for some relevant materials. Platinum had long been used to make seals to soda-lime glass, their Temperature Coefficients of Expansion (TCE) being close enough at 9.1×10^{-6} and 9.2×10^{-6} (Corning G8, "Soft Glass"). The thermal match between *Pyrex* (3.2×10^{-6}, "Hard Glass"), a borosilicate commonly used for chemical glassware, and tungsten (4.7×10^{-6}) is not ideal, but vacuum-tight joints are usually possible with wires under 1.5 mm diameter. If the wire is larger, a bead of *Nonex* glass (3.6×10^{-6}) was fused to the wire first and then this was joined to the *Pyrex* wall. Generally, metals with high TCE, like copper, cannot be joined to Hard Glasses but in the 1920s, Housekeeper (1923) showed that certain designs of joint could form satisfactory seals with copper. In one such case, the wires were flattened where they passed through the glass, a design often visible in incandescent light bulbs. Also in the 1920s, the American company Westinghouse developed iron-nickel-cobalt alloys that were well-matched to certain borosilicate glasses. *Fernico* is shown in Figure 1.3, but *Kovar* is now better known and became the standard alloy for making metal packages with glass feed-throughs, such as those shown in Figure 1.5(a). Figure 1.4 shows a hermetically-packaged device with metal-in-glass feed-throughs; it is a military device which illustrates the most advanced technology of that time.

1.6.4 Glass ceramics and solder glasses

The range of electrical conductors which could be used in TCE-matched feed-throughs was greatly increased by the development of glass-ceramics, which started from the discovery by Stookey at the Corning Glass Works just after the Second World War. He found that glasses in which a small amount of gold had been precipitated could then be converted to ceramic (devitrified) in a relatively short heat treatment, the gold acting as nucleation sites. Subsequent research (McMillan 1979) found that these *glass-ceramics* could be stronger than the original glass and have higher softening temperatures, allowing further high-temperature processing without distortion. Furthermore, these materials, which are often lithium-silicate mixtures with one other oxide, such as Pb, Zn or Al, can have TCE values from more than 15×10^{-6} to -3.9×10^{-6}, a range which includes all the metallic conductors that are of interest for implanted devices.

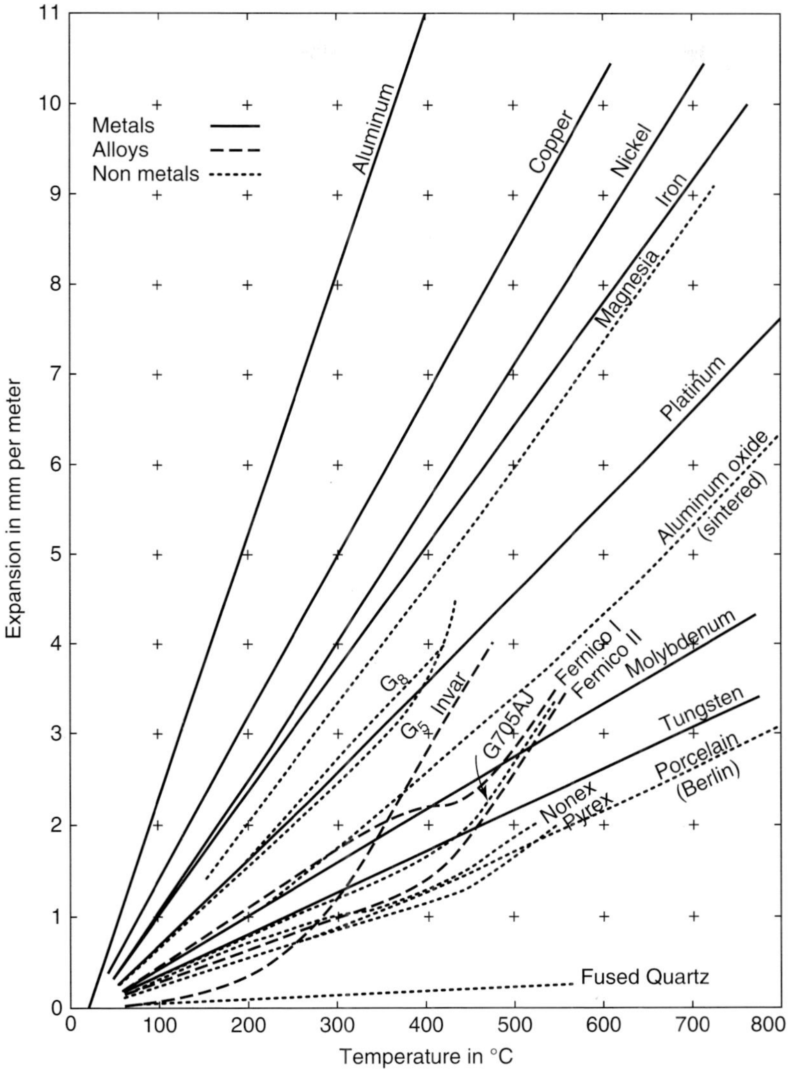

Figure 1.3 Thermal expansion curves for some metals, some special alloys, some ceramics and a few glasses, taken from Strong (1938). The broken lines are the insulators.

So-called solder glasses have been used for much longer. These are glasses that can be worked at moderate temperature, often about 500°C, and can be used for joining metal or ceramic or indeed other glasses. A package formed in this way is shown in Figure 1.5(b). They also need to have matched TCE, and since TCE and softening temperature are generally inversely related, it follows that they must be unusual glasses. They are sometimes not silicate but borate glasses, particularly lead borate. They may remain entirely vitreous or they may devitrify after making the joint; the

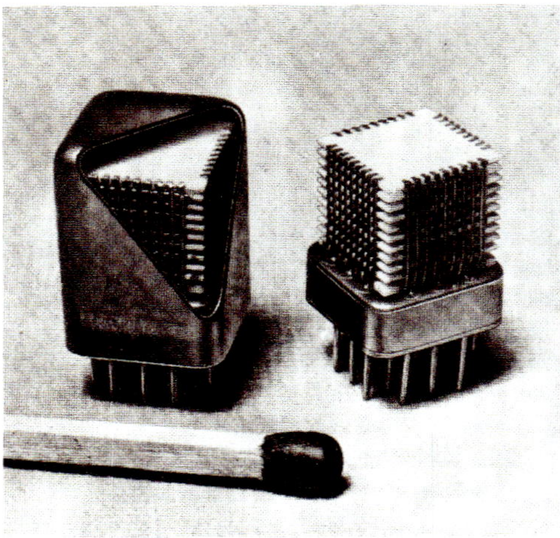

Figure 1.4 High-reliability microelectronic module (aerospace) taken from Manfield (1969). The package has metal-in-glass feed-throughs and is sealed by electron beam welding.

(a) Kovar package with metal-in-glass feed-throughs. The Kovar is gold-plated; lid usually resistance-welded to wall.

(b) Flat-Pack with alumina base and Kovar lead frame and seal ring. The side wall is solder glass. The metal beams of the lead frame pass through the side wall.

(c) Ceramic dual-in-line package made by co-firing three layers of alumina with tungsten metallisation, then brazing the lead frame outside and gold plating.

Figure 1.5 Three types of package that were available in the 1970s, showing that several technologies were available that might be suitable for hermetic packages for implants.

latter are sometimes called solder glass ceramics. Since TCE and solubility tend to be correlated, in some applications solder glasses may dissolve too quickly if exposed to moisture. A comprehensive review of solder glasses was published by Frieser (1975). They are significant materials for implant designers because they may allow hermetic seals to be formed that are not electrical conductors and therefore will not affect magnetic fields.

1.6.5 Ceramics

An alternative approach to making vacuum-tight joints between conductors and insulators is to metallise ceramics. The first method of metallisation was developed in

Germany during the Second World War: a molybdenum layer is applied by heating the ceramic to around 1500°C in a reducing atmosphere. The molybdenum starts as a powder, often mixed with some manganese. After metallising high-alumina ceramic in this way, a brazed joint can be made to the metal using Ag/Cu eutectic after plating the moly-manganese with nickel. This process was widely used in the ceramics industry by 1970 (Twentyman 1975).

As transistors became available in the 1950s, there was widespread interest in how to reduce the size of electronic assemblies and in 1958, the US Army set an objective to reduce sizes of military radio equipment ten-fold. The result has been described by Stetson (1986), who was one of the team at RCA to work on the contract. They made two breakthroughs: Gyurk developed the method for laminating metallised *green* (pre-sintered) ceramic sheets to make ceramic chip capacitors (1959) and Stetson himself invented the method for making three-dimensional circuits by punching holes to form vias (vertical connection between layers of metallisation) before metallising and laminating the green sheets (1960). With tungsten metallisation and alumina ceramic, the structure can be tough and impermeable to water vapour and gases, and was used to form high-reliability hermetic packages, such as that shown in Figure 1.5(c). In such a structure, the feed-throughs from the cavity inside to the connections outside can be part of the interconnection for the internal components. This is known as co-fired ceramic technology.

1.6.6 Microcircuit technologies

The microelectronics industry wanted to be able make multilayer circuits at lower temperatures than the moly-manganese process, because such a high processing temperature was neither convenient nor appropriate for forming resistors, capacitors or inductors. A technology that is usually called "thick film" was developed in which the metals were noble, so that they could be heated in air without oxidising, and these were fused to a ceramic substrate using a glass frit (powder). This technology originated from the need to make radio fuses for mortar shells in the Second World War, the circuit having to withstand the forces of acceleration at 10 000 g (Stetson 1986). The metal and glass mixture was screen printed with an organic vehicle which was then burnt off before sintering the mixture in place, usually at about 900°C. Dielectric layers of glass or devitrifying glass and higher resistivity materials could also be deposited so that by repeated printing and firing, a circuit could be built up. This process became attractive with the invention of transistors and later integrated circuits, as dice could be connected to the thick film circuit, achieving small, light and reliable devices. The technology was being developed in the 1960s. It was not originally a method for making seals but nevertheless, a soldered joint made to thick film metallisation may be vacuum-tight.

For completeness, we will mention the third method of metallisation which is thin film. Where thick films are typically 20 microns thick, thin film metallisation is under 1 micron. Early thin films were usually deposited by evaporation or sputtering in a vacuum. These films were not used for making seals, being too thin for brazing or soldering and generally not having adequate adhesion to the substrate.

1.6.7 Leak testing

These new glass and ceramic technologies for the formation of feed-throughs in hermetic packages could only be shown to be effective in a practical time if some method of measuring the gas leak rate was available. Fortunately, a method became available by this time; it was the use of mass spectrometers tuned to helium, generally known as *Helium Leak Testing*. Investigations in the 19th century into currents (rays) in discharge (vacuum) tubes found that there were both anode rays (ions) and cathode rays (electrons). At the end of the century, the cathode rays were more mysterious, because no particles smaller than atoms were expected. In 1886, Goldstein showed that the anode rays could be deflected by a magnetic field and by 1913, J.J. Thomson had shown that neon appeared to be composed of two isotopes of different mass. In 1919, his student F.W. Aston (1919), described the first mass spectrometer and listed the isotopes present in six elements, leading to instruments for analysis of elements. During the Second World War, the Manhattan Project required a method to show that containment vessels were gas-tight, to ensure that uranium hexafluoride would not escape, and a special mass spectrometer was designed (Nerkan 1991). This instrument was tuned to helium, because atoms of that gas are very small, penetrating the smallest crack, and its concentration in the atmosphere is low, so the background signal is small. It was described by Nier *et al.* in 1947 and industrial instruments were produced immediately after the war for the electrical industry. According to Hilleret (1999), sensitivity has increased from 10^{-7} Pa.m^3.s^{-1} in 1946 to 10^{-10} in 1970 and 10^{-13} in 1999[6].

To summarise, by 1970 many electrical components such as diodes, transistors and electrolytic capacitors were mass produced in hermetic enclosures. Glasses and glass-ceramics were available which allowed matched thermal expansion to all technologically-useful metals and alloys so that metallic packages could be made with metal-in-insulator feed-throughs. Furthermore, packages which were largely ceramic were available in several configurations, some of which had large cavities like pacemaker packages. Helium leak testing was already a standard method for quality assurance, and the relationship between helium leakage from the package and water vapour ingress into the package was known (Davey 1975).

1.7 ENCAPSULATION (ELECTRICAL INSULATION)

1.7.1 Insulation

Early in the 18th century, Gray discovered that materials can be classified as electrical conductors or electrical insulators (Mendenhall 1895) and this led to new demonstrations of electrical effects. Gray himself showed that a human body, a boy suspended by silk threads, could be charged with a friction generator and later discharged by drawing a spark. Even more spectacular were demonstrations by Nollet that electricity could be passed by chains of up to 600 people (200 monks on one occasion) before

[6] 1 Pa.m^3.s^{-1} is approximately 10 at.cc/s, which has been the familiar unit of leak rate.

allowing a spark jump to the ground (Fara 2002). Friction machines were used to generate electricity throughout that century and some technical progress was made. It is well-known that Franklin was active in electrical science, showing that lightning is an electrical discharge and making a practical electrostatic motor.

The quality of insulation became a critical matter early in the 19th century, following Oersted's discovery that an electric current can cause a magnetic needle to deflect (1820). This led would-be inventors in many countries to try to develop electric telegraph systems. The system that became most popular was that of Morse. For his first experimental telegraph line, from Washington to Baltimore, he buried the wire (Mendenhall 1895), but he abandoned this method because too much current leaked through the insulation and so he changed to the system common in Europe, which had first been used by O'Shaughnessy in Calcutta in 1839, of suspending the wires from insulators on poles. This avoided the need for good insulation of the wires.

1.7.2 Underwater insulation

The desire to lay telegraph cables under rivers meant that the need for better insulation could no longer be avoided. Fibres including cotton and hemp, impregnated with tar and other materials, were tested but none lasted longer than a few days. Fortunately, a British surgeon, stationed in Singapore, experimented with a material made from the sap of the Gutta Percha tree, which he found suitable for making splints. He sent specimens to London in 1822, where its properties of elasticity and good insulation were soon recognized. Unlike natural latex, this rubbery material is thermo-softening and becomes sticky when hot, so it could be extruded around the wires to form an insulating sheath (Routledge 1881). This timely discovery allowed a successful telegraph cable to be laid across the Hudson River in 1848 and the English Channel in 1850. Each copper wire was coated in Gutta Percha, before these wires were spun into a rope impregnated with tarred hemp and then protected by outer wires of iron. The first cable to be laid across the Atlantic did not last long because its insulation was damaged by high voltages (500 V) being applied, but it demonstrated the feasibility of the idea in 1858 and the first successful cable operated after 1865. A description of this cable technology from 1915 describes the copper core being coated with what we might call an adhesion promoter called *Chatterton's Compound* before the Gutta Percha extrusion (Anon 1915). The operating voltage was limited to 60.

Here we see the essentials of a successful underwater insulator or encapsulant: it must be flexible (like rubber) so that it can deform as necessary in service without rupture; it must have a sufficient volume resistivity; and it must adhere to the conductor so that water, which will diffuse through the permeable encapsulant, cannot form a layer of moisture at the interface. Any moisture allows corrosion of metals, dissolution of non-metals like solder glasses, and leakage currents to flow between conductors at different potentials.

1.7.3 Silicones

Major advances in insulation materials were made by industrial chemists at the American companies, Corning Glass and General Electric; the story has been very well told by one of the major contributors (Rochow 1987). There were parallel developments behind the Iron Curtain. The problem was to try to find insulation that could withstand higher temperatures, above 125°C, allowing better electrical machinery. It was understood that it is the carbon chains in natural rubbers which were susceptible to oxidation. Kipping had shown that the silicon analogues of alcohols (silanols) are much less stable than their carbon counterparts, so that, for example:

$$2\,R_3SiOH \rightarrow R_3SiOSiR_3 + H_2O \tag{1.1}$$

condensation occurred, forming longer molecules. Di-silanols, molecules with two –OH groups, are capable of polymerisation and the –Si-O-Si–, siloxane structure, was often a product in silicon chemistry. When Franklin-Hyde at Corning was working on flexible binders that hold glass fibres together at temperatures over 125°C, he remembered that Kipping had described glue-like substances that formed from di-phenyl di-silanols. He succeeded in synthesizing silicones with phenyl and ethyl side groups. When heated to 200°C, the ethyl groups were oxidised, cross-linking and curing the material to form a resin varnish that withstood 180°C. This discovery led to the formation of the Dow Corning company in 1943. Rochow made two major contributions. First, he foresaw that the most promising structure for high service temperature would be one that minimised the number of carbon atoms on the silicon-oxygen backbone, which suggested a material that had never previously been synthesised, di-methyl

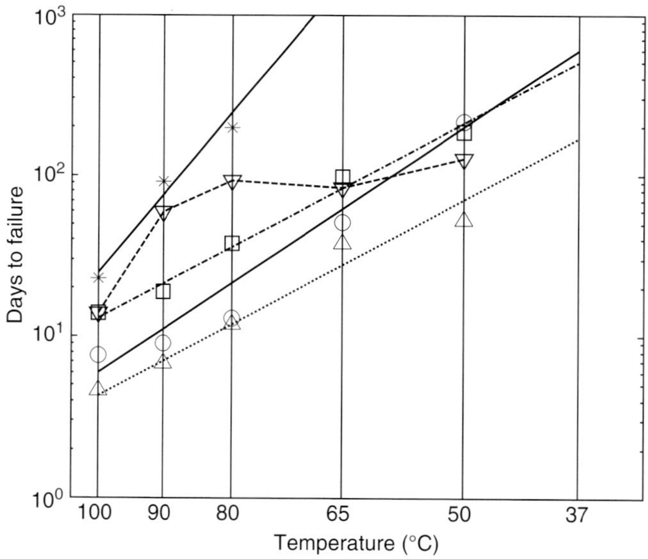

Figure 1.6 Arrhenius plot for accelerated adhesive life tests, replotted from Donaldson (1982). Adhesive is Dow Corning 3140, an alcohol-evolving RTV. The substrates are alumina (*), tin-lead solder (□), kovar (∇), copper-nickel alloy (○), oxidised titanium (Δ). Bonds to alumina are easily superior to all others.

silicone. Initially he synthesised this material using the expensive Grignard process, but later he invented a direct method of making chlorosilanes directly from silicon which was then hydrolysed to form the silanol that polymerised by condensation to form the poly di-methyl siloxane (PDMS or silicone). Direct synthesis is the basis of the modern silicone industry. Box 1.3 describes further processing to form practically useful silicone rubbers.

Box 1.3 Making practical silicones

Two major steps were necessary to obtain useful silicone rubbers. Methods had to be found to achieve very long chains, but with enough cross-links to hold them together. Two common methods are:

(i) The siloxane chains include some vinyl groups and some hydrogen in place of methyl groups. An addition reaction can occur between the Si-H and Si-CH=CH_2 to form a cross-link of Si-CH_2-CH_2-Si . This is catalysed by platinum and takes place with heating to moderate temperatures.

(ii) *Room Temperature Vulcanizing* (RTV) is an older method in which there are hydroxyl side groups and acyloxy side groups. When exposed to moist air, the water allows cross-linking:

$$
\begin{aligned}
&-\mathrm{Si(CH_3)_2{-}OH} + \mathrm{Ac{-}O{-}Si(CH_3)_2}- \\
&\rightarrow -\mathrm{Si(CH_3)_2{-}O} - \mathrm{Si(CH_3)_2} - + \mathrm{AcOH}
\end{aligned}
\tag{1.2}
$$

in this case, evolving acetic acid AcOH, but some evolve alcohols (Arkles and Redinger 1983).

The second major step was the incorporation of a filler to which the chains bond. Most commonly this is achieved with fumed silica, which has a similar atomic structure to the PDMS. However, the silica would soon be covered with water molecules, making the surface unavailable for bonding, so it is pre-exposed at high temperature to small cyclic siloxanes. It is these two steps that give us the rubbers that are strong, can stretch by 300%, and are stable above autoclave temperatures (134°C).

This inertness and purity made these attractive implantable materials (Box 1.5), but perhaps their most useful feature of the Room Temperature Vulcanizing (RTV) silicone (Box 1.3) is one that is hardly acknowledged by the manufacturers – long-term adhesion in the presence of water, as shown in Figure 1.6 (Gajewski 1983). The mechanism of cross-linking in the RTV types always involves bond formation with a silanol group on one chain and it seems likely that the hydroxyl group can form bonds to the surface of oxides, including glass, silica and ceramics. For example, bonding to silicate glass:

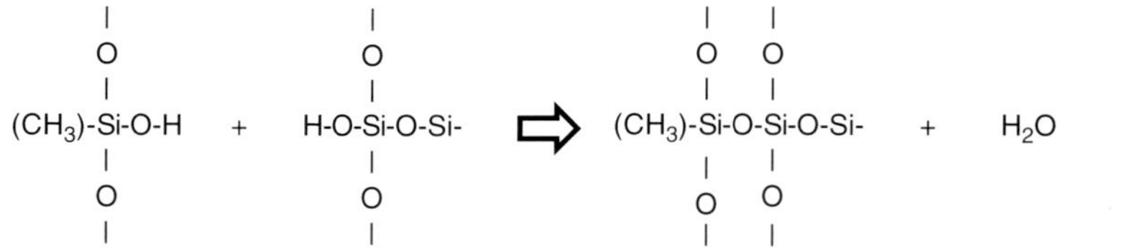

Siloxane chain   Hydroxyl on glass   Chemical bond   Condensed water

or on a metal surface:

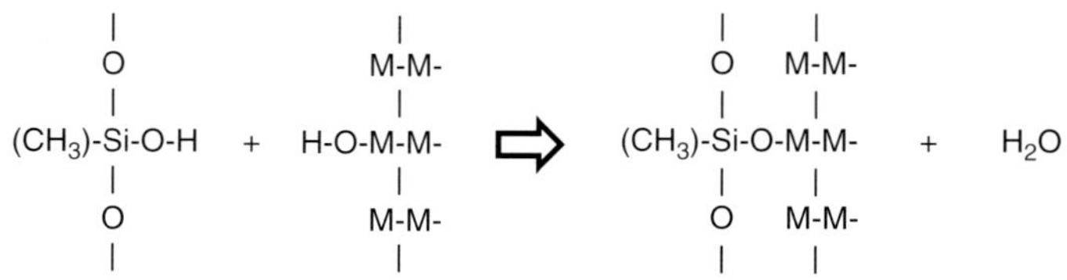

Siloxane chain   Hydroxyl on metal   Chemical bond   Condensed water

The nature of the bond between the oxygen and the metal has been a matter of scientific enquiry (Pluedermann 1982). This is fundamental to neural prostheses, because the insulation relies on the durability of the adhesive bonds.

1.7.4 Primers

The stability of the bond formed when the silanol group meets an oxide surface has had other applications besides the bonding of silicone polymers to other materials. The problem of making strong composites by including glass fibres in a resin matrix presented itself after the Second World War. In 1947, a report to the US Navy showed that the application of a coating of triethoxy silane on glass fibres before impregnating them in polyester resin, dramatically reduced the effect of boiling water on the strength of the composite (Pluedderman 1982). Glass fibres were treated by silanes that bonded to the hydrophilic surface of the glass and presented organic groups to which the resin could bond.

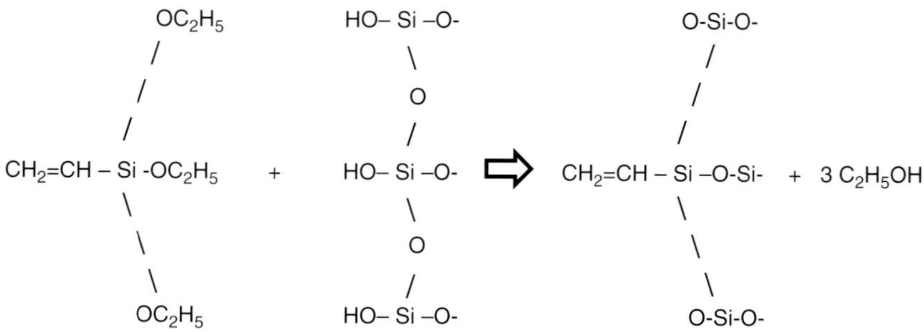

Vinyl triethoxy silane   Glass fibre   Vinyl coupling group   Ethanol

The remaining double bond in the vinyl groups allows it to be incorporated into the polymer as it cures against the surface. This is the treatment that allows the long life of fibreglass boats in water and printed circuit boards in moist environments (Rochow 1987). This research was done in the 1960s in the USA. There are many silane coupling agents from which primers can be chosen to improve bonds between organic polymers (e.g. encapsulants), and the parts to which they should bond (Plueddermann 1982). This is relevant to high-modulus encapsulants (Box 1.4).

Box 1.4 High-modulus encapsulants

The use of low-modulus silicone rubbers makes the shape of the parts to be insulated and the control of the encapsulation process less critical. That simplification reduces the amount of development needed and is particularly helpful for experimental devices being made quickly without much engineering effort. High-modulus encapsulants like epoxy resins are normally used for pacemakers and structurally similar devices (Box 1.2). The resin forms the socket for the electrode lead, as well as insulating the connections at the feed-throughs and the wires running from there to the socket. The encapsulant must be stiff to support these conductors and prevent fracture due to forces applied after implantation (Figure Box 4a and 4b). The design and the encapsulation process should be developed so that nowhere during manufacture or service is the interfacial stress higher than the adhesive strength. As the adhesive strength declines as the joint is hydrolysed, the considerations should include the possible use of primers to diminish the rate of hydrolysis.

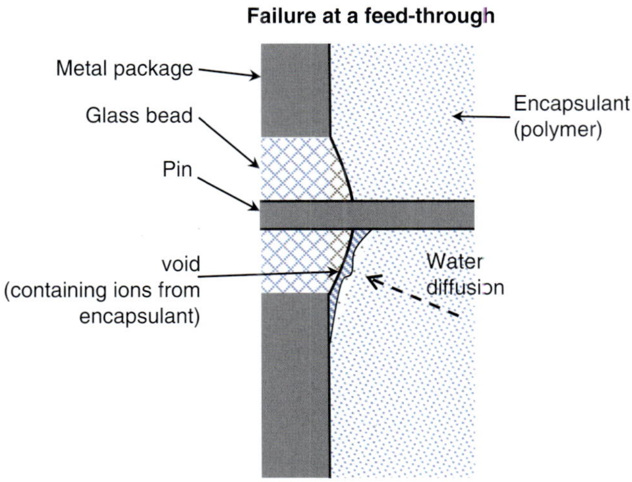

Figure Box 4a The void on the surface of the glass bead of the feed-through fills with water that diffuses through the encapsulant. Soluble salts that are residues left after inadequate cleaning, or that are present in the encapsulant, dissolve in the water and increase its conductivity. Failures like this are often made visible because of corrosion products from the metals.

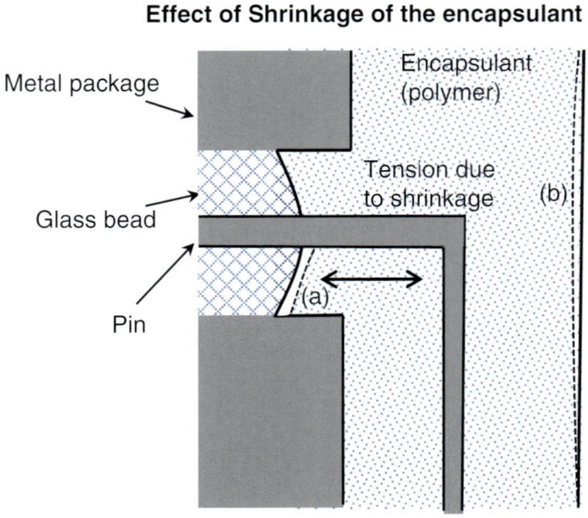

Figure Box 4b This diagram illustrates the disadvantage of using a high-modulus encapsulant if it will have any cause to shrink, for example due to cooling after cure. With this feed-through, which has a recessed glass bead, shrinkage sets up tension in the direction shown, tending to cause adhesion failure (a). This is exactly where adhesion is vital to prevent electrical leakage. Soft encapsulants are more likely to be able to relieve the tension by distortion at a free surface (b).

Box 1.5 Other important favourable properties of silicone rubbers as encapsulants

Besides the hydrolytic stability of their adhesive joints, many silicone rubbers have several other important properties.

(i) *Low Ionic Content:* Perhaps, because of the way they are manufactured from silicon via gaseous methyl silicon chlorides in the Direct Synthesis process, silicones have little contamination. Of particular importance is the low contamination by salts that would ionise so that concentrations of common ions like Na^+ and Cl^- are usually around 1 p.p.m. The epoxies used in earlier implants contained much more Na^+ and Cl^- than this, which probably partly explains why they failed so readily. Figure Box 4b shows the usual failure mode in early pacemakers with epoxy encapsulant. A void would form at the interface between the encapsulant and the package or component. The void might have been due to an absence of adhesion after curing the encapsulation, or was perhaps a bubble; alternatively, although there was initial adhesion, the stresses built up as the implant cooled after curing exceeded the strength of the adhesive joint and the materials parted. Either way, after implantation, water vapour will diffuse through the encapsulant and condense in the void. If that liquid is pure water, which has a very high resistivity, the current flowing through it will be so small that the function of the implant is not compromised. Gas is likely to evolve at the conductors that have become electrodes under the encapsulant but if the current is small, the gassing rate may be slow enough for the gas to diffuse away harmlessly. If, on the other hand, the encapsulant has a relatively high concentration of salt impurities, some of these ions will dissolve in the water film, increasing the leakage current and the gassing rate. Rapid gas evolution will create high pressure, tending to lift off the encapsulant, and typically this tears the encapsulant away from the surface, extending the initial void. This type of runaway failure is exacerbated by encapsulants with high levels of ionic impurity and this has been one advantage of using silicones.

(ii) *Low Modulus and Low Cure Temperature:* The volume of the encapsulant will change: if it is cured at elevated temperature and then cools; if it is sterilised by autoclave; and when water diffuses into the encapsulant after implantation. This strain builds up stresses according to the modulus of the encapsulant and the shapes involved. Figure Box 4b shows an example where this effect caused failures on the glass bead of the feed-through, just where insulation is most important. This type of feed-through with deeply-recessed beads makes successful encapsulation very difficult. In general, the outside of the package should be convex, as far as possible, so that tension in the encapsulant presses the encapsulant against the surface. Although the bulk modulus of silicone rubbers is high, their low Young's Modulus, compared to resins like epoxy, means that distortion of the shape can usually relieve the low internal pressure due to volume shrinkage (Donaldson 1977).

(iii) *Transparency:* In order to inspect the encapsulation, it is convenient if the material is transparent. Some silicones are nearly ideal in this respect; for example, Nusil MED-6215, but this is an unfilled rubber and its extension-to-failure is only 80%. Most rubbers are filled and will not fail until hundreds of percent extension, so they are tougher and therefore more suitable for most implanted devices. This improvement is achieved by including silica filler, but this carries the disadvantage that the material is slightly cloudy so less detail can be seen by inspection.

1.8 EARLY IMPLANTED DEVICES

There were some early pacemakers that, like the animal stimulator of Louks (1933), had a coil into which current was induced, connected directly to the stimulating electrodes[7], but tuning the primary and secondary coils improves efficiency and allows

[7] See Exhibit A6002.73 in the Medical Collection, Science Museum, London.

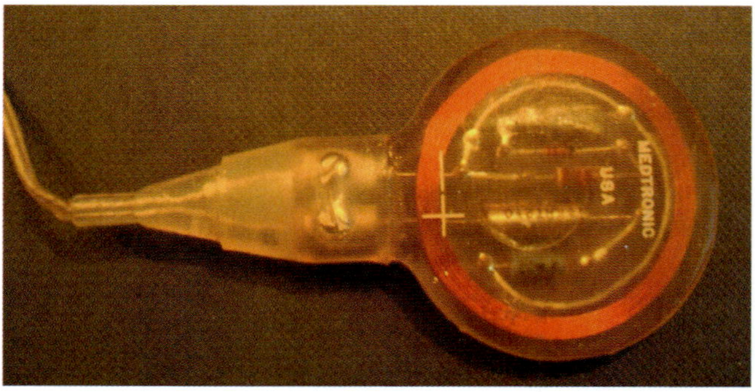

Figure 1.7 Dorsal column stimulator made by Medtronic from 1968–1978.

a smaller size coil if a radio frequency carrier is used. Such an inductively-coupled tuned-coil stimulator is shown in Figure 1.7.

Inductive-powering is advantageous because no battery is required in the implant, which means that it may be smaller in volume, and its life will not be limited by the battery capacity. Devices can be designed to which power and stimulation commands are sent together over one pair of coils. These have been very successfully used, particularly in cochlear implants for the deaf. A review of this technology was presented by Patrick *et al.* (1990). The small size of modern cochlear implants has allowed them to be implanted into pre-lingual infants enabling them to learn spoken language. However, continuous inductive powering does require a power transmitter to be held in place during use: this may be inconvenient or seen as unattractive.

Pacemakers[8] must be continuously active for life support and from the beginning it was seen as necessary that the implant should be battery-powered. But this requirement presented designers with a difficult challenge, because the electromotive force of the battery was continuously present, unlike inductive stimulators, like that shown in Figure 1.7, in which pulses are usually less than 1 ms duration with a low duty cycle (perhaps 1/50) and there is no voltage difference to drive destructive reactions at all while the device is not in use (except galvanic potentials between metal couples). In the earliest pacemakers, all the components were encapsulated in polymer (Figure 1.1 and Chardack *et al.* 1960). Figure 1.8 shows an intermediate design in which the electronic circuit was fabricated in thin film and then sealed in a metal enclosure with glass feed-throughs. At that time, the battery could not be included in the sealed package because these cells produced gas and could reach a pressure that might cause an explosion. Greatbatch solved this problem by developing the lithium iodide cell that did not gas, was inherently reliable, and could itself be hermetically-sealed (Greatbatch 1984). This led to the modern form of the implanted device that has replaced all others for pacemakers and is predominant for most other bionic devices that are implanted, the titanium package package with glass feed-throughs, containing the electronics

[8]History of Pacemakers, undated.

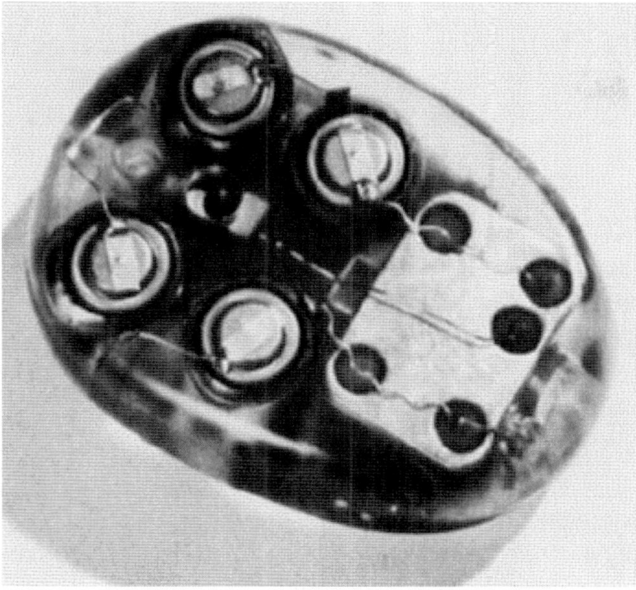

Figure 1.8 Intermediate design of pacemaker from Devices Implants Ltd. The hermetically-sealed package is on the right and the battery on the left, all encapsulated in epoxy resin (Kenny 1969). Figure with permission from the Annals of the New York Academy of Sciences.

and the battery, with some external insulation where the cable or connector is joined to the feed-throughs (Box 1.2). For stimulators such as this, the battery is inside the hermetic package and only pulses appear at the output cables which is much less demanding on the encapsulantion.

1.9 AFTERWORD

The development of implantable devices for medical application is at the intersection of neuroscience, clinical science and engineering. Although some justification for attempting new experimental procedures has been based on experiments on animals, supplemented by anatomical studies on human cadavers and experiments on human volunteers using imaging techniques, much of the justification has derived from observations made during treatment, some fortuitous, of patients who are injured or undergoing some surgical operation.

The treatment of patients by electricity is very ancient and in principle neuroprosthetic devices might have been attempted after Galvani and after Faraday's discovery of electromagnetic induction (Louks 1933), but implantation would not have been practical until after anaesthesia allowed prolonged surgery without extreme pain, and

Table 1.1 Packaging Methods discussed at Stanford Meeting in 1979

	Substrate	Package	Package Seal	Encapsulant
1	Thick Film Hybrid	Machinable ceramic	Epoxy	Not described
2	PCB	Copper	Solder	Not described
3	PCB	None	None	Wax
4	PCB	Stainless steel	Weld	Not described
5	Thick Film Hybrid	Kovar	Weld	Not described
6		Tantalum	Weld	Not described
7	Thick Film Hybrid	Ceramic	Epoxy	Epoxy
8	Thick Film Hybrid	Titanium	Weld	Parylene
9	Thin Film Hybrid	Glass	Solder	Epoxy
10	None	None	None	Silicone
11	None	None	None	Epoxy

aseptic techniques brought the risk of surgery down to an acceptable level. These techniques are the great contributions from the 19th century, but neuroscience was not then sufficiently advanced.

The key that opened the door for neuroprosthetic devices was the transistor and its arrival occurred when much more was known about the brain and nervous system than in the 19th century. We have shown that there were also many other lines of technological development which have been essential to success: in vacuum science, metallurgy, glass, ceramic and polymer technology and improved battery technology. Much of this know-how came from innovations in the USA, often driven by the Second World War or the Cold War. Nevertheless, in the 1970s, it was far from clear how to make reliable implants and researchers tried all sorts of methods, with far greater variety than we see now. For example, in a workshop held in Stanford in 1979 called "Implantable Transducers and Systems: Packaging Methods and Testing Criteria", the speakers described the types of implant construction shown in Table 1.1.

Dr Robert White from the host group described the major difficulties at that time (Hambrecht and Reswick, in McNeal 1977). "The packaging … and specifically the achievement of reliability against insulation breakdown, fluid seepage and lead breakage is perhaps the most widespread and common problem." However, after pacemaker batteries could be included inside the hermetic packages and with good choice of materials, design and quality control, the form of the pacemaker became established and this has been copied in most other types of device that have followed (Box 1.2). The industry has achieved significant reduction in size and increase in functionality as integrated circuits, programmable devices and now remote communication have been introduced. During this time, the form of the packaging has changed little, but that conventional design is now often increasingly inappropriate as new types of device have to be very small with many electrodes (e.g. retina stimulators). New types of hermetic micropackages are needed to fit in such small spaces and it may be that it is better to dispense with the package entirely and rely on encapsulation of the integrated circuit (Vanhoestenburghe and Donaldson 2013). These new requirements should encourage innovation using available technology, some of which is new, such as Diamond-like Carbon films, but much is original, as described above, from the prehistoric period of bionics.

Box 1.6 The Medical Research Council, Neurological Prostheses Unit, 1968 –1992

One of us (G.S.B.) designed and built an implant to stimulate the visual cortex and this was implanted in a blind volunteer in 1967. Although the device was not reliable, it established the possibility that blind people might have implants to restore some sense of sight (Brindley and Lewin 1968). There are two bases for the visual prosthesis: the less important is the mapping of the visual field on the occipital cortex. This was discovered and its fine detail well established during the First World War by recording the visual field losses from wounds to the occipital poles of the hemispheres, mainly by the British neurologist Gordon Holmes. The more important base for the visual prosthesis was the electrical stimulation of the occipital cortex during operations to treat epilepsy, which occurred as a late effect of such wounds. This was done by German surgeons during the first few years after the First World War. Similar stimulation was done by Penfield in Montreal in the 1930s and 1940s, with apparently conflicting results. My decision to make the first visual prosthesis depended absolutely on my judging that the descriptions given by Foerster (1929) and Krause and Schum (1931) were accurate, and those given by Penfield substantially inaccurate.

I learned only in 1969 that the idea of stimulating the occipital cortex to provide artificial vision had already been suggested by Krieg (1953). I presented two Communications to the Physiological Society in 1964, one on the number of stimulation channels that should be necessary for reading (Brindley 1964a), and the other on the design of an array of radio-frequency transmitters and receivers that could be used to excite the implanted electrodes (Figure Box 6) (Brindley 1964b). The first implantation was done in 1967 (Brindley and Lewin 1967, 1968).

The British Medical Research Council established a research unit in 1968 so that further applications of the technology and methods could be developed, and new applications explored. The Chief Engineer from foundation to closure in 1992 was P.E.K. Donaldson, father of the first author.

A large number of methods were explored by Brindley and his many collaborators, which were reported for the 4-yearly reviews. It is interesting to see how many of these ideas were far-sighted and which have been found ineffective or been superseded.

Because of their historical interest, G.S.B. reviewed the 26 projects in the table to consider the extent to which each was based on clinical observation. The majority actually had no basis on clinical observation, but the following are either exceptions or are worth further description:

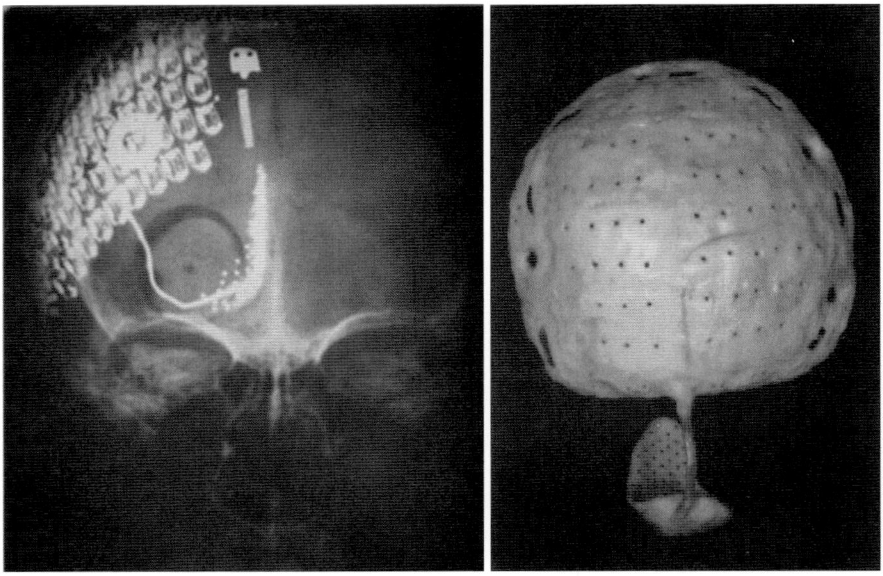

Figure Box 6 Eighty-channel visual prosthesis, implanted in a blind volunteer in 1967. The extracranial part of the device had 80 radio frequency receivers, each connected to one of the electrodes on the intracranial silicone cap that was implanted over the right occipital cortex. Thirty-nine channels produced phosphenes after implantation. The receivers and the electrodes are visible in the X-ray image (Brindley and Lewin 1968). Each receiver in the implant was encapsulated in epoxy resin before being inserted into the silicone cap. Failures were probably due to wire breakage since failures were never partial. Biocompatibility had previously been tested in baboons.

Table 1.2 Implanted Devices reported from the MRC Neurological Prostheses Unit

	Project	1974	1978	1982	1987
1	Artificial visual pathway (occipital cortex)	✓	✓	✓	✓
2	Auditory-tactile signalling	✓	✓		
3	Implantable bladder and sphincter controller (S3 –S4)	✓	✓	✓	✓
4	Artificial motor pathway (nowadays called a Brain-Computer Interface)	✓	✓	✓	✓
5	Artificial auditory pathway (cortex)	✓	✓		
6	Carotid sinus stimulator for hypertension	✓	✓	✓	
7	Denervated muscle stimulation	✓	✓		
8	Artificial sphincter (stress incontinence)	✓	✓	✓	✓
9	Flexible implantable lens to allow accommodation		✓		

Table 1.2 (***Continued***)

	Project	1974	1978	1982	1987
10	Cerebellar stimulator (epilepsy)		✓	✓	
11	Posterior column stimulator (multiple sclerosis)		✓		
12	Deep brain stimulation (pain relief)		✓		
13	Deep brain stimulation (auditory prosthesis)		✓		
14	Deep brain stimulation (obesity or anorexia nervosa)		✓	✓	
15	Stimulation for paraplegic walking		✓	✓	✓
16	Pancreatic stimulator (diabetes)			✓	
17	Stimulator for grasp			✓	
18	Intracranial pressure sensors			✓	✓
19	Conditional pudendal nerve stimulator (detrusor instability)			✓	✓
20	Intercostal nerve stimulator (tetraplegia)				✓
21	Artificial anal sphincter				✓
22	Deep brain stimulator for tremor (in MS)				✓
23	Hypogastric plexus stimulation for ejaculation (SCI)				✓
24	Cannula and reservoir for semen (SCI)				✓
25	Cochlear implant				✓
26	Facial nerve stimulator (UMN facial palsy)				✓

Not all of these of these projects are bionic; numbers 8, 9, 21 and 24 were not electrical in any way.

1. Clinical observation, in conjunction with post-mortem examination, showed that the retina was mapped on the visual cortex, and by 1918 the map was well-established and accurate. Without the map, Foerster and Krause would probably not have operated on their patients with occipital epileptic foci, and if they did operate, would not have stimulated. So for the visual prosthesis, three things were absolutely necessary: good clinical observation, post-mortem examination of the same patients, and the initiative taken independently by two German surgeons in different centres to stimulate the occipital lobes before removing them.
2. Clinical observation told us that people with defective hearing often do better with a tactile supplement, for example being able to touch the speaker's larynx. The details of the device depended slightly on clinical observation.
3. It is a clinical observation that a person with the spinal cord cut through at any level above S3 has no bladder control. This could also be deduced from the anatomy, but the deduction could, by an extreme sceptic, be deemed uncertain, whereas the observation, made by the patient himself, is certain. The design of the device owes nothing to clinical observation.

4 and 15. This project was about two kinds of implant: leg-muscle activators controlled by switches worked by the patient's fingers, and similar activators controlled by signals taken from the leg area of the motor cortex. The biological knowledge needed for the design of the leg-muscle activators is anatomical, and was derived about 96% from dissection, 2% from experiments on corpses, and 2% from experiments on live human volunteers. Decisions on how to use them were based mainly on clinical observation.

10. This project had, in my view, absolutely no scientific justification. A respected American neurosurgeon, Irving Cooper, decided on theoretical grounds that stimulation of the cerebellar cortex was likely to suppress epilepsy. He tried it, uncontrolled, on seven patients, and reported good results. The King's College Hospital neurologists read his paper, and asked me if I would help them try to replicate it. All the relevant clinicians concurred that the risks were very small, so I checked that the MRC Unit staff had no objections, and then consented. I based this project only on Cooper's clinical observations, which were weak because uncontrolled, but not negligible. I dismissed his original theoretical basis as entirely worthless. We made three stimulators. They were all implanted, and all three patients thought their fits had become fewer. Only one patient would consent to a blind trial. The trial showed that there was no effect whatever on the frequency of fits. The patient nevertheless continued to believe that the implant was doing her good, and continued to use it.

11. This project was also, in my opinion, based on dubious clinical observation, with no scientific justification whatever. Posterior column stimulators for pain relief work very well, but this was not for pain relief. We abandoned the project before we had wasted much time on it.

12. This project was based on a chance observation by an American neurosurgeon during stereotaxic surgery. The observation was wholly unexpected, and could well be classified as a clinical observation.

18. It was chiefly clinical observation that convinced neurosurgeons of the need to monitor intracranial pressure. The *design* of our sensors was influenced only slightly by clinical observation.

20. Needed the clinical observation that when both phrenic nerves are injured, the intercostal muscles, perhaps with some help from the sternomastoids, trapezius and platysma, can give adequate breathing, at least for a few hours.

26. It was from clinical observation that I decided that our patient was likely to gain benefit from a facial nerve stimulator. Clinical observation influenced the *design* only in that I respected the patient's wish that the stimulator be inconspicuous, and consulted him about details.

REFERENCES

Anon (1915) History of the Atlantic Cable and Undersea Communications. http://www.atlantic-cable.com/

Anon, History of Pacemakers; http://www.biotele.com/pacemakers.htm

Arkles, B. and Redinger, P. (1983) Silicones in biomedical applications, in *Biocompatible Polymers, Metals and Composites*, Chapter 32, Szycher, M. (ed.), Technomic Publishing, Lancaster PA.

Aston, F.W. (1919) A positive ray spectrograph. *Philosop. Mag.* **38**: 707–714.

Birmingham, K., Gradinaru, V., Anikeeva, P., Grill, W.M., Pikov, V. et al. (2014) Bioelectronic medicines: a research roadmap. *Nature Rev. Drug Disc.*, **13**: 399–400.

Brindley, G.S. (1964a) The number of information channels needed for efficient reading. *J. Physiol.*, **177**: 44.

Brindley, G.S. (1964b) Transmission of electrical stimuli along many independent channels through a fairly small area of intact skin. *J. Physiol.*, **177**: 44–46.

Brindley, G.S. and Lewin, W.S. (1967) The visual sensations produced by electrical stimulation of the medial occipital cortex. *J. Physiol.*, **194**: 54–55.

Brindley, G.S. and Lewin, W.S. (1968) The sensations produced by electrical stimulation of the visual cortex. *J. Physiol.*, **196**: 479–493.

Chardack, W.M., Gage, A.A. and Greatbatch, W. (1960) A transistorized self-contained implantable pacemaker for the long-term correction of complete heart block. *Surgery*, **48**(4): 643–654.

Colas, A. and Curtis, J. (2004) Silicone biomaterials: history and chemistry, in *Biomaterials Science: An Introduction to Materials in Medicine*, 2nd Edition. Ratner, B.D., Hoffman, A.S., Schoen, F.J. and Lemons, J.E. (eds), Elsevier, San Diego and London.

Cumberbatch, E.P. (1929) *Essentials of Medical Electricity*, 6th Edition, L. Henry Kimpton, London.

Davey, J.G. (1975) Model calculations for maximum allowable leak rates of hermetic packages. *J. Vac. Sci. Tech.*, **12**(1): 423–429.

Donaldson, P.E.K. and Davies, J.G. (1973) Microelectronic devices for surgical implantation. *J. Inst. Elec. Radio Eng.*, **43** (1/2): 125–132.

Donaldson, P.E.K. (1977) Silicone rubber adhesives as encapsulants for microelectronic implants: Effects of high electric fields and tensile stress. *Med. Biol. Eng. Comput.*, **15**: 712–715.

Donaldson, P.E.K. (1982) The underwater life of joints between some adherend and adhesive materials useful in neurological prosthesis-making. *Proc. BES/PRI Meeting*, Durham.

Erlanger, J. and Gasser, H. (1930) The action potential in fibres of slow conduction in spinal roots and somatic nerves. *Am. J. Physiol.*, **92**: 43–82.

Famm, K., Litt, B., Tracey, K.J., Boyden, E.S. and Slaoui, M. (2013) A jump-start for electroceuticals, *Nature*, **496**: 159–161.

Fara, P. (2002) *An Entertainment for Angels*. Icon, Cambridge, UK.

Foerster, O.J. (1929) Beiträge zur Pathophysiologie der Sehbahn und der Sehsphäre. *J. Psychol. Neurol. (Leipzig)*, **39**: 463–485.

Frieser, R.G. (1975) A review of solder glasses. *Electrocomp. Sci. Tech.*, **2**: 163–199.

Gajewski, H.M. (1983) Synthesis, characterisation and fabrication of high-purity biomedical grade silicone rubbers, in *Biocompatible Polymers, Metals and Composites*, Chapter 31, Szycher, M. (ed.), Technomic Publishing, Lancaster PA.

Gasser, H. and Newcomer, H.S. (1921) Physiological action currents in the phrenic nerve: An application of the thermionic tube to nerve physiology. *Am. J. Physiol.*, **57**(1): 1–16.

Glenn, W.W.L., Holcomb, W.G., Hogan, J., Matsuto, I., Gee, J.B.L. *et al.* (1973) Diaphragm pacing by radiofrequency transmission in the treatment of chronic ventilator insufficiency. *J. Cardiovasc. Surg.*, **66**: 505.

Greatbatch, W. (1984) Implantable power sources – a review. *J. Med. Eng. Tech.*, **8**(2): 56–63.

Haase, W. (1937) "Rostfreier" Stahl in der Chirurgie. *Archiv für orthopädische Chirurgie*, **37**: 606–611.

Hilleret, N. (1999) Leak Detection. http://cds.cern.ch/record/455564/files/p203.pdf

Horsley, V. and Clarke, R.H. (1908) The structure and functions of the cerebellum examined by a new method. *Brain*, **31**: 45–124.

Housekeeper, W.G. (1923) The art of sealing base metals through glass. *The Journal of the American Institute of Electrical Engineers*, **42**: 954.

Kearns, W.H. (Ed.) (1984) *Welding Handbook: Metals and their Weldability*, 7th Edition, vol. **4**. American Welding Society (Pub).

Kellaway, P. (1946) The part played by electric fish in the early history of bioelectricity and electrotherapy. *Bull. Hist. Med.*, **20**: 112–122.

Kenny, J. (1969) Experience in the development of a thin-film pacemaker. *Ann. NY Acad. Sci.*, **X**: 886–890.

Koopman, F.A. *et al.* (2012) Pilot study of stimulation of the cholinergic anti-inflammatory pathway with an implantable vagus nerve stimulation device in patients with rheumatoid arthritis. *Arth. Rheum.*, **64** (suppl.): abstr. 451.

Krause, F. and Schum, H. (1931) *Neue Deutsche Chirurgie*, Stuttgart, Enke, vol. **49a**: 482–486.

Krieg, W.J.S. (1953) *Functional Neuroanatomy*, 2nd Edition. Blakiston, Evanston IL, pp 207–208.

Lane, A. (1895) Some remarks on the treatment of fractures. *Brit. Ed. J.*, **1**(1790): 861–863.

Limousin, P., Pollak, P., Benazzouz, A., Hoffmann, D., Le Bas, J.F. *et al.* (1995) Effect of Parkinsonian signs and symptoms of bilateral subthalamic nucleus stimulation. *Lancet*, **345**: 91–95.

Louks, R.B. (1933) Preliminary report of a technique for stimulation of tissue beneath the integument. *J. Comp. Psychol.*, **16**: 439–444.

Macfarlane, A. and Martin, G. (2004) A world of glass. *Science*, **305**: 1407–1408.

McMillan, P.W. (1979) *Glass Ceramics*, 2nd Edition. Academic Press, London.

McNeal, D.R. (1977) *2000 years of electrical stimulation, in Functional Electrical Stimulation*, Hambreckt, F.T. and Reswick, J.B. (eds), Dekker.

Manfield, H.G. (1969) Packaging techniques for electronic systems, in *Microcircuits and their Applications*, Gore, W. (ed.), Iliffe, London.

Matteucci, C. (1842) Sur une phénomène physiologique produite par les muscles en contraction. *Compt. Rend. Acad. Sc.*, Paris 4: 797.

Melzack, R. (1973) *The Puzzle of Pain*. Penguin, UK and Australia.

Mendenhall, T.C. (1895) *A Century of Electricity*, Houghton, Mifflin and Co., Boston and New York.

Nerken, A. (1991) History of helium leak detection. *J. Vac. Sci. Tech. A*, **9**(3): 2036–2038.

Newton, I. (1730) *Opticks*, 4th Edition, Qu. 12.

Nier, A.O., Stevens, C.M., Hustrulid, A. and Abbott, T.A. (1947) Mass spectrometer for leak detection. *J. App. Phys.*, **18**: 30–33.

Patrick, J.F., Seligman, P.M., Money, D.K. and Kuzma, J.A. (1990) *Cochlear Prostheses* Clark, G.M., Tong, Y.C. and Patrick, J.F. (eds), Churchill Livingston, Edinburgh, pp 99–124.

Plueddemann, E.P. (1982) *Silane Coupling Agents*. Plenum, New York.

Pool, J.L. (1954) Psychosurgery in older people. *J. Am. Geriat. Soc.*, **2**: 456–466.

Raslan, A.M., McCartney, S. and Burchiel, K.J. (2007) Management of chronic pain: Cerebral neuromodulatory and neuroablative approaches. *Acta Neurochir. Suppl.*, **97**(2): 17–26.

Rawlings, C., Rossitch, E. and Nashold, B.S. (1992) The history of neurosurgical procedures for the relief of pain. *Surg. Neurol.*, **38**: 454–463.

Redhead, P.A. Undated. History of Vacuum Devices. http://www.cientificosaficionados.com/libros/CERN/vacio23-CERN.pdf

Rochow, E.G. (1987) *Silicon and Silicones*. Springer-Verlag, Berlin-Heidelberg.

Routledge, R. (1881) *Discoveries and Inventions of the Nineteenth Century*. George Routledge & Sons, London, pp 517–519.

Sakas, D.E., Panourias, I.G., Simpson, B.A. and Krames, E.S. (2007) An introduction to operative neuromodulation and functional neuroprosthetics, the new frontiers of clinical neuroscience and biotechnology. *Acta Neurochir. Suppl.*, **97**(1): 3–10.

Smith-Petersen, M.N. (1939) Arthroplasty of the hip: A new method. *J. Bone Jt. Surg.*, **30B**: 59.

Stetson, H.W. (1986) Multilayer ceramic technology, in *High Technology Ceramics: Past, Present and Future: The Nature of Innovation and Change in Ceramic Technology*, American Ceramics Soc., Proc. 88th Annual Meeting, pp 307–322.

Strong, J. (1938) *Procedures in Experimental Physics*. Prentice-Hall, New York.

Sweet, W.H. and Wepsic, J.G. (1968) Treatment of chronic pain by stimulation of fibers of primary afferent neuron. *Trans. Am. Neurol. Assoc.*, **93**: 103–105.

The History of Welding. http://www.weldinghistory.org/whfolder/folder/wh1900.html (accessed August 2014).

The Parkinson's Appeal. http://www.parkinsonsappeal.com/dbs/dbshistory.html (accessed February 2014).

Twentyman, M.E. (1975) High-temperature metallizing. *J. Mat. Sci.*, **10**: 765–776.

Vanhoestenberghe, A. and Donaldson, N. (2013) Corrosion and lifetime predictions in implantable electronic devices. *J. Neural. Eng*, **10**(3); DOI: 10.1088/1741-2560/10/3/031002.

Venable, C.S. and Stuck, W.G. (1938) Electrolysis controlling factor in the use of metals in treating fractures. *J. Am. Med. Ass.*, **111**: 1349–1352.

Volta, A. (1800) On electricity excited by the mere contact of conducting substances of different kinds. *Phil. Trans.* 1800–1814, **1**: 27–29.

Williams, D.F. and Roaf, R. (1973) *Implants in Surgery*. Saunders and Co, London.

2

DEVELOPMENT OF STABLE LONG-TERM ELECTRODE TISSUE INTERFACES FOR RECORDING AND STIMULATION

JENS SCHOUENBORG

Neuronano Research Center, Experimental Medical Science and the Nanometerconsortium, Lund University, Lund, Sweden

2.1 INTRODUCTION

Brain Computer interfaces (BCIs), that are both structurally and functionally stable inside the brain over long periods of time and that have minimal effects on the physiological conditions of the neural tissue, would provide extremely powerful research and clinical tools. A key to achieve this goal is to overcome the foreign body reaction and consequent loss of neurons adjacent to the implanted BCI, which eventually may render the BCI useless. It is also necessary to accomplish sufficiently stable contact with individual neurons to permit long-term studies of, for example, learning mechanisms and to link data from the BCI to anatomical atlases. This chapter reviews the state-of–the-art of neural interfaces and the concepts behind recent research and development of biocompatible BCIs.

Neurobionics: The Biomedical Engineering of Neural Prostheses, First Edition.
Edited by Robert K. Shepherd.

2.2 TISSUE RESPONSES IN THE BRAIN TO AN IMPLANTED FOREIGN BODY

2.2.1 Acute tissue responses

Common to all implants in the brain and spinal cord is that the surgical procedures during implantation cause trauma to the tissue including local bleeding, edema and rupture of the brain blood barrier and local neuronal damage (Figure 2.1). The magnitude of the trauma depends on a number of factors, such as the size and shape of the implant and speed of insertion [1]. The initial trauma, in turn, triggers an acute inflammatory tissue response including, for example, astrocyte proliferation, microglia (the principal phagocytic cell type in the brain that takes care of local debris), migration towards the implant and angiogenesis that typically lasts for 7–10

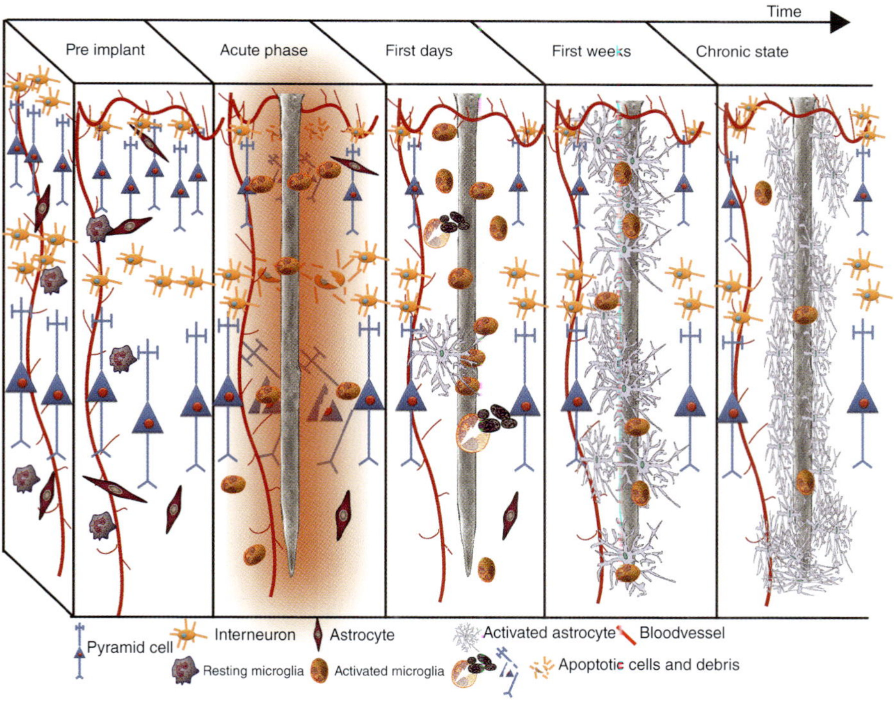

Figure 2.1 Schematic of the normal brain tissue response to the implantation of an electrode. From left to right: pre-implantation, acute to chronic phases after implantation. During the implantation, neurons, glia cells and blood vessels along and adjacent to the insertion track line are mechanically injured. As a consequence, local bleeding (illustrated in red) and edema are common. Within the next few days, an acute local tissue response to the injured cells and debris is initiated. This acute inflammatory tissue response includes activation and migration of microglia toward the implant (which starts within minutes after the injury) and proliferation of astrocytes in the tissue nearby the implant as well as neuronal death. During the subsequent weeks, an astrocytic scar that encapsulates the implanted electrode is formed, whereas many microglia cells undergo apoptosis. The number of neurons adjacent to the electrode is also significantly reduced.

days [2–6]. This acute response is a normal pathophysiological response to the surgery with a purpose to restore normal tissue conditions.

2.2.2 Chronic tissue responses

Despite the use of biocompatible surface materials on the implants, the acute tissue response is usually followed by a chronic tissue response that may even escalate with time. The chronic reaction is largely constituted by astrocytes, which encapsulate the implant in a tightly woven sheet (sometimes referred to as "membrana gliae limitans accessoria" [7]. It is commonly assumed that this tissue response, which occupies a space 20–100 um from the implant, can lead to functional "encapsulation" of electrodes and a reduction in electrode function. Moreover, current electrode arrays also cause a chronic microglia reaction [3, 6, 8] that, together with the astrocytic proliferation, may displace neurons away from the electrodes causing smaller amplitudes from recorded units. It is also conceivable that, for example, cytokines released from microglia can impair or kill the neurons nearby [9–12]. Indeed, the neuronal density in the tissue adjacent (50 um from the border of the electrode array) to the implanted electrode array is typically reported to be reduced by between 20 and 60% as compared to normal brain tissue [3].

2.2.3 On the importance of physiological conditions

The brain and spinal cord are normally floating in cerebrospinal fluid, mechanically protected by the skull or vertebrae and kept in a highly regulated milieu by the blood brain barrier. It is therefore not surprising that normal information processing and learning in the brain and spinal cord is critically dependent on maintained physiological conditions. Indeed, deviations from the normal state, such as changes in exocrine and endocrine hormone levels [13], infections, anaesthesia, sedation or brain trauma [14] can drastically change the ongoing information processing and impair learning mechanisms. Unfortunately, much of our present knowledge on how neuronal networks process, store and recall information are based on conditions that are far from physiological, for example from electrophysiological *in vivo* studies in anaesthetized animals, decerebrated spinal animals, *in vitro* tissue slices that are disconnected from the rest of the nervous system or from tissue slices/cultures. Consequently, we are far from understanding the dynamic information processing in neural networks in physiological conditions such as during normal behavior or sleep, plastic network processing underlying, for instance learning and recall of memory [15], alterations in these systems during development of chronic pain [16, 17] or during neurodegenerative disease such as Alzheimer's or Parkinson's diseases.

Electrodes that can reside in the brain for long periods of time and stably communicate with multiple individual neurons without causing significant disruption of normal physiological conditions, would thus offer an invaluable tool to understand fundamental brain functions and to therapeutically interact with the brain. Moreover, stable long-term connections with neurons would be a considerable advantage in the control of prosthetic devices. For example, the stability of the neuronal recordings significantly affects the training time spent by the patient to learn to control prosthetic devices. To fully exploit the potential of chronically implanted electrodes, it is thus crucial to elucidate how to create electrodes that has a minimal influence on the adjacent neural tissue, while providing stable electrophysiological contact with the local neural population.

2.3 BRAIN COMPUTER INTERFACES (BCI) – STATE-OF-THE-ART

A number of implantable micro-electrode arrays have been developed for use as neural interfaces for BCIs. Perhaps the most well-known are the Utah electrode array [18–20], the "Michigan electrode array" and its later versions [21–23], the tetrode [12], the cone electrode [25–27] and plain wire constructions [28, 29]. These electrode constructions have permitted the recordings of numerous neurons in awake animals. As shown in the 1960s by Fetz and colleagues [30], the brain can learn to use implanted electrodes to control external devices, demonstrating a considerable plastic capacity. These pioneering studies have been followed up by demonstrations that rats and monkeys can learn to use implanted electrodes in the brain to control robotic arms [28, 29] and, perhaps most importantly, that quadriplegic humans can learn to use neural recordings to communicate with computers [27, 31] (Chapter 9). However, as mentioned above, current BCI electrodes cause significant tissue responses, loss of nearby neurons and consequently significant reorganization of the neural circuits being targeted [3, 6]. This confounds an analysis of the information processing in the brain during physiological conditions and during disease processes. Moreover, the quality of recording from current BCIs usually deteriorates over time, although there are examples of very long-lasting recordings in the literature [32]. A related concern is that neuronal recordings are usually not stable over time, that is, they fluctuate considerably or are lost altogether. Hence, it is usually not clear whether the same neurons are studied from day to day. Few studies have therefore attempted recordings of the response patterns in the same neurons over long periods of time precluding detailed studies of changes in neuronal networks.

From a clinical perspective, the focus has been on stimulation of nervous tissue, for example Deep Brain Stimulation (DBS) and Spinal Cord Stimulation (SCS) [33, 34], rather than recording. Currently, relatively large electrodes (in the mm scale) are used, with consequent large spread of current in the tissue, resulting in a lack of specific stimulation effects and often marked side effects that may differ depending on the precise target (Chapter 8). For example DBS of the Subthalamic nucleus can cause unwanted mood changes, hallucinations, speech difficulties and impair certain aspects of cognitive processing [35]. Also, stimulation intensity often needs to be increased over time to reach the desired therapeutic effect leading to an increased occurrence of side effects. It appears likely that the deterioration of electrode function is due to a significant tissue response that over time causes an encapsulation of the implanted electrodes and a reduction in the neural tissue adjacent to the probe. Indeed, a recent metastudy based on post-mortem studies indicate the presence of substantial glial scarring around DBS electrodes [36]. From a practical point of view, the large currents used also require replacement or frequent recharging of batteries. Another problem arising from the large spread of current is that the mechanisms of action become very difficult to elucidate. In fact, there is still no clear consensus as to the mechanisms underlying the DBS effects in Parkinson's disease or the SCS effects on pain [33, 35, 37].

2.4 BIOCOMPATIBILITY OF BCI – ON THE IMPORTANCE OF MECHANICAL COMPLIANCE

Biocompatibility of a neural interface is determined not only by the material it is made of, but is also influenced by its anchorage, size, shape and mechanical properties, such as flexibility and specific weight. Most electrodes used today are rather rigid to be implantable and are anchored to the skull or the skeleton. However, during normal life, the brain and spinal cord exhibit considerable movement relative to the cranium and vertebrae, respectively, due to respiration, arterial pulsation and forces caused by the acceleration and rotation of the body [38–40]. The resulting microforces between anchored rigid electrodes and nearby tissue can thus be expected to irritate and if large enough also injure nearby cells with consequent long-term activation of glia cells such as astrocytes or microglia. Fluctuating microforces between an electrode and tissue can also lead to instable recordings or loss of unit recordings. From studies using sharp and rigid micro-electrodes in acute experiments, it is well known that in order to obtain stable extracellular recordings, the electrode should not be positioned too close to the cell membrane, since local tissue movements caused by blood pressure changes and respiration may easily damage the neuron despite precautions to stabilize the skull or vertebrae. Hence, it can be expected that if the electrode is not stable in the tissue only distant neurons will eventually be recorded from, consequently with much smaller spike amplitudes. An unstable electrode position also precludes possibilities for chronic intracellular recordings.

To address these issues, we have performed several studies to systematically analyze the effects of different features (anchoring, size, density and flexibility) of the probes on neurons and glia cells in long-term studies. Implantable probes made of biocompatible materials differing in one of the mentioned aspects were used. In a 12-week follow-up study, we found significant tissue displacement and alteration in neuronal morphology adjacent to stiff probes anchored in the skull [41]. These were more pronounced in the anterior-posterior axis than in the medio-lateral axis, reflecting that the implanted probes during daily life move relative to the tissue primarily along this axis [38–40]. Conspicuously, neurons around the probes were abnormally flat, most likely due to the continuous movements of the probes (Figure 2.2). Such neurons are not likely to function normally, albeit apparently staying alive. Moreover, the tissue responses did not decline between the 6th and 12th week post-implantation. By contrast, electrodes that were not tethered to the skull, and thus could move with the brain, showed much less tissue response and no obvious derangement of

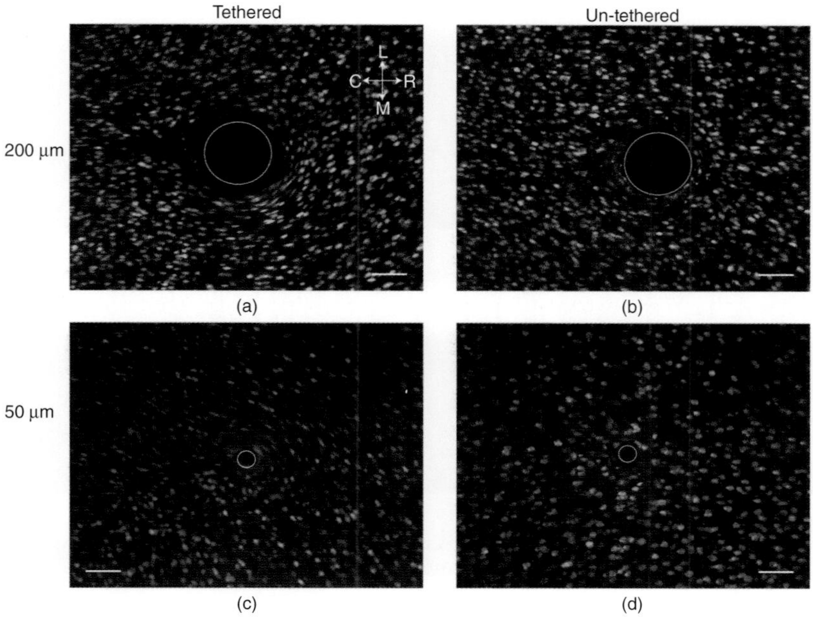

Figure 2.2 Morphological alterations in neural morphology after 12 weeks implantation in the rat cortex cerebri depends on probe size and anchoring. Tethered (left) and untethered to the skull (right) with two different probe sizes (top 200 μm, bottom 50 μm). Neurons were stained for NeuN (from Thelin *et al.* 2010 [41] with permission). Note the flat shape of neurons adjacent to the tethered probes.

neuron morphology nearby. The tissue responses were also more intense for larger electrodes. Recently, we investigated the impact of density (specific weight) by using probes with the same shape, dimension and surface material but differing in density in a 6-week study [42, 43]. These probes were not tethered to the skull. Despite using a rather large size, probes with a density close to that of the tissue was found to produce much less astrocytic reaction and microglia reaction than probes with more normal electrode densities. In fact, there was sometimes no discernable astrocytic reaction at all around the density matched probes (Figure 2.3). Moreover, neuronal density adjacent to the density matched probes tended to be less reduced. These experiments demonstrated that probe density should be tissue matched and provided evidence that microforces between tissue and implanted probe due to differences in inertia is a key trigger for long-term glial activation. To investigate the importance of flexibility, we constructed flat probes that were 8000 times more flexible in one dimension than in the other dimensions and took advantage of the fact that the rat brain movements are predominantly in the anterio-posterior axis during daily life [44]. By using this approach, we avoided confounding the analysis by differences in materials, size, weight and shapes. The result demonstrated that the astrocytic proliferation, but not microglial reaction, was reduced when the probes were oriented with

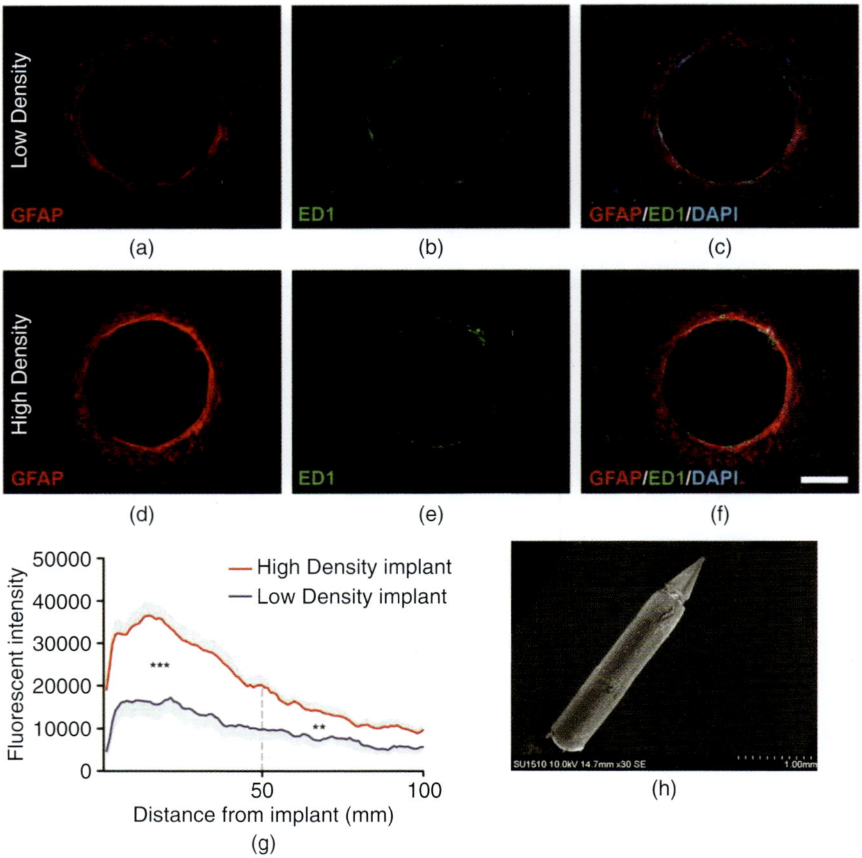

Figure 2.3 Comparison of glial reaction to rigid probes differing in density. The probes were not tethered to the cranium during the implantation time. The probes were explanted 6 weeks after implantation in rat cortex. (a–c) tissue reaction to low density probe, and (d–f) tissue reaction to high density probe. Scale bar: 200 um. Astrocytes were stained for GFAP and Microglia for ED1, cell nuclei was stained for DAPI. (g) GFAP staining plotted against distance from border of implant (in um). (h) Shape of probe used. The probes were 500 um in diameter and 3.5 mm long. Both low (tissue matched) and high density probes (density similar to platinum electrodes) had the same shape and were coated with Parylene C. Note the conspicuous difference in astrocyte proliferation between the two probes. While there was no difference in ED1 in the tissue sections, an examination of microglia attached to the probes after explantation showed a significantly larger number of microglia attached to the high density probes (not shown) (from Lind *et al*. 2013 [42] with permission).

the flexible axis in the anterior-posterior axis, thus for the first time providing direct evidence that flexibility is an important aspect at least for the astrocytic reaction.

A complexity with the microforces occurring in the tissue is that they are not always uniform. For example, blood pressure changes will cause a propagating wave of tissue movements. The tissue adjacent to an electrode, that is rigid in its longitudinal axis, will thus move relative to the electrode if it is located close to a large vessel. Hence, electrodes should ideally be flexible/stretchable also along their main axis. To avoid slow drift in location of the electrodes in the tissue, caused by for example, the action of the heart and respiration, electrodes presumably also need to be anchored in or near the target tissue. However, when used in the clinic, this anchoring must not cause extensive damage if the electrode array has to be explanted.

In conclusion, to achieve stable electrode positions in soft tissue, electrode arrays need to be as mechanically compliant with the tissue as possible, truly floating and appropriately anchored. Our electrode development has been focused on finding solutions to these requirements.

2.5 NOVEL ELECTRODE CONSTRUCTS AND IMPLANTATION PROCEDURES

A major problem that has prevented the development of ultra-thin and highly flexible electrodes in the past is how to implant them with precision into soft tissue. Traditionally, relatively thin and straight electrodes (e.g. tungsten wires with a diameter 35–50 um) have been guided by a stiff cannula. However, the cannula itself must be larger than the electrodes and upon implantation therefore causes more tissue destruction than the electrode itself. Moreover, when withdrawing the guiding cannula, the thin electrodes may easily move in the tissue due to adhesion to the guiding cannula, resulting in a lack of control of the recording/stimulation sites. Such an arrangement also complicates the procedure of connecting multiple electrodes outside the skull.

2.5.1 Methods to implant ultraflexible electrodes

To solve the problem of how to implant ultra-flexible electrodes, we developed a technique using bundles of straight electrode leads held together with a non-dissolvable glue. Collectively, this assembly provided the necessary stiffness for insertion into the brain. Each lead is equipped with an ultra-thin electrode protruding backwards, like a barb [45]. This simple construct, termed the "umbrella electrode", permitted for the first time very thin electrodes to be dragged into the tissue (as opposed to being pushed into the tissue, which easily results in deviations from the intended track). By mounting the electrode protrusions such that they are directed in all dimensions and at different positions along the electrode bundle, a three-dimensional electrode bundle can also be achieved. In addition, by pulling the electrode bundle slightly backwards after it has been positioned close to the target tissue, the ultra-thin protrusions can be made to unfold inside the target tissue.

The protrusions themselves function as active electrodes (at the uninsulated tips) and will also have an anchoring effect on the electrode in the tissue, thereby stabilizing the electrode tips with respect to target neurons. A related type of electrode array with protrusions containing the active sites, termed "μ-foil" was developed in parallel (46,

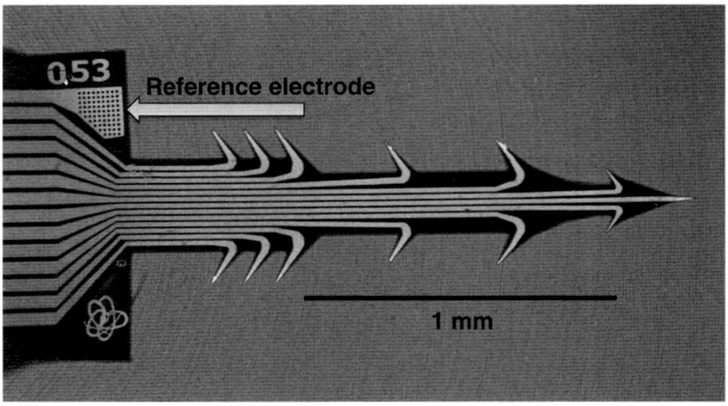

Figure 2.4 A scanning electron microscopy (SEM) image of a novel polymer-based electrode array termed "u-foil electrode array" equipped with recording sites on protrusions [46]. The backbone of the electrode array is made of a thin film (<10 um) of a photo-structurable polyimide (Durimide 7505, FujiFilm Belgium) that serves to insulate the array of thin (<150 nm) gold leads. The electrode array is equipped with protrusions on which the individual recording sites are located at respective tips. The protrusions are directed backwards and serve the dual role to provide recording sites in less compromised tissue (i.e. at a distance from the major bulk of the array) and to anchor the electrode array in the tissue.

Figure 2.4). This latter electrode construct is manufactured using photolithographical methods and can therefore potentially be produced in large quantities. The current μ-foil electrode is very thin (<10 um in thickness), but flat. Its construction is primarily flexible in one dimension, although the protrusions, if made of flexible material, may flex in two dimensions. Notably, the length of the protrusions of the umbrella electrode and the μ-foil electrode can be made such that their tips will be outside the major tissue response caused by the shank of the electrode, hence connecting with neurons in less compromised tissue. By stacking μ-foil electrode arrays, the number of protruding electrodes can be increased manifold. Moreover, it is relatively easy to tailor the dimensions of the protrusion and their spacing with great precision to fit the cytoarchitecture of the target tissue. This construction can also be used as a convenient platform for testing of different surface modifications. Initial tests of this construction show promising results as regards signal-to-noise level of recorded units, but it remains to be tested in long-term studies.

2.5.2 Surface configurations

The surface configuration of the electrodes is another parameter of interest. Johansson *et al.* demonstrated [47] that probes with a porous surface implanted in the sciatic nerve resulted in less tissue response than implants with a smooth surface. It is likely that a porous surface provides improved anchoring in the tissue, thereby reducing the tissue response to micro-motion. In this context, it may be worth noting, however, that the specific weight of the implant has not been adequately considered before [42, 43]. Hence, if the electrode arrays are truly floating in the tissue, there will be less need for anchoring. Consequently, the previous conclusion that rough surfaces provide less tissue responses than smooth surfaces may have been confounded by effects of micro-forces caused by differences in inertia between implant and tissue. However, *in vitro* studies, in which the substrate is not subject to movement, indicate that neurons tend to avoid smooth surfaces and adhere to and grow along rough or nanostructured surfaces [48]. The mechanisms of interactions between neurons and surface structures (also including the insulated lead) are still poorly understood and clearly warrant further studies.

2.5.3 Matrix embedded electrodes

While the umbrella electrode is equipped with very thin protrusions, its shank is relatively rigid to permit implantation. To circumvent the tissue problems inherent to rigid implants, we developed a new technology where the individual ultra-thin conductive leads in the shank are embedded in a rigid dissolvable/degradable matrix that provides the necessary stiffness during implantation [49]. Collagen-based gelatin was our first choice as embedding material, as it is a well-known biocompatible and hemostatic material. In the first study, we found that it is possible to insert multiple ultra-thin (7 and 12 um in diameter) electrodes through a small opening in the arachnoidea of the cerebral cortex, and spread out the electrodes in the deep parts of the cortex (50, Figure 2.5). The same study also indicated that gelatin reduces tissue responses following implantation. In fact, implanted pure gelatin shaped as needles could not be

Figure 2.5 Gelatine-embedded electrode bundle inserted through a narrow opening and spread out in rat cortex to target a larger cortical volume. Each platinum lead has a diameter of 12 um and a 4 um thick layer of Parylene C insulation. A thick cortical section containing the entire implanted electrode bundle was clarified with methyl salicylate. In this process, lipids are extracted so that the brain tissue becomes transparent (from Lind *et al.* 2010 [50] with permission).

traced 12 weeks post-implantation, whereas a stab wound caused by a metal needle with the same dimensions and insertion depth, still produced a clear tissue response at that time point. Likewise, gelatin embedded wire bundles tended to show less tissue response than wire bundles coated with a thin layer of gelatin. In a recent study, we confirmed that gelatin reduces the microglia reaction (but not the astrocytic reaction) by comparing the tissue reaction to gelatin coated flat probes with uncoated probes [44]. Importantly, there was minimal loss of neurons around the gelatin coated probe, while the neuron density was significantly reduced around the uncoated probe [44]. To our knowledge, this is the first time normal neuronal densities have been found adjacent to implanted probes in the brain. The mechanisms underlying the beneficial effects of gelatin remain to be elucidated, but it is likely that its hemostatic properties play a significant role. Nevertheless, this finding is very encouraging, since it indicates that the goal of accomplishing physiological conditions around the electrode array is indeed attainable.

When implanting bundles of electrodes in superficially located tissue, one support matrix may be sufficient, but for implantation in deep tissue it can be advantageous to use materials that dissolve with different time courses, such that the support matrix coating the proximal parts of the bundle, so that the part nearest to the micromanipulator remains supportive for a longer time. After the distal part of the matrix is dissolved it is possible to spread out the electrodes in the target tissue by pushing the electrodes further into the tissue. This simple principle was recently confirmed using gelatin-embedded thin electrode bundles (Figure 2.6, platinum wires, diameter 20 um including insulation) implanted into the area of the sub-thalamic nucleus [51]. In this study, a three-step process was used in which the bundle was first introduced to a pre-target area above the target, then after a pause of 10 min the bundle

Figure 2.6 Gelatin-embedded bundle of electrodes for deep brain recording/stimulation. Platinum wires, 12 um in diameter and insulated with Parylene C. The distal part of the gelatin-embedded probe is shown enlarged to the right. Bar indicates 500 um.

was then further advanced and the thin electrodes spread out in the area of the subthalamic nucleus. Neuronal recordings over a period of 8 weeks confirmed that the position of the electrode tips was relatively stable, with only marginal changes in the neural signal-to-noise over time. Some drift in unit amplitudes was observed over time in some channels, indicating the need for improved anchoring in or close to the target tissue. It should be noted that both the umbrella and gelatin-embedded electrode arrays can be inserted through relatively small holes, thus limiting tissue injuries along the insertion track, while at the same time targeting a larger tissue volume using the unfolding mechanism. From a clinical perspective, selection of electrodes producing desirable results from such a cluster of electrodes and discarding those that do not, may offer substantially improved stimulation specificity as compared to conventional techniques for DBS. While the individual thin electrode cannot safely eject as high an amount of current as large electrodes, this problem can be circumvented by using many electrodes and by expanding their respective active electrode area, for example by de-insulating a larger part of the electrode leads or by adding multiple protrusions.

The embedding technique also offers the potential for the development of electrodes that are highly flexible and stretchable [52, 53]. Recently, techniques for producing ultra-thin wavy, and therefore stretchable, gold leads equipped with protrusions were developed in our laboratory (Figure 2.7). The gold leads were cut out from 4 um thin gold leaf using laser milling, embedded in a gelatin-based matrix and implanted in the rat cortex [54]. The electrodes were insulated (4 um insulation) with Parylene C, except for at their active sites. This array of individually three-dimensional flexible electrodes embedded in a hard matrix of gelatin can be introduced into the tissue without substantial deformation during the implantation procedure. In initial experiments, lasting 3 weeks, we have confirmed that high-quality (>150 uV amplitudes) cortical neuronal recordings can be obtained with this type of electrode array. Electrodes that are flexible in all three dimensions may be particularly useful in tissues, exhibiting large movements during daily life such as the spinal cord, peripheral nerves or tissue that is close to major blood vessels.

2.5.4 Electrode arrays encorporating drugs

The dissolvable matrix used to embed the ultra-thin electrodes can also serve as a vehicle for drugs, viral vectors, etc. to be implanted together with the electrodes [55]. This has the potential to improve implantation safety, but also offers a novel method to analyze the effects of local drug release. For example, antibiotics may be included at strategic points along the implant to reduce the risk of infection after the implantation. Another strategy may be to incorporate anti-inflammatory drugs. Successful attempts to reduce acute inflammation in this way have been reported [56, 57]. However, it should be kept in mind that, as mentioned above, inflammatory responses may be beneficial. This may be particularly true if the inflammation is due to a maintained irritation caused by continuously changing microforces between electrodes and tissue. In this case, a tissue response resulting in encapsulation of the implant

Figure 2.7 Electrode array composed of ultra-thin electrodes that are individually flexible in three dimensions. Left, electrode array shown next to a British penny for comparison of size. Right, electrode array embedded and configurationally locked in hard gelatin for implantation.

may be necessary to protect the brain tissue (see above). For this reason, there is a need to evaluate the long-term consequences of anti-inflammatory drug strategies using probes that are as mechanically compliant with the tissue as possible.

2.6 CONCLUDING REMARKS

In conclusion, our development and research focusing on how to minimize the foreign body reactions and loss of neurons around the implant has resulted in identification of several key features that electrodes need to fulfill and the development of novel ultra-thin/elastic and flexible constructs that fulfil these requirements. By using hard but dissolvable matrix materials, arrays of ultra-flexible electrodes can be implanted. Importantly, gelatin-based embedding materials also appear to protect the neurons close to the BCI. Hence, a combination of mechanically compliant neural interfaces and neuroprotective matrix materials that also serve to preserve the electrode array configuration during implantation are likely to provide new opportunities for long-term high resolution studies under *physiological conditions* and to distinguish *pathological* alterations in network functions that occur during, for instance, neurodegenerative diseases. It should be kept in mind, however, that there is still a need for a careful, and therefore time-consuming, evaluation of the long-term impact of different types of electrode architectures, surface structures, embedding media and local administration of drugs, on adjacent glia and neurons and other tissue constituents such as blood vessels and the blood brain barrier. From a clinical perspective, the prospect of improving stimulation specificity and thus reduction of side effects, by selection from a surplus of ultra-thin biocompatible electrodes spread out in the target tissue is of considerable interest, but needs to be evaluated with respect to safety and functionality in long-term studies.

ACKNOWLEDGEMENTS

This work was supported by grants from the Knut and Alice Wallenbergs Foundation, the Swedish Research Council (project grant 01013 and Linneus grant) and the Medical faculty of Lund University. The author thanks Professor Martin Garwicz for constructive criticism on the manuscript and Dr Jonas Thelin and Dr Cecilia E. Linsmeier for help with the illustrations.

Declaration of conflict of interest: The author is a minority shareholder in Neuronano Inc. Sweden, who owns the patents cited in the reference list.

REFERENCES

1. Welkenhuysen, M., Andrei, A., Ameye, L., Eberle, W. and Nuttin, B. (2011) Effect of insertion speed on tissue response and insertion mechanics of a chronically implanted silicon-based neural probe. *IEEE Trans. Biomed. Eng.*, **58**(11): 3250–3259.
2. Winslow, B.D. and Tresco, P.A. (2010) Quantitative analysis of the tissue response to chronically implanted microwire electrodes in rat cortex. *Biomaterials*, **31**(7): 1558–1567.
3. Biran, R., Martin, D.C. and Tresco, P.A. (2005) Neuronal cell loss accompanies the brain tissue response to chronically implanted silicon microelectrode arrays. *Exp. Neurol.*, **195**(1): 115–126.
4. Skousen, J.L., Merriam, M.E. Sr, Srivannavit, O., Perlin, G., Wise, K.D. and Tresco, P.A. (2011) Reducing surface area while maintaining implant penetrating profile lowers the brain foreign body response to chronically implanted planar silicon microelectrode arrays. *Progr. Brain Res.*, **194**: 167–180.
5. Purcell, E.K., Thompson, D.E., Ludwig, K.A. and Kipke, D.R. (2009). Flavopiridol reduces the impedance of neural prostheses *in vivo* without affecting recording quality. *J. Neurosci. Meth.*, **183**(2): 149–157.
6. Polikov, V.S., Tresco, P.A. and Reichert, W.M. (2005) Response of brain tissue to chronically implanted neural electrodes. *J. Neurosci. Meth.*, **148**(1): 1–18.
7. Hanisch, U.K. (2002) Microglia as a source and target of cytokines. *Glia*, **40**: 140–155.
8. Mathewson, A.J. and Berry, M. (1985) Observations on the astrocyte response to a cerebral stab wound in adult rats. *Brain Res.*, **327**(1–2): 61–9.
9. Szarowski, D.H., Andersen, M.D., Retterer, S., Spence, A.J., Isaacson, M. et al. (2003) Brain responses to micro-machine silicon devices. *Brain Res.*, **9839**(1–2): 23–25.
10. Streit, W., Mrak, R. and Griffin, W.S. (2004) Microglia and neuroinflammation: A pathological perspective. *J. Neuroinflammation*, **1**: 14.
11. Verkhratsky, A. and Butt, A. (2007) *General Pathophysiology of Glia, Glial Neurobiology*. John Wiley & Sons Ltd., Chichester UK, pp 153–165.
12. Skousen, J.L., Bridge, M.J. and Tresco, P.A. (2015) A strategy to passively reduce neuroinflammation surrounding devices implanted chronically in brain tissue by manipulating device surface permeability. *Biomaterials*, **36**: 33–43.
13. Haas, B.W. and Canli, T. (2008) Emotional memory function, personality structure and psychopathology: A neural system approach to the identification of vulnerability markers. *Brain Res. Rev.*, **58**: 71–84.

14. Binder, L. (1986) Persistent symptoms after mild head injury: A review of the post-concussive syndrome. *J. Clin. Exp. Neuropsychol.*, **8**: 323–346.
15. Schouenborg, J. (2008) Action-based sensory encoding in spinal sensorimotor circuits. *Brain Res. Rev.*, **57**(1): 111–117.
16. Dolgin, E. (2010) Animalgesic effects. *Nature Med.*, **16**(11): 1237–1241.
17. Woolf, C.J. (2010) Overcoming obstacles to developing new analgesics. *Nature Med.*, **16**(11): 1241–1247.
18. Campbell, P.K., Jones, K.E. Huber, R.J., Horch, K.W. and Normann, R.A. (1991) A silicon based, three-dimensional neural interface: manufacturing processes for an intracortical electrode array. *IEEE Trans. Biomed. Eng.*, **38**: 758.
19. Schwartz, A.B. (2004) Cortical neural prosthetics. *Ann. Rev. Neurosci.*, **27**: 487–507.
20. Parker, R.A., Davis, T.S., House, P.A., Normann, R.A. and Greger, B. (2011) The functional consequences of chronic, physiologically effective intracortical micro-stimulation. *Progr. Brain Res.*, **194**: 145–165.
21. Drake, K.L., Wise, K.D., Farraye, J., Anderson, D.J. and BeMent, S.L. (1988) Performance of planar multisite microprobes in recording extracellular single-unit intracortical activity. *IEEE Trans. Biomed. Eng.*, **35**: 719–732.
22. Williams, J.C., Rennaker, R.L. and Kipke, D.R. (1999) Long-term neural recording characteristics of wire microelectrode arrays implanted in cerebral cortex. *Brain Res. Protocols*, **4**: 303.
23. Kipke, D.R., Vetter, R.J., Williams, J.C. and Hetke, J.F. (2003) Silicon-substrate intracortical microelectrode arrays for long-term recording of neuronal spike activity. *Cerebral Cortex. IEEE Trans. Neural Syst. Rehab. Eng.*, **11**: 151.
24. Jog, M.S., Connolly, C.I., Kubota, Y., Iyengar, D.R. et al. (2002) Tetrode technology: advances in implantable hardware, neuroimaging, and data analysis techniques. *J. Neurosci. Meth.*, **117**: 141–152.
25. Kennedy, P.R. (1989) The cone electrode: a long-term electrode that records from neurites grown onto its recording surface. *J. Neurosci. Meth.*, **29**: 181–193.
26. Kennedy, P.R., Mirra, S.S. and Bakay, R.A.E. (1992) The cone electrode: ultrastructural studies following long-term recording in rat and monkey cortex. *Neurosci. Lett.*, **142**: 89–94.
27. Kennedy, P.R., Bakay, R.A.E., Moore, M.M., Adams, K. et al. (2000) Direct control of a computer from the human central nervous system. *Rehab. Eng. IEEE Trans.*, **8**: 198–202.
28. Chapin, J.K., Moxon, K.A., Markowitz, R.S. and Nicolelis, M.A. (1999) Real-time control of a robot arm using simultaneously recorded neurons in the motor cortex. *Nature Neurosci.*, **2**(7): 664–670.
29. Carmena, J.M., Lebedev, M.A., Crist, R.E., O'Doherty, J.E., Santucci, D.M. et al. (2003) Learning to control a brain-machine interface for reaching and grasping by primates. *PLoSBiol.*, **1**(2): 193–208.
30. Fetz, E.E. (1969) Operant conditioning of cortical unit activity. *Science*, **163**: 955–958.
31. Hatsopoulos, N.G., Mukand, J.A., Polykoff, G., Friehs, G.M. and Donoghue, J.P. (2005) Cortically controlled brain-machine interface. *IEEE Eng. Med. and Biol., 27th Ann. Conf.*, pp 7660–7663.
32. Tresco, P.A. and Gerhardt, G.A. (2008) The biotic-abiotic interface, in *Brain Computer Interfaces*, Berger, T.W. et al. (eds), Springer Science + Business Media BV, pp 31–45.

33. Benabid, A.L., Torres, N., Moro, C., Aksenova, T., Costecalde, T. et al. (2011) Deep brain stimulation: BMI at large, where are we going to? *Prog. Brain. Res.*, **194**: 71–82.

34. Tierney, T.S., Sankar, T. and Lozano, A.M. (2011) Deep brain stimulation: emerging indications. *Progr. Brain Res.*, **194**: 83–95.

35. DiLorenzo, D.J., Jankovic, J., Simpson, R.K., Takei, H. and Powell, S.Z. (2014) Neurohistopathological findings at the electrode–tissue interface in long-term deep brain stimulation: systematic literature review, case report, and assessment of stimulation threshold safety. *Neuromodulation*, **17**: 405–418.

36. Perlmutter, J.S. and Mink, J.W. (2006) Deep brain stimulation. *Ann. Rev. Neurosci.*, **26**: 229–257.

37. Zhang, T.C., Janik, J.J. and Grill, W.M. (2014) Mechanisms and models of spinal cord stimulation for the treatment of neuropathic pain. *Brain Res.*, **1569**: 19–31.

38. Thiagarajah, J.R., Papadopoulos, M.C. and Verkman, A.S. (2005) Noninvasive early detection of brain edema in mice by near-infrared light scattering. *J. Neurosci. Res.*, **80**(2): 293–299.

39. Poncelet, B.P., Wedeen, V.J., Weisskoff, R.M. and Cohen, M.S. (1992) Brain parenchyma motion: measurement with cine echo-planar MR imaging. *Radiology*, **185**(3): 645–51.

40. Chen, T., Xue, Z., Wang, C., Qu, Z., Wong, K.K. and Wong, S.T. (2010) Motion artifact correction of multi-photon imaging of awake mice models using speed embedded HMM. *Med. Image Comput. Assist. Interv.*, **13**(3): 473–480.

41. Thelin, J., Jörntell, H., Psouni, E., Garwicz, M., Schouenborg, J. et al. (2011) Implant size and fixation mode strongly influence tissue reactions in the CNS. *PLoS ONE*, **6**(1): e16267.

42. Lind, G., Eriksson Linsmeier, C. and Schouenborg, J. (2013) The density difference between tissue and neural probes is a key factor for glial scarring. *Sci. Rep.*, **3**: Article number: 2942DOI:doi:10.1038/srep02942.

43. Schouenborg, J. (2014) Displacement resistant microelectrode, microelectrode bundle and microelectrode array. US patent 8,751,014 B2.

44. Köhler, P., Wolff, A., Ejserholm, F., Wallman, L., Schouenborg, J. and Linsmeier, C.E. (2015) Influence of probe flexibility and gelatin embedding on neuronal density and glial responses to brain implants. *PLoS One*, **10**(3): e0119340.

45. Schouenborg, J. (2013) Electrode bundle. US patent US8457762 (B2), June.

46. Ejserholm, F., Köhler, P., Granmo, M., Schouenborg, J., Bengtsson, M. and Wallman, L. (2014) μ-Foil polymer electrode array for intracortical neural recordings. *J. Transl. Eng. Health Med.*, June 10, **2**: Digital Object Identifier 10.1109/JTEHM.2014.2326859.

47. Johansson, F., Wallman, L., Danielsen, N., Schouenborg, J. and Kanje, M. (2009) Porous silicon as a potential electrode material in a nerve repair setting: tissue reactions. *Acta Biomateria*, **5**(6): 2230–2237.

48. Kanje, M. and Johansson, P.F. (2011) Nanomodified surfaces and neurite outgrowth. *Progr. Brain Res.*, **194**: 253–262.

49. Schouenborg, J. (2012) Dissociating multi channel electrode. US patent US8306632 B2), November.

50. Lind, G., Linsmeier, C.E., Thelin, J. and Schouenborg, J. (2010) Gelatine-embedded electrodes – a novel biocompatible vehicle allowing implantation of highly flexible microelectrodes. *J. Neural. Eng.*, **7**(4): 046005.

51. Mohammed, M., Etemadi, L., Ekstrand, J., Friberg, A., Granmo, M. et al. (2013) *Matrix embedded ultrathin microelectrodes for multisite deep brain stimulation and recording*. Abstract presented at INS Conference, Berlin.
52. Witteveen, J.A., Suyatin, D.B., Gällentoft, L., Schouenborg, J., Danielsen, N. and Prinz, C.N. (2010) Gelatin/glycerol coating to preserve mechanically compliant nanowire electrodes from damage during brain implantation. *J. Vac. Sci. Technol.*, **28**(6): C6K13–C6K16.
53. Schouenborg, J. (2013) Medical electrode, electrode bundle and electrode bundle array. US patent US 8,386,006 B2, February.
54. Agorelius, J., Tsanakalis, F., Friberg, A., Thorbergsson, P.T., Pettersson, L.M.E. and Schouenborg, J. (2015) An array of highly flexible electrodes with a tailored configuration locked by gelatin during implantation-initial evaluation in cortex cerebri of awake rats. *Front Neurosci.*, **9**: 331.
55. Ek, F., Danielsen, N., Linsmeier, C.E., Petersson, P. and Schouenborg, J. (2015) Microelectrode and multiple microelectrodes comprising means for releasing drugs into the tissue. US patent US 8,954,142 B2, February.
56. Zhong, Y. and Bellamkonda, R.V. (2007) Dexamethasone-coated neural probes elicit attenuated inflammatory response and neuronal loss compared to uncoated neural probes. *Brain Res.*, **1148**: 15–27.
57. Mercanzini, A., Reddy, S., Velluto, D., Colin, P., Maillard, A. et al. (2011) Controlled released from nanoparticle embedded coatings reduce the tissue reaction to neuroprostheses. *J. Controlled Release*, **145**: 196–202.

3

ELECTROCHEMICAL PRINCIPLES OF SAFE CHARGE INJECTION

STUART F. COGAN

Department of Bioengineering, University of Texas at Dallas, Richardson, TX, USA

DAVID J. GARRETT

Department of Physics, The University of Melbourne, Parkville and The Bionics Institute, East Melbourne, Victoria, Australia

RYLIE A. GREEN

Graduate School of Biomedical Engineering, UNSW, Sydney, NSW, Australia

3.1 INTRODUCTION

The criteria for electrochemically safe charge injection are that the reactions mediating transfer of charge across the electrode-tissue interface:

1. do not induce corrosion or degradation of the electrode sufficient to compromise its electrical and mechanical function
2. do not produce chemical species that are toxic to cells; and
3. that the electric field in the tissue adjacent to the electrode does not induce electroporation of cells that leads to damage of excitable tissue.

Our approach to items 1 and 2 is to require that the electrode potential during stimulation pulsing, as well as in the period between pulses, is not polarized to a level sufficient to induce electrode or tissue damage by electrochemical processes. For

Neurobionics: The Biomedical Engineering of Neural Prostheses, First Edition.
Edited by Robert K. Shepherd.

implantable devices that may need to work continuously for many years over many millions of pulse cycles, even very small amounts of damage will eventually be problematic. This does not mean that all charge transfer reactions must be reversible, rather that irreversible reactions that do occur must not induce unacceptable levels of electrode or tissue damage. Similarly, if the current density at an electrode induces irreversible electroporation of nearby cells, any cell damage and consequent cellular response must be sufficiently contained that the therapeutic or functional response to chronic stimulation is preserved.

3.2 CHARGE INJECTION REQUIREMENTS

Charge injection for neural stimulation is usually defined by the charge delivered in the leading phase of a stimulation pulse in units of microcoulombs/phase and the charge per unit geometric surface area (GSA) of the stimulating electrode in units of microcoulombs/square centimeter. Proper selection of other stimulation parameters, such as the pulse frequency, pulse width and the duty cycle, is also important for obtaining the desired functional response and ensuring that the stimulation is delivered without electrode corrosion or tissue damage.

3.2.1 Stimulation levels for functional responses

The range of stimulation intensities employed in clinical neural stimulation devices is shown in Figure 3.1 on a log-log plot of charge density versus charge per phase [1]. The data shown include deep brain stimulation (DBS), vagus nerve stimulation (VNS), cochlear nucleus stimulation, and retinal stimulation as well as other more investigative therapies. Results from animal studies [2–10] reporting tissue damaging and non-damaging levels of stimulation are included. From the perspective of charge injection reactions at the electrode-tissue interface, these are all planar macroelectrodes. That is, transport of reactants and counter ions to and from the electrode surface is governed by linear rather than hemispherical diffusion or migration. For microelectrodes, somewhat arbitrarily having a GSA $<10\ 000\ m^2$, threshold stimulation intensities for neural activation depend on the charge/phase. Excitable tissue more distant from the electrode than a few times the largest exposed electrode dimension is in effect stimulated from a point source and the shape of the electrode and charge density are less important. For this reason, stimulation thresholds for eliciting a physiological response are usually given as a charge/phase. A compendium of representative stimulation intensities, for which there is histological evidence of damage or absence of tissue damage, is provided in Table 3.1.

3.2.2 Tissue damage thresholds

For stimulation with macroelectrodes, tissue damage thresholds are often described by the Shannon equation [3.1], which defines the boundary between tissue damaging

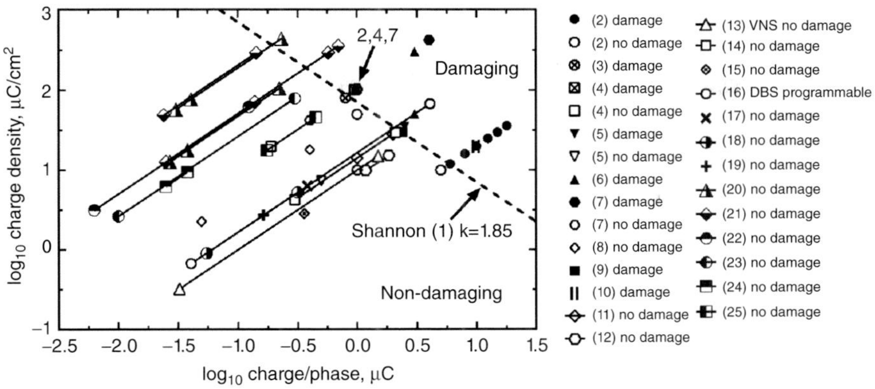

Figure 3.1 Damaging and non-damaging levels of stimulation are shown as a function of pulse charge density and charge/phase for platinum electrodes. The boundary between damaging and non-damaging stimulus intensity is delineated by the Shannon line [1] with k = 1.85. These data include acute animal studies of tissue damage thresholds and studies in humans for which no tissue damage was reported. The absence of reported damage does not exclude damage and the occurrence or absence of tissue damage can depend on a variety of factors including pulse frequency, duty cycle and the type of tissue being stimulated. ([1–25]).

Table 3.1 Representative charge/phase (μC/ph) and charge density (μC/cm²) values for damaging and non-damaging levels of stimulation with planar platinum electrodes.

damaging			non-damaging		
μC/ph	μC/cm²	ref.	μC/ph	μC/cm²	ref.
0.8	80	[3]	0.01	2.6	[23]
1	100	[3]	0.05	2.3	[8]
1	100	[2]	0.19	20	[4]
6	12	[2]	0.3	79	[23]
8	16	[2]	0.4	18	[8]
10	20	[2]	0.4	42	[7]
12	25	[2]	1	10	[2]
15	30	[2]	1	50	[2]
18	36	[2]	5	10	[2]

and non-damaging levels of electrical stimulation reported by McCreery *et al.* [2] on a log charge density (D) versus log charge/phase (Q) plot as a line with the equation:

$$\log(D) = k - \log(Q) \tag{3.1}$$

where k is an adjustable parameter typically chosen between 1.5 and 2.0. While Equation [3.1] was derived from a limited dataset, other reports of tissue damage with similar macroelectrodes fit the Shannon equation quite well. It should be appreciated, however, that the Shannon equation does not account for current density, pulse frequency or duty cycle and is based entirely on data obtained with planar surface electrodes used to stimulate cortical neurons. These studies were also limited in duration and their applicability to chronic stimulation has to be inferred. From the Shannon equation there has emerged a threshold charge density limit of 30 $\mu C/cm^2$ for tissue damage. As discussed by Kuncel and Grill [26], such a limit emerges by extending the charge density/charge per phase line for a typical macroelectrode GSA (0.06 cm^2 in the case of Medtronic DBS electrodes, in their example) to the Shannon line for $k \approx 1.75$. However, damaging stimulation protocols with charge densities as low as 12 $\mu C/cm^2$ have been reported [2] and this emphasizes the importance of both charge density and charge per phase when stimulating with large-area electrodes. As should become evident, it is possible to stimulate below 30 $\mu C/cm^2$ and damage tissue and also possible to stimulate above 40 $\mu C/cm^2$ and not damage tissue [7].

Stimulation electrodes must be capable of safely delivering charge to tissue at levels that are above the functional thresholds. The charge transfer mechanisms available for delivering this supra-threshold charge and the electrochemical limits to achieving reversible charge injection and avoiding reactions that degrade the electrode or damage tissue are discussed in the following section.

3.2.3 Charge injection processes

From the perspective of neural stimulation, reversible charge injection can be defined as that which avoids generating noxious products at the electrode-tissue interface that either diffuse away from the interface or cause degradation of the electrode. If reaction products are consumed at the interface by the use of charge-balanced waveforms or react chemically to form non-harmful species, then the generation of these products, which is strictly irreversible, may be acceptable. An example of the latter would be the generation of OH^- or H^+ ions, which are typically consumed by the physiological buffer before their accumulation causes a damaging pH change. There are two categories of charge transfer at the electrode tissue interface: capacitive charge transfer by double-layer charging, and Faradaic charge transfer in which species are oxidized or reduced.

3.2.4 Capacitive charge injection

The double-layer at the electrode-electrolyte interface has been described in detail in various texts [27]. In a simplified representation, the double-layer comprises an inner layer of absorbed water molecules and specifically absorbed ions, and an outer layer of hydrated ions which form a diffuse layer. The total thickness of the double layer in physiological saline (~0.1 M NaCl) is less than 10 nm. Besides the electrolyte composition, the capacitance of the double layer (C_{dl}) depends on the electrode material and potential, although the potential dependence is difficult to model theoretically.

The double-layer can be modeled as a parallel-plate capacitor:

$$C_{dl} = \frac{\varepsilon_0 \varepsilon_r ESA}{t} \tag{3.2}$$

where ε_o and ε_r are the permittivity of free space and the dielectric constant of the absorbed water layer (inner layer), respectively, *ESA* is the electrochemical surface area of the electrode, and *t* the thickness of the inner layer ($\varepsilon_o = 8.85 \times 10^{-12}$ $C^2/N\text{-}m^2$ in S.I. units). The dielectric constant of the primary layer of absorbed water at the surface of a metal is about 6. This is much lower than the dielectric constant of liquid water due to the highly aligned dipole orientation of absorbed water. However, there is a significant contribution to the capacitance from the less aligned water molecules in the outer layer, for which the estimated average dielectric constant is about 40. The double-layer capacitance is the series sum of these two capacitances. Equation [3.2] asserts that C_{dl} is independent of the potential applied to the electrode. This is not observed experimentally and reflects the far greater complexity of the interface. However, Equation [3.2] does predict that a larger C_{dl} can be obtained by increasing the ESA of an electrode and by increasing the dielectric constant at the interface. Both approaches are effective in increasing the available capacitance of an electrode for charge-injection. A striking example of increasing charge capacity by increasing ESA is that of porous titanium nitride, sometimes called fractal TiN [28].

Incorporating a thin dielectric film at the interface has the potential to increase the charge that can be stored across the electrode-electrolyte interface by increasing the dielectric constant and by increasing the breakdown voltage at the interface. The most common example of this approach is the Ta_2O_5/Ta capacitor electrode in which the Ta_2O_5 is formed on a tantalum substrate by anodization [29]. Anodically formed Ta_2O_5 has a dielectric constant of about 25, somewhat less than the hydrated metal-water interface. To maximize the available charge storage, it is necessary to create a porous electrode having a high ratio of electrochemical to geometric surface area and to operate the electrode at a positive bias that is a significant fraction of the voltage used to anodize the tantalum; on the order of 5 V. Under optimal charge injection conditions and at an anodic bias of about 5–8 V, Ta_2O_5/Ta capacitor electrodes can inject about 150–250 $\mu C/cm^2$ in a cathodal-first current pulse [30]. Maintaining the anodic bias of an electrode results in a small leakage current that is usually less than about 10 nA per nF of electrode capacitance. These leakage currents are about the same magnitude as those required to maintain a positive bias with activated iridium oxide films (AIROFs). These bias levels have not been found to damage neural tissue [31]. Shepherd *et al.* [32], however, have shown that direct current (DC) leakage currents of more than 400 nA can cause tissue damage and adopted a maximum safe level of 100 nA.

The primary disadvantages of capacitor electrodes are the difficulty in creating a sufficiently high ESA/GSA ratio to achieve useful levels of charge-injection, which is particularly true for microelectrodes, and the limitations imposed by pore resistance in electrodes with a high ESA. The primary use of capacitor electrodes for clinical devices is that of a charge storage electrode in the BION [33]. Charge storage at the

Ta/Ta_2O_5 interface is discharged through an iridium oxide electrode to produce a stimulus current pulse.

Dielectric coatings of TiO_2 and barium titanate, which have higher dielectric constants than Ta_2O_5, have also been investigated for capacitor electrodes [30]. These coatings suffer from the same disadvantages as Ta_2O_5/Ta electrodes and also exhibit higher leakage currents under bias. At present, there is little interest in these dielectrics for capacitor electrodes.

3.2.5 Faradaic charge injection

By definition, Faradaic charge injection involves the reduction and oxidation reactions, that is, an electron transfer reaction mediates the change from electronic to ionic conduction across the electrode–tissue interface. These redox reactions occur at the surface of the electrode and may involve the electrode material itself or a coating on the electrode. Reduction and oxidation of species in the electrolyte around the electrode is almost always irreversible for electrodes implanted in tissue and generally should be avoided. For platinum and platinum-iridium alloy electrodes, the primary reversible charge injection reactions involve surface confined species:

$$\mathrm{Pt + H_2O \Longleftrightarrow PtO + 2H^+ + 2e^-} \quad \text{oxide formation and reduction, and}$$
$$\mathrm{Pt + H_2O + e^- \Longleftrightarrow Pt\text{-}H + OH^-} \quad \text{H-atom plating.}$$

These reactions form an adsorbed species on the surface of the platinum and are often referred to as pseudocapacitive, since their electrochemical characteristics are similar to those of a capacitor. However, these reactions are electrochemical, as opposed to electrostatic, as in the case of double-layer charge storage, and when the consumption or generation of H^+ and OH^- ions occurs. Therefore charge injection with platinum electrodes, even if reversible by the definition in Section 3.2, is accompanied by pH changes.

Since the reversible reactions on platinum are limited to a monolayer on the electrode surface, the ability of platinum electrodes to inject charge without altering the adjacent electrolyte is limited. In early investigations of reversible charge reactions on platinum, a total available charge of approximately 350 $\mu C/cm^2$ was estimate for the reactions above [34]. This total was bounded by the potentials for reduction and oxidation of water on the platinum. It is important to emphasize that this total charge cannot be accessed during stimulation, because the starting potential of the platinum electrode *in vivo* is about midway between the water electrolysis potentials (the water window). In addition, current density dependent overpotentials further limit the charge that can be injected without the electrode polarization exceeding the water window, and measurements with expected physiologically relevant stimulation pulses suggest a charge density limit of about 35–150 $\mu C/cm^2$, depending on pulse width [35–37]. Higher charge injection densities (~300 $\mu C/cm^2$) can be obtained for cathodal-first pulsing by holding the platinum electrode at a positive potential between pulses, but this does require a continuous anodic current in the interpulse period to maintain the bias. These charge densities also include the double-layer contribution which occurs at all potentials. While the surface-bound reactions on platinum are reversible, other reactions must be occurring since platinum dissolution, at least in saline, has been observed at pulsed charge densities as low as 20–50 μC/cm^2. Protein addition to saline electrolytes greatly reduces or even eliminates platinum dissolution [38–40]. Oxygen reduction also occurs readily on platinum and has the potential to generate highly reactive and tissue-damaging species [41], although estimates of the contribution of oxygen reduction to charge injection during pulsing indicate that this is a very minor reaction, presumably due to the rate limiting diffusion of oxygen to the electrode [42].

Platinum has proved to be a versatile and safe stimulation electrode in a variety of neural applications, including cochlear implantation for deafness (Chapter 7), VNS for epilepsy, and DBS for movement disorders (Chapter 8), spinal cord stimulation for pain management, and subdural cortical stimulation for epilepsy. All these applications involve electrodes with a comparatively large surface area – greater than 0.06 cm^2. A notable limitation of platinum is its lack of suitability in microelectrode stimulation. With electrodes having a GSA of $0.5–3 \times 10^{-5}$ cm^2 (500–3000 μm^2), typical for microelectrode applications, the minimum charge densities to reach functional threshold are on the order of 100 μC/cm^2 and higher charge densities will likely be necessary in clinical practice [43, 44]. Efforts to increase the charge injection capacity of platinum by surface roughening to increase the ESA have produced coatings capable of injecting about 1 mC/cm^2 in saline, but *in vivo* measurements suggest a much reduced charge capacity of <100 μC/cm^2 in chronic implantations [45].

To overcome the low charge capacity of platinum, iridium oxide was introduced as a neural stimulation electrode coating [46, 47]. Iridium oxide is a Faradaic electrode that injects charge via reversible redox reactions between Ir^{3+}/Ir^{4+}. Whether the counterion is H^+ or OH^- is uncertain, but measurements suggest that it is OH^- at neutral pH [48]. Iridium oxide is formed by a variety of techniques including activation of iridium metal [47], thermal decomposition of iridium salt solutions [49], electrodeposition [50] and sputtering [51]. Due to the three-dimensional nature of the iridium oxide coating, compared with the surface monolayer platinum, there intrinsic charge storage capacity of an iridium oxide electrode might range from 20–60 mC/cm^2, versus 1–2 mC/cm^2 for platinum, and has a charge-injection capacity of up to 4–8 mC/cm^2, depending on the stimulation waveform [51]. AIROFs implanted in cat cortex have been observed to fail by delamination if pulsed negative of the water reduction potential [52]. Oxygen reduction appears negligible on iridium oxide [42] although, like platinum electrodes, the charge injection capacity appears to decrease very significantly for chronically implanted electrodes *in vivo* [53].

3.2.6 Stimulation waveforms

Electrical stimulation can be delivered using voltage-controlled or current-controlled pulses. Voltage control has been used extensively in DBS and in spinal cord stimulation. Voltage control is easier to implement electronically than current-controlled stimulation. However, it has the disadvantage that the charge injection and the electric field in the tissue depend on the electrode impedance, which varies considerably between patients and also varies over time, resulting in uncertainty in the quantity of charge injected [26]. For constant current pulsing, the injected charge per phase is readily calculated as the product of the current and pulse width and the charge density determined by dividing by the electrode GSA. Since charge density is the primary pulse parameter determining the severity of stimulation with respect to electrode stability and the reversibility of charge injection reactions, and the combination of charge/phase and charge density are often used to assess the likelihood of tissue damage, it is usually much easier to assess the safety of constant current pulsing compared with constant voltage pulsing. This is particularly true for stimulation involving large driving voltages. It is assumed that the majority of the voltage drop occurs in the tissue and is presumed to be non-harmful. However, because of the large driving voltage, there is always some uncertainty as to where the voltage drop occurs and if the drop occurs across the electrode-tissue interface, damaging electrochemical actions may be elicited.

From the perspective of electrochemical safety, the design of pulse generators and selection of stimulus parameters has focused mostly on minimizing net DC current flow during stimulation, and ensuring that any imbalance in the charge between the cathodal and anodal phases of a stimulus pulse is prevented from accumulating and polarizing the electrode to harmful potentials [54]. The manner in which a small imbalance in charge within a single biphasic current pulse can accumulate during pulsing and polarize the electrode to damaging potentials has been described by Merrill *et al.* [55]. If the imbalance in the waveform is modest, a small shift in the interpulse potential of the electrode should occur in the direction of the imbalance and this potential displacement should drive a counter-balancing reaction that just offsets the imbalance in the biphasic pulse. In effect, this establishes a new steady state interpulse potential that reflects the imbalance. Voltage-controlled pulse waveforms offer no control over the charge balance between phases and some method of compensating for charge imbalance during the pulse is essential.

Avoiding net DC current flow is often achieved by placing an isolation capacitor in series with the output of the stimulator. To avoid polarization due to charge imbalance, a stimulation electrode will often be electrically shorted to the metal case of the implanted pulse generator (IPG) or the electrode will be shorted to the other electrodes in a multi-electrode array. When shorting to the case of the pulse generator, the objective is to establish a galvanic couple between the electrode and the much larger metal case. The mixed-potential of this galvanic couple is dominated by the metal case. Since the *in vivo* equilibrium potential of titanium, a typical IPG case material, is about 0.0 versus Ag|AgCl, the stimulation electrode is quickly polarized to the same potential. When there is concern that shorting to the titanium case alone will be insufficient to prevent polarization, a platinum coating may be applied to the case. The platinum also adopts an *in vivo* potential of about 0.0 V and can provide high galvanic currents and will more rapidly establish a 0.0 V potential at the stimulation electrode.

A common strategy for achieving charge-balanced pulses in neural stimulation applications is to employ a biphasic current pulse in which the charge in the leading and trialing phases of the pulse are equal and opposite. Combining some form of interpulse potential control, as described above, with a charge balanced biphasic pulse usually ensures that the E_{ipp} of the electrode remains close to its *in vivo* equilibrium value and well within a potential range that avoids electrode or tissue damage. However, this is not a sufficient condition for safe stimulation. Large potential excursions driven by high current densities during a pulse, including any charge-balancing second phase pulses, can also be harmful to the electrode and tissue. A variety of different stimulation waveforms have been employed and the rationale and details of these waveforms are discussed by Merrill *et al.* [55] and Cogan [56].

3.2.7 Voltage transient analysis

The voltage excursion of an electrode during a stimulation pulse can be used to estimate the polarization of the electrode and thus what reactions are thermodynamically possible. For most electrode materials, the acceptable polarization range is defined by the potentials at which water is reduced or oxidized. The potential range between the reduction and oxidation limits is frequently called the *water window*. Provided the polarization of the stimulation electrode during a pulse remains positive of the water reduction potential and negative of the water oxidation potential, that is within the water window, the stimulus intensity, as defined by the charge density in the pulse, is often considered "safe." There are several limitations to this approach. First, due to the high current density at the electrode during a stimulation pulse, resistive and capacitive voltage drops in the tissue, as well as activation and concentration overpotentials, make it difficult to determine the true polarization of the electrode [56]. Failure to accurately account for these overpotentials may result in an over-estimation of the polarization across the electrode–tissue interface. In addition, electrochemical reactions are usually kinetically limited and polarization beyond the thermodynamic potential is necessary to achieve a significant reaction rate. This is why different electrode materials do not necessarily exhibit the same water oxidation and reduction potentials. A compendium of water electrolysis potentials for different electrode materials is provided in Table 3.2.

3.2.7.1 Biphasic constant-current pulsing Biphasic constant-current pulsing is commonly employed for neural stimulation and usually, although not always, the current magnitude and pulse width for the cathodal and anodal phase will be equal. The polarization of an electrode can be evaluated by analyzing the voltage transient at the electrode in response to the current pulse. These measurements are usually made in buffered saline using a three-electrode configuration in which the voltage transient is measured with a non-current carrying reference electrode in response to current imposed between the stimulation electrode and a large area counter electrode. This mode of stimulus delivery, in which current flows from a stimulation electrode to a remote counter electrode, is termed monopolar. In clinical practice, the large area monopolar return electrode would be the metal can of IPG. It is equally possible to

Table 3.2 Water reduction and oxidation potentials on electrode materials at pH 7 vs. Ag|AgCl.

Electrode	Reduction, V	Oxidation, V	Reference
Pt and PtIr-alloys	–0.6	0.8	[35]
Activated iridium oxide (AIROF)	–0.6	0.8	[57, 58]
Thermal iridium oxide (TIROF)	–0.6	0.8 V	[49]
Sputtered iridium oxide (SIROF)	–0.6	0.8 V	[51]
Titanium nitride	–0.9	0.9	[28]
n-doped UNCD	–1.5	1.2	[59]
PEDOT	–0.9	0.6	[60]

make voltage transient measurements at an electrode that is acting as one member of a bipolar pair in which current is delivered between the two electrodes rather than a remote counter electrode. However, if the bipolar pair are in close proximity, such that their respective electric fields or counter ion concentration gradients in the tissue or electrolyte overlap, it becomes more difficult to unambiguously determine the potential across the electrode-electrolyte interface for the individual electrodes.

An example of a voltage transient in response to a biphasic current pulse is shown in Figure 3.2 for an AIROF electrode with a GSA of 1000 μm^2. The measurement was made in buffered saline (pH 7.4) using a current of 100 μA and half-phase pulse width of 400 μs to delivery 4 nC/phase pulses with a charge density of 4 mC/cm^2. The maximum cathodic and anodic potential excursions, E_{mc} and E_{ma}, are –0.42 V and 0.6 V versus Ag|AgCl, respectively, and within the water electrolysis limits for AIROF (Table 3.2). The interpulse potential of the AIROF in this example has been biased to 0.4 V (Ag|AgCl) to increase the charge injection capacity [57]. A voltage transient response from the same AIROF microelectrode polarized to potentials beyond the water electrolysis limits by increasing the pulse width from 400 μs to 600 μs is shown in Figure 3.3. The charge per phase is 70 nC/ph and the charge density 7 mC/cm^2, resulting in an E_{mc} of –1.4 V and E_{ma} of 1.2 V (Ag|AgCl), well beyond the reduction and oxidation limits of water on AIROF. The inflection in the voltage transient response indicated by the red arrow is typical for iridium oxide films pulsed beyond the water limits. The inflection indicates the recruitment of additional reactions at the electrode beyond reversible reduction and oxidation of the Ir^{3+}/Ir^{4+} redox couple. These reactions will include reduction and oxidation of water, but may also include other reactions [55]. It is important to note that when the charge injected at an electrode has a significant irreversible component, overpotentials at the electrode are difficult to accurately determine and the electrode-tissue interfacial polarization may not be accurately determined by the simple procedure of equating the electrode voltage in the interphase period with the interfacial polarization.

During constant current pulsing, the driving force for charge delivery is the voltage output of the stimulator. This voltage output adjusts continuously during the course of a stimulus pulse to maintain a constant current at the electrode. In clinical devices, this driving voltage is measured between the outputs of the stimulator and includes

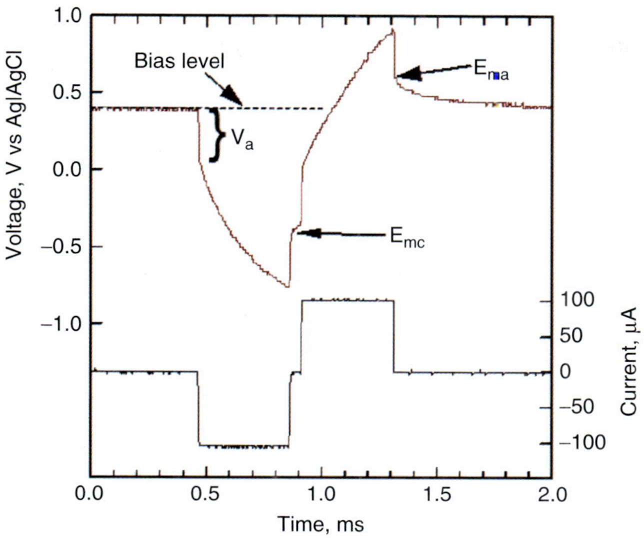

Figure 3.2 Voltage transient response of an AIROF microelectrode in response to biphasic 100 µA, 400 µs current pulsing at 50 pps. The electrode was maintained at a positive bias of 0.4 V in the interpulse period. E_{mc} maximum cathodal potential excursion (–0.42 V vs. Ag|AgCl); E_{ma} maximum anodal potential excursion (0.6 V vs. Ag|AgCl), and V_a access voltage (0.4 V vs. Ag|AgCl).

voltage drops in the implanted leads, in the tissue between the stimulation and return electrode, and the voltage drop across the electrode-tissue interface at the stimulation and return electrodes. If the return electrode has a large surface area relative to the stimulation electrode (>10 × surface area), as is the case if the metal can of an IPG is used as the return, then the voltage drop at the return electrode–tissue interface is often ignored. If polarization at the return electrode is a concern or must be minimized, it is common practice to use a platinum coating on the IPG, which is usually fabricated from titanium. For stimulation using one or more *stimulation electrodes* as the return, as is the case with bipolar or tripolar electrode configurations, the voltage drop at the return electrode–tissue interface can be significant and potentially damaging, particularly for the anodal-first member of a bipolar electrode pair. In most studies of charge injection in physiological saline or in many animal studies, an independent, non-current carrying reference electrode is employed for measuring the voltage transient at the stimulation electrode. In this case, polarization at the return or counter-electrode is not recorded by the measurement.

3.2.7.2 Imbalanced pulses The need to maintain charge balance by matching the charge per phase in the cathodal and anodal phases of a biphasic current pulse has been discussed previously and strategies for minimizing deleterious polarization of an electrode due to this imbalance described. Measurements in physiological saline using deliberately unbalanced biphasic current pulses demonstrate the consequences of charge imbalance. The voltage transients measured on a 400 µm diameter SIROF

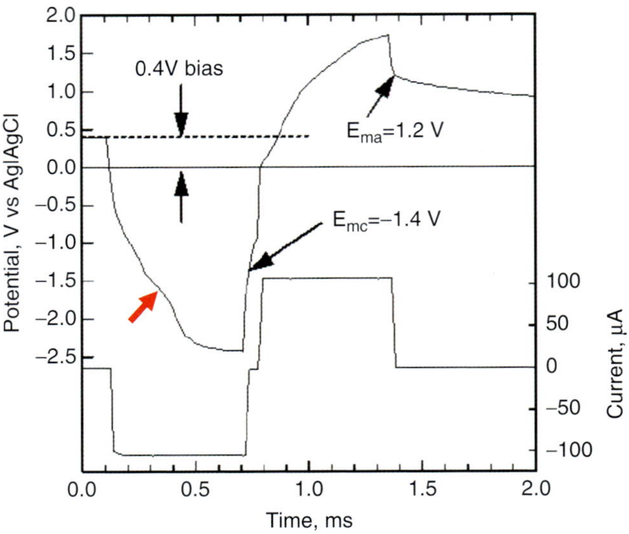

Figure 3.3 Voltage transient response of an over-pulsed AIROF microelectrode in response to biphasic 100 μA, 600 μs current pulsing at 50 pps. The electrode was maintained at a positive bias of 0.4 V in the interpulse period. E_{mc} maximum cathodal potential excursion (–1.4 V vs. Ag|AgCl); E_{ma} maximum anodal potential excursion (1.2 V vs. Ag|AgCl), and Va access voltage (0.4 V vs. Ag|AgCl).

electrode are shown in Figure 3.4 for a well-balanced biphasic current pulse (400 μs pulse width, 150 μA per phase, 150 Hz; 60 nC/phase, 48 μC/cm^2) and for two unbalanced waveforms in which the anodic current is decreased to 125 μA and 120 μA while the cathodic current is maintained at 150 μA. The current pulses were delivered without a series DC isolation capacitor and with the electrode electrically floating during the interpulse period. There is a modest change in the voltage transient when the anodic current is reduced to 125 μA (17% imbalance) with a slight negative shift in the electrode potential. The maximum cathodal and anodal potential excursions and the interpulse potential remain well within the limits for water electrolysis. However, an additional imbalance to an anodic current of 120 μA (20% imbalance) results in rapid polarization of the electrode to water reduction potentials during the cathodal pulse and a shift in interpulse potential from 0 V to –0.3 V (Ag|AgCl). The distorted transient in the cathodal phase is due to at least large pH changes and possibly gas evolution.

It is worth noting that the boundary between damaging and non-damaging levels of charge imbalance can be quite abrupt. The effect of pulse frequency and anodal versus cathodal imbalance are shown in Figure 3.5. These data show that at 150 Hz, the cathodal and anodal phases needed to be charge matched to 80–87% to avoid damaging electrode polarization, regardless of which phase has the larger charge. At 100 Hz the range is 73–78% and at 50 Hz the range is about 45% for lower cathodal charge and about 65% for lower anodal charge. These large charge imbalances can be sustained without excessive electrode polarization because of chemical reactions

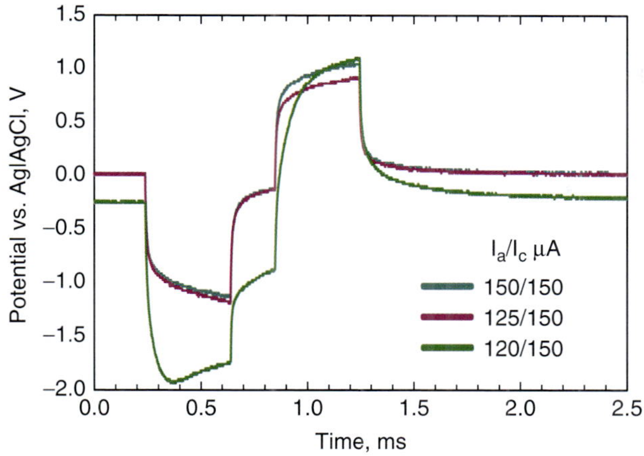

Figure 3.4 Voltage transients for an SIROF electrode pulsed wih a balanced (150/150 mA) and two unbalanced biphasic current wveforms (125/150 μA and 120/150 μA). Polarization of the SIROF electrode to potentials negative of the water reduction potential occurs abruptedly as the imbalance is increased from 25 μA to 30 μA.

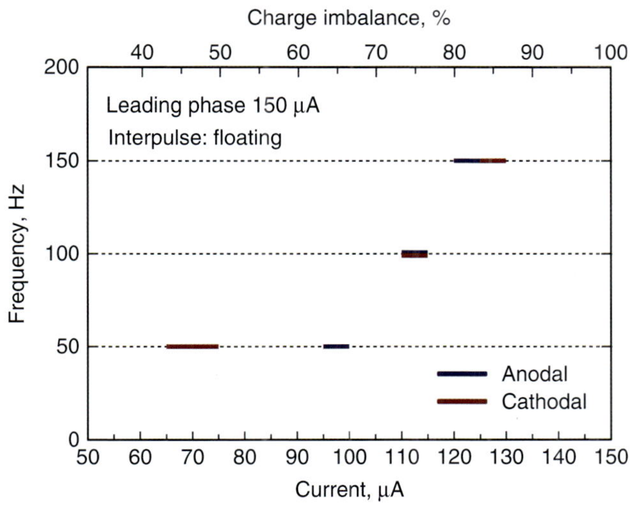

Figure 3.5 The effect of pulse frequency on the polarization for an SIROF electrode subjected to unbalanced biphasic pulsing. Each line represents the range of currents over which the polarization of the SIROF was observed to transition from potentials within the water window to potentials at which water electrolysis occurs. Data for both an anodal and cathodal imbalance are shown.

that occur on the electrode in the interpulse period. These reactions tend to drive the potential of the electrode to its equilibrium value for the medium. Oxygen reduction on a platinum electrode pulsed with a larger charge in the cathodal half-phase would be an example of a reaction that tends to restore the equilibrium potential of the electrode. The actions of buffers will also tend to restore the equilibrium potential of an electrode if H_3O^+ and OH^- ions are involved in maintaining the equilibrium potential, such as with iridium oxide electrodes. The large negative shift in the interpulse potential, evident in Figure 3.4 for the SIROF voltage transient with a 20% imbalance (120 μA anodal/150 μA cathodal), acts as a driving force to restore the equilibrium potential in the interpulse period.

As would be expected, higher pulse frequencies exacerbate the effects of charge imbalance. In addition, it is likely that these SIROF electrodes, used to generate the data in Figures 3.4 and 3.5, tolerated these large imbalances, at least in part due the low charge injection levels of 48 $\mu C/cm^2$, which is a small fraction of the charge injection capacity of SIROF under these pulse conditions. While these data suggest that electrodes might tolerate a significant charge imbalance during biphasic pulsing, they were acquired in buffered saline over short periods of time and care should still be taken to avoid such large imbalances in animal or clinical settings.

In early work, the use of a small charge imbalance had been proposed as a means of increasing the charge injection capacity of stainless steel electrodes [61]. The leading cathodal phase of a biphasic pulse in delivered with approximately 1% more charge than the trailing anodal phase. It was presumed that some Faradaic reaction products in the leading phase would diffuse too far from the electrode to be oxidized during the anodal pulse and therefore the electrode would be polarized positive of its equilibrium potential during the anodal phase, raising the possibility of electrode corrosion. Although stainless steel electrodes are rarely used for chronic neural stimulation and the strategy of a deliberate charge imbalance to increase charge injection capacity is not employed, the underlying principal for achieving reversibility is important. The fundamental idea behind the approach by McHardy *et al.* [61] is that the electrode potential at the end of the imbalance biphasic pulse, when the imposed current is zero, is the same as the potential just before the onset of the current pulse. Since the electrode potential before and after the pulse is unchanged, the equilibrium state of the electrode–tissue interface is maintained. While this means that the potential of the electrode in the interpulse period remains within safe electrochemical limits, it does not exclude harmful electrochemical potentials during the pulse when the electrode is polarized at high current densities.

The strategies described previously of electrically shorting the stimulation electrode in the interpulse period to the metal case of an IPG or to other electrodes, either unpulsed or pulsed with the opposite current polarity, all attempt to establish an interpulse potential that is consistently close to the equilibrium potential of the electrode. It is also possible to actively control the potential of a stimulation electrode in the interpulse period. This strategy is shown in Figure 3.6, in which a monophasic current pulse is used for stimulation. At the end of the current pulse, the stimulator switches from current control to voltage control and maintains the potential of the stimulation electrode at a pre-determined potential with respect to an implanted reference

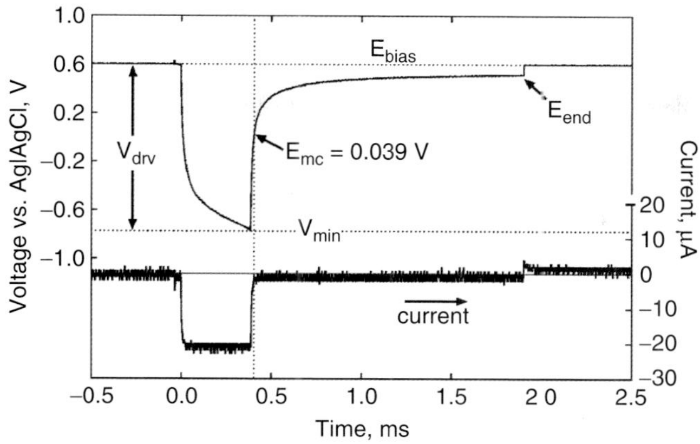

Figure 3.6 Voltage transient in response to a monophasic cathodal current pulse with charge-balance achieved by controlling the interpulse potential (E_{bias}) of the electrode to 0.6 V (vs. Ag|AgCl) in the interpulse period.

electrode. In Figure 3.6, the potential interpulse potential is labeled E_{bias} to indicate that it is actively controlled. This iridium oxide electrode, E_{bias} was maintained at 0.6 V versus Ag|AgCl to increase the charge injection capacity of the electrode. A period of about 1.5 ms is allowed between the end of the cathodal pulse and the onset of voltage control. This point is denoted by E_{end} in Figure 3.6 and a small anodic current transient can be observed as the stimulator re-establishes the bias potential.

In the approach by Troyk *et al.* [62], the magnitude of the restoring anodic current is limited so that the electrode is not polarized positively of the water oxidation potential (0.8 V vs. Ag|AgCl on this iridium oxide electrode). The electrode potential is thus maintained at a constant and safe value within the interpulse period. The reference electrode could be the metal case of the pulse generator or a dedicated electrode which would typically be included as part of an electrode array. Electrode materials such as PtIr, AIROF and sputtered iridium oxide (SIROF) have been used as implanted reference electrodes. Although precise equilibrium potentials are not readily established in the body for these materials, experience has shown that they are reasonably predictable and close to 0 V versus Ag|AgCl, making them suitable for the purpose of maintaining a safe interpulse potential. A consequence of maintaining the interpulse potential at a non-equilibrium value, as is the case with this positively biased iridium oxide electrode, is the need to maintain a net DC current in the interpulse period. The magnitude of this current is small, on the order of a few micro-amperes per square centimeter of electrode area, and has not been associated with tissue or electrode damage [31, 62]. However, this approach precludes the use of a DC isolation capacitor to protect the electrode and tissue in the event of an unexpectedly large or sustained current, which might arise from an electrical circuit fault in the stimulator for example, and places an additional burden on stimulator design for safety.

3.3 ELECTRODE MATERIALS

Many reviews provide lists of electrode materials suitable for chronic recording and stimulation [55, 56]. In common clinical practice, this list is limited to platinum and its alloys with iridium, porous titanium nitride and, to a lesser extent, iridium oxide, and some stainless steels. Emerging electrode coatings based on intrinsically conducting polymers, carbon nanotubes (CNTs), doped ultra-nano-crystalline diamond and graphene are discussed in later sections of this chapter. Highlights of the properties of those more conventional electrode materials are provided below.

3.3.1 Non-noble metal electrodes

The non-noble metal electrodes include stainless steels such as the ferrous-based 316 LVM alloy, Elgiloy, and the refractory metals titanium and tantalum. Stainless steel has been employed extensively as an intramuscular electrode for phrenic nerve stimulation, limb animation in quadriplegia and as an extraneural recording and stimulation electrode for footdrop in stroke patients [63–65]. These electrode materials injection charge primarily by double-layer capacitance. Charge-injection via oxidation and reduction of the surface of ferrous alloys and CoCr-based alloys is unlikely to be chemically reversible and some dissolution is often observed during charge-injection. To emphasize the irreversibility of surface reactions involving the passive layers on these alloys, extensive corrosion of MP35N, a CoCr alloy, was observed at the modest charge injection level of 20 $\mu C/cm^2$ *in vitro* [66]. Similar levels of charge injection are observed with titanium and tantalum, unless they are employed as capacitor electrodes with an anodically oxidized surface (Section 3.1.4).

3.3.2 Noble metals

Noble metals for stimulation electrodes include Pt and PtIr-alloys, with recent interest in palladium as a lower-cost alternative. Early studies of the charge injection capacity of Pt electrodes reported reversible limits of about 300 $\mu C/cm^2$ [34]. However, this limit represents the total charge available from the water oxidation potential to the water reduction potential. This total charge is not accessible during a charge injection pulse, unless the Pt electrode is biased to 0.8 V (Ag|AgCl) in the interpulse period. In addition, the measurements by Brummer and Turner [34] were made with pulse widths that are longer than is typically used in clinical practice. Later studies with platinum microelectrodes revised the charge injection capacity for platinum to about 70 $\mu C/cm^2$ in a 200 μs pulse [35]. More recent measurements by Leung *et al.* [37] further revised the charge injection capacity for large-area Pt electrodes to about 50 $\mu C/cm^2$. Charge injection capacities, specifically charge per unit GSA, are expected to decrease with increasing size of a stimulation electrode due to the more severe constraints on transport in the electrolyte for large electrodes [56, 67].

3.3.3 High surface area capacitor electrodes

The basic properties of capacitor electrodes have been discussed in Section 3.1.4. It is possible to increase charge injection capacity by increasing electrochemical surface area, for a fixed geometric surface area, by creating an electrode with a highly porous surface. The best example of this strategy is fractal titanium nitride (f-TiN), which exhibits charge injection capacities on the order of 1 mC/cm^2 [28]. The limitation of this approach is the resistance of the electrolyte within the pore of the electrode. This pore resistance combined with the intrinsic double-layer capacitance of the electrode material results in a time-constant that limits the charge that can be accessed during a pulse; when the time constant is comparable to or longer than the pulse width used in the stimulation pulse, not all of the total charge can be accessed [56]. This limitation applies to all electrodes that employ porosity as a means of enhanced charge-injection capacity.

3.3.4 Three-dimensional noble metal oxide films

Noble metal oxide electrodes are based primarily on iridium oxide. Thin films of iridium oxide are mixed-conductors, that is, they conduct both electrons and ions. The mixed conductivity permits access to the Ir^{3+}/Ir^{4+} reduction-oxidation (redox) couple within the three-dimensional structure of the oxide film, which is on the order of 100–500 nm thick. The total charge available from the Ir^{3+}/Ir^{4+} redox couple is on the order of 20–40 mC/cm^2, as measured by slow sweep rate cyclic voltammetry. During current pulsing, about 10–20% of this capacity can be accessed during a short-duration stimulation pulse resulting in charge injection capacities of 0.5–8 mC/cm^2, depending on pulse width and whether the iridium oxide is positively biased during pulsing [57, 68]. Other mixed-conductor oxides including ruthenium oxide and rhodium oxide have been investigated, as well as mixtures of these oxides with iridium oxide, but none have achieved the combination of charge injection capacity and stability obtained with iridium oxide. Iridium oxide films can be deposited by sputtering from an iridium metal or metal oxide target (SIROF), thermal decomposition of an iridium salt solution (TIROF), electrodeposition (EIROF), or electrochemical activation of iridium metal (AIROF) [47, 49–51].

3.4 FACTORS INFLUENCING ELECTRODE REVERSIBILITY

3.4.1 *In vivo* versus saline charge injection limits

We have described how charge injection limits in buffered saline are determined by measuring electrode polarization in response to a current pulse and comparing the maximum polarization to water reduction and oxidation potentials, after appropriate adjustment for iR drops and overpotentials. The same strategy has been employed *in vivo* for acutely and chronically implanted electrodes. These *in vivo* measurements of charge injection capacity are significantly lower than those measured in saline [37, 45, 53, 69]. A comparison of saline and *in vivo* measurements in three animal models is provided in Table 3.3.

Observations from *in vivo* studies using AIROF microelectrodes demonstrate that physical electrode damage is possible when these electrodes are pulsed to potentials

Table 3.3 Comparison of saline and *in vivo* charge injection capacities ($\mu C/cm^2$).

Electrode	Pre-Implant Saline	*In Vivo*	Explanted Saline	Ratio[a]%	Time	Placement	Ref.
Porous Platinum	845	178	870	21	45 days	Rabbit suprachoroidal	[45]
Platinum	34–54	4–17	Not reported	12–31%	Acute to 3 months	Cat suprachoroidal	[37]
AIROF	1330	140	Recovered[b]	11	acute	Zebra finch brain	[69]
SIROF	1410	660	Not reported	47	154 days	Cat cortex	[53]

a – $(Q_{in\ vivo}/Q_{saline})\bullet 100\%$; the *in vivo* charge injection capacity as a percent of the pre-implant charge capacity measured in a saline electrolyte
b – pre-implant transients recovered, (Q_{inj} not reported)

more negative than that for reduction of water (~ –0.6 V vs. Ag|AgCl) [52, 70]. While the response of other electrode materials is uncertain, it is obviously prudent to avoid polarization to potentials at which water electrolysis occurs and, if such excursions cannot be avoided, the *in vivo* stability of the electrode and the possibility of tissue damage should be carefully evaluated. An interesting consideration in the *in vivo* evaluation of charge injection limits is the effect of tissue capacitance on the voltage waveform measured during constant current pulsing. Modeling studies suggest the impact of tissue capacitance can be significant at short pulse widths and high current amplitudes, and *in vivo* measurements of charge injection capacity at pulse widths shorter than about 200 μs may be affected by tissue capacitance [58, 70].

3.4.2 Degradation mechanisms and irreversible reactions

3.4.2.1 Metal dissolution Platinum dissolution in saline electrolytes has been observed at charge densities well below charge-injection limits for avoiding water electrolysis [39]. Observations that the presence of proteins in the saline greatly reduces or even eliminates dissolution during pulsing, suggest that dissolution *in vivo* will be greatly reduced [38, 71]. This appears to be the case, with Pt or PtIr-alloy electrodes providing chronic stimulation of neural and cardiac tissue in many applications without obvious deterioration of electrodes due to dissolution [40].

3.4.2.2 Oxygen reduction Oxygen reduction is a chemically irreversible reaction that results in the formation of reactive species that can be damaging to tissue. Due to the relatively low tissue oxygen concentration and the limited diffusion rate of the neutral oxygen molecule to an electrode, oxygen reduction is not a likely major charge injection mechanism. Measurements by Cogan *et al.* [42] suggest that less than 7% of the charge injected during *in vivo* pulsing with a platinum electrode would involve oxygen reduction. This value was even lower on iridium oxide electrodes, which has a higher overpotential for oxygen reduction than platinum. A similarly low level of oxygen reduction is expected on TiN, which also has a significant overpotential for oxygen reduction.

3.4.2.3 Delamination Delamination of AIROF from pulsing more negatively than the water reduction potential has been reported [52]. Severe damage was observed on explanted AIROF electrodes examined by scanning electron microcopy (SEM) and deposition of thin layers of AIROF was observed in cortical tissue around the electrode. Chronic *in vivo* studies of iridium oxide electrodes suggest that delamination is avoidable, provided the charge injection limits of the electrode are not exceeded [31]. While the extent to which charge injection limits for different electrode materials can be exceeded during pulsing is not known, the observations with AIROF electrodes suggest that charge injection limits defined by water electrolysis should be exceeded with caution.

3.5 EMERGING ELECTRODE MATERIALS

3.5.1 Intrinsically conductive polymers

Research conducted from the 1970s onwards earned Alan Heeger, Alan McDiarmid and Hideki Shirakawa the 2000 Nobel Prize in Chemistry for the discovery and development of conductive polymers (CPs). CPs, including polypyrrole (PPy) and poly(3,4-ethylenedioxythiophene) (PEDOT), show favorable properties as an alternative to metallic electrodes [72–75]. The conjugated backbone of these polymers enables charge transfer through both capacitive and Faradaic mechanisms. As with most other coating technologies, the major contributing factor to increased charge transfer capacity is the substantially higher surface area provided by a CP. In Figure 3.7, scanning electrode microscopy (SEM) images show the comparison of a bare Pt electrode to one coated with PEDOT. Specifically, in this image a small dopant ion, paratoluene sulfonate (pTS), creates a highly roughened electrode surface morphology [73], shown at 1000× magnification in Figure 3.7(c). The typical surface morphology of a CP, produced from either PPy or PEDOT, has been

(A) (B) (C)

Figure 3.7 SEM images of (A) bare Pt electrodes, (B) coated with PEDOT/pTS at 200× magnification, and (C) PEDOT/pTS at 1000× magnification (reproduced with permissions from [78]), shows the nodular structure of a typical CP film.

described as highly nodular, with electrical studies being indirectly used to estimate an increase in charge transfer area of the order of one magnitude, when electrodes are operated in saline [76–78]. However, there is minimal data specifying the absolute increase in real charge transfer area determined from direct methods, such as profilometry, atomic force microscopy or electrochemistry. Most CP studies for applications in neural prostheses have focused on PEDOT, as it has been shown that the backbone chemistry of PEDOT is more stable than that of PPy [79]. Although there are a wide variety of potential CP chemistries in addition to PPy and PEDOT, they have been largely discounted due to instability issues or the requirement for harsh solvents to enable effective fabrication and processing [80].

All CPs require a doping ion which is incorporated during fabrication. When efficiently doped, the ions are distributed along the polymer backbone, bending the CP around them, to create continuous bipolarons, which enable electron transport through the CP [81]. In this way, CPs can transfer charge from an electrode to an electrolyte by relatively fast capacitive charging mechanisms, using their enhanced surface area to enable faster charge transfer per geometric area than a typical metal [78]. However, CPs also employ Faradaic mechanisms to transfer charge through oxidation and reduction reactions [82]. Reduction of the polymer shifts the charge along the conjugated backbone and drives the dopant ion out of the CP matrix. However, this is dependent on both the mobility of the dopant ion within the CP and the time for which the voltage or current driven stimulus is applied. In most commercial neural prostheses, the driving stimulus is applied for less than 1 ms [83] and diffusion inhibition of the doping ion out of the CP matrix prevents this from being the primary mechanism of charge transfer. If the doping ion is completely immobile within the CP matrix, Fardaic charge transfer can occur through diffusion of cations into the CP coating. Electrochemical analysis of CPs by cyclic voltammetry can reveal the potential for charge transfer via oxidation and reduction reactions. This characterization technique applies a voltage ramp on the order of seconds, which enables ions the time required to diffuse out or into the CP. This yields a charge storage capacity, which enables comparison of the charge transfer capability of electrode materials, but should not be mistaken for their ability to transduce charge as part of a neurobionics device which is driven by biphasic stimuli. Figure 3.8 shows the potential increase in charge storage capacity provided by a CP coating.

The most efficiently doped CPs are produced by electrodeposition, which also enables direct coating of electrode sites during fabrication without the need for further processing. The *in vitro* charge storage capacity of CPs produced in this way is extremely high, having been reported for PEDOT doped with pTS as around 130 mC/cm^2 [78, 84]. When less mobile dopants are used to produce CPs, such as the polymeric poly(styrene sulfonate) (PSS), the surface roughness of the CP coating is reduced and the charge storage capacity is similarly lowered. Baek *et al.* report that PEDOT/PSS produced at the same thickness of the PEDOT/pTS cited above, has a charge storage capacity of 100 mC/cm^2 when tested in physiological saline [84]. Perchlorate has also been investigated as a doping ion for CPs, which is highly effective in providing a high charge transfer area, with the charge storage capacity reported as 150 mC/cm^2 [84, 85]. However, perchlorate ions remain highly mobile within

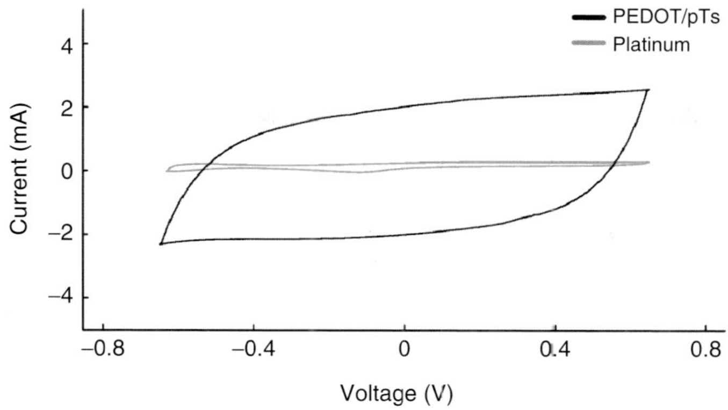

Figure 3.8 Cyclic voltammetry showing voltage response of Pt compared to PEDOT/pTS. Charge storage capacity is the integration of the current with respect to time. These curves were obtained in saline versus Ag/AgCl reference electrode with the voltage ramped at 150 mV/s.

the formed CP and have been shown to present biological limitations due to dopant toxicity when coatings are produced at a thickness of over 0.5 μm [84].

The highly roughened surface, combined with the capacity to pass both electronic and ionic charge, has resulted in CPs having substantially lower impedance than Pt at frequencies below 1 kHz. At frequencies between 0.1 and 10 Hz, a PEDOT coating can reduce the impedance of a flat metallic electrode by up to three orders of magnitude [72, 84]. The phase angle across the same range is similarly reduced from close to 80° for metallic electrodes to less than 20° for CP coated electrodes [78, 84]. As such, when the biphasic charge injection limit is assessed for these coatings, this low impedance behavior facilitates an *in vitro* charge injection limit which ranges from 0.8 to 4 mC/cm^2, dependent on both the coating chemistry and stimulation waveform. However, it is known that upon implantation, the highly roughened organic polymer is significantly blocked by proteins, which reduces the electrical benefit provided. *In vitro*, a typical PEDOT coating can reduce the voltage excursion at a Pt electrode interface by an order of magnitude, but *in vivo* this reduction is limited to roughly 50% [78]. In addition, while continued activation of a metallic electrode usually conditions the surface and reduces impedance over time, the same behavior is not observed for PEDOT coatings [77]. While still promising, these coatings can exhibit delamination [73, 86], resulting in a total loss of their electrochemical advantage.

Despite the limitations of CP coatings, they are proposed to have some advantages over other inorganic coating technologies in both the ability to modify the mechanical stiffness of the electrode and to provide biological cues at the neural interface. It is thought that the lower stiffness of these polymers [87] offers a more compliant electrode interface that may limit micro-trauma [88] (Chapter 2). In addition, the polymer can be easily functionalized to both present covalently bound biological cues and deliver reagents such as anti-inflammatories or neurotrophic factors [74, 89–91]. While it has been shown that these molecules have benefit *in vitro* and a short-term benefit *in vivo*, the inclusion of biomolecules within CPs has a detrimental effect on mechanical properties, reducing coating coherency and exacerbating problems with delamination [91–93].

To address the mechanical limitations of CPs, Green *et al.* recently combined PEDOT with a poly(vinyl alcohol) hydrogel to create a conductive hydrogel that showed greatly improved mechanical stability [94]. This composite approach has dual benefits as modification of hydrogel chemistries with biofunctional groups is a well-established field of research, enabling polymers to be designed which incorporate tailored proteins that can be presented at the surface of the electrode or released into the tissue [95, 96]. As such, the hydrogel component and hence the conductive hydrogel coating can be easily functionalized to include biochemical molecules, which promote cell attachment and neural cell regeneration without impacting on the mechanical properties of the polymer hybrid [97]. Conductive hydrogels have been evaluated *in vitro* and found to be both stable under electrical stimulation and further improve the charge transfer capacity of an electrode [98]. Kim *et al.* have also produced PEDOT within alginate hydrogel and found that the composite PEDOT-alginate had an impedance two orders of magnitude lower than the PEDOT coating alone at 1 KHz [99]. This additional improvement in electrical performance has been attributed to the ion sequestering behavior of hydrogels, which provide a ready supply of mobile ions at the charge transfer interface due to interactions with the charged hydrogel backbone. Preliminary *in vivo* studies using composites of CPs and hydrogels as recording electrodes reflect these improved charge transfer properties, indicated by higher signal-to-noise ratios when compared to uncoated controls [100]. In these studies, the impact of the softer interface was also demonstrated by reduced neural cell death at the electrode site [100]. While conductive hydrogels have demonstrated beneficial properties *in vitro* and for sub-chronic *in vivo* recording applications, their performance under chronic implant conditions with active electrical stimulation remains an area for future studies.

3.5.2 Carbon nanotubes and conductive diamond

Since their 1991 popularization by Iijima *et al.* [101], CNTs have been regarded as a biomaterial with many potential clinical applications. Over the intervening 20+ years, research into biological applications of CNTs has been intense. The various types of CNTs exist in the same size domain and as many of the fundamental elements of biology. Single-walled CNTs range from 1–5 nm in diameter, similar to the smallest building blocks of life such as proteins, DNA and glycoproteins such as fibronectin. Multi-walled CNTs can be as large as hundreds of nanometers in diameter, inhabiting the size domain of axons, neuritis, dendrites and the like. The very high mechanical strength and flexibility of CNTs has led to extensive research using CNTs as platforms for tissue and nerve regeneration [102]. Considering the high electrical conductivity of CNTs, it is no great surprise that a wide range of neural interfacing applications have been proposed.

For neural stimulation applications, CNTs provide a number of specific advantages. Almost all stimulating electrodes fabricated from CNTs are films containing many CNTs. They can be mats of curly, intertwined CNTs (such as those shown in

Figures 3.9 (c) and (d)) or forests of CNTs with some degree of alignment, typically vertical to the growth substrate [103, 104]. Such electrodes can feature very high surface-to-geometric area ratios and can exhibit very high capacitance and low impedance, even for electrodes with very small GSA. Specific capacitance can vary widely depending primarily on the nature and density of the nanotubes, but values of 2000–4000 μF cm^{-2} for planar electrodes are commonplace, in particular for dense vertically aligned CNT forests [105]. CNT electrodes operate by capacitive charging. The high capacitance values are achieved by virtue of extremely high real surface areas. Keefer *et al.* demonstrated that gold electrodes coated with a mat of multi-walled CNTs required a threshold voltage pulse of 195 mV compared to

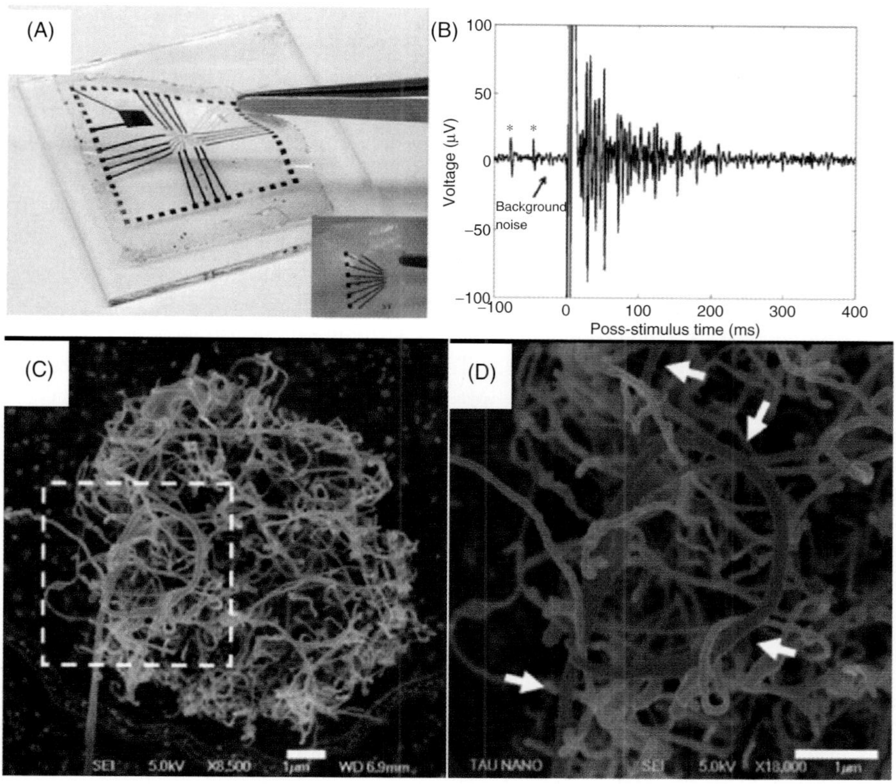

Figure 3.9 Electrodes fabricated from CNTs. (A) A CNT microelectrode array (MEA), fan out wires and contact pads [109]. (B) Patch clamp recordings made from ganglion cells in an excised chick retina. Recordings were made following stimulation with the MEA shown in (A). The small arrow in (B) indicates the artifact coincident with the stimulation pulse. Following are a train of natural retinal ganglion depolarization spikes. (C) An SEM image of curly intertwined CNTs upon which rat neurons have been cultured [111]. (D) Higher magnification of the section indicated in (C) by the dashed lines. The fiber shown colored in blue and indicated by arrows is a neuronal process that has grown into and become intertwined with the nanotubes.

725 mV for the gold electrode alone, to elicit a response from cultured neurons [106]. The flexibility and small size of CNTs makes them an attractive option for implantable electrode applications, where mechanically stiff electrodes can cause chronic tissue inflammation and fibrotic response [107, 108]. Recent work has demonstrated that MEAs of CNT can be fabricated and used to efficiently stimulate retinal ganglion cells in excised chick retina [109]. The MEAs described in that paper are shown in Figure 3.9(a) and recordings from adjacent CNT electrodes cells close to the stimulation site are depicted in Figure 3.9(b). De Asis *et al.* demonstrated that vertically-aligned CNTs were highly effective in stimulating neurons in hippocampal brain slices taken from rats [110].

Despite the apparently outstanding properties of CNT electrodes, their adoption into real-world devices has been slow. This is also reflected in the low number of *in vivo* experiments that have been conducted with these materials. Kolarcik *et al.* conducted the only chronic *in vivo* implantation study focusing on stimulation currently published. The study was conducted in rat cortex using PEDOT/CNT coated electrodes. In that study, stimulation efficacy was not assessed, the authors instead using the high surface area of the CNT electrodes to deliver a precoated anti-inflammatory drug, thus reducing tissue damage following a long-term stimulation protocol [112]. Samba *et al.* also employ PEDOT/CNT electrodes to stimulate retinal ganglion cells in excised rat retina, requiring less than 200 mV voltage pulse amplitude compared to 600 mV for TiN electrodes to generate a response [113].

The lack of *in vivo* results from CNT electrodes may be driven by lingering doubts regarding the toxicity of carbon nantoubes, in particular their fibrous nature and morphological similarity to asbestos fibers. Toxicity of CNTs is dictated by the size and shape of the CNTs and whether they are surface confined or not. Long straight CNT fibers have been shown to cause frustrated phagocytosis, the same phenomenon that leads to asbestosis; however, short or bundled CNTs appear to be well tolerated [114]. CNTs bound to polymer films caused negligible inflammation implanted into the back muscle of guinea pigs [115], and rat neurons grown on CNT films have been shown to send out neural processes that become intertwined with the CNTs and remain apparently healthy [111] (Figures 3.9 c and d).

Conductive diamond electrodes are fundamentally a different proposition to CNTs. Diamond is extremely hard with the consequence that it is more likely to cause mechanical damage and chronic inflammation if implanted in soft tissue [107, 108]. However, there are no toxicity concerns with any of the forms of diamond which invariably exhibit negligible toxic, fibrotic or inflammatory response *in vitro* or *in vivo* [116–118]. Diamond is also extremely chemically and biochemically inert, meaning it is very unlikely that electrodes will dissolve, degrade or give rise to toxic leachate products.

For the purposed of device fabrication, diamond is most commonly prepared by microwave plasma chemical vapor deposition (MPCVD), a process whereby nanodiamond seed crystals are enlarged in a carbon containing plasma. Preparations may be in the form of large single crystals if the seeds are dispersed, but more commonly the growing seeds coalesce into films of polycrystalline (micron sized), nanocrystalline (~50–100 nm) or ultra-nanocrystalline (3–5 nm) diamonds. The most common form of conductive diamond is boron-doped diamond (BDD), which can be prepared in any of the previously listed forms. BDD crystals or films, almost without exception, exhibit very low electrochemical double layer capacitance values and are therefore not suitable for neural stimulation. They are, however, an attractive prospect for neural recording applications [119] where low capacitance is desirable. The only form of diamond that exhibits double layer capacitance values high enough for neural stimulation is nitrogen-doped ultra-nanocrystalline diamond (N-UNCD). Nitrogen, when present in the MPCVD plasma during UNCD growth is not in fact included as a dopant in these films. Instead it appears to alter the structure of the grain boundaries between the nanodiamonds, making them wider and more graphitic. The electrical conductivity in N-UNCD films is thought to be via these graphitic grain boundaries [120]. N-UNCD films grow with a high degree of nano- and microscale roughness, which undoubtedly contributes to the high capacitance by increasing the ESA of the electrode.

However, of equal or perhaps greater importance appears to be the nature of the chemical termination at the electrode surface. Garrett *et al.* [59] showed that moderately high (179 μC cm^{-2}) charge injection capacities could be measured *in vitro*, but only after electrochemical oxidation of the N-UNCD film surface. Analysis of the oxidized surface by X-ray photoelectron spectroscopy indicated an increase in oxygen content consistent with functionalization of the surface. Similar increases in the electrochemical capacitance of CNT electrodes has been observed after oxidation by heat and acid treatment [121]. Though not entirely clear, it appears that increased capacitance through oxygen functionalization is correlated with increased hydrophillicity. Increased hydrophillicity alters the nature of the contact between the electrode and the electrolyte, possibly affecting the dielectric constant of the adsorbed water inner layer. There is also a possibility that the oxidized layer may act as a dielectric increasing capacitance by a similar mechanism to Ta_2O_5 layers deposited on tantalum, as discussed in Section 3.3.1. Hadjinicolaou *et al.* [122] demonstrated effective stimulation of rat ganglion cells in excised retina using single electrochemically-oxidized N-UNCD microelectrodes. In that paper, charge injection capacities of 300 μC cm^{-2} were measured.

Following recent advances in the ability to reliably fabricate and make electrical contact to diamond, an all-diamond electrode array for use in a retinal prosthesis for the blind has been proposed [116, 122]. An artist's impression of the diamond retinal prosthesis is shown in Figure 3.10. Garrett *et al.* more recently demonstrated that N-UNCD and other conducting and non-conducting forms of diamond caused less chronic immunohistological response than medical grade silicone when implanted in the back muscle of a guinea pig [123]. Early *in vivo* experiments using a 120×120 μm square N-UNCD electrode reveal that, under ideal conditions, threshold currents of 10 μA for a 500 μs per phase stimulus at the retina were sufficient to generate a corresponding response in the visual cortex of a cat. A 12-week implantation study comparing 700 μm diameter discs of N-UNCD, platinum and laser roughened platinum implanted behind the retina, revealed that the impedance of the N-UNCD electrodes was lower than the platinum varieties throughout the study and required similar current amplitudes to generate a cortical response [124]. Bergonzo *et al.* developed a

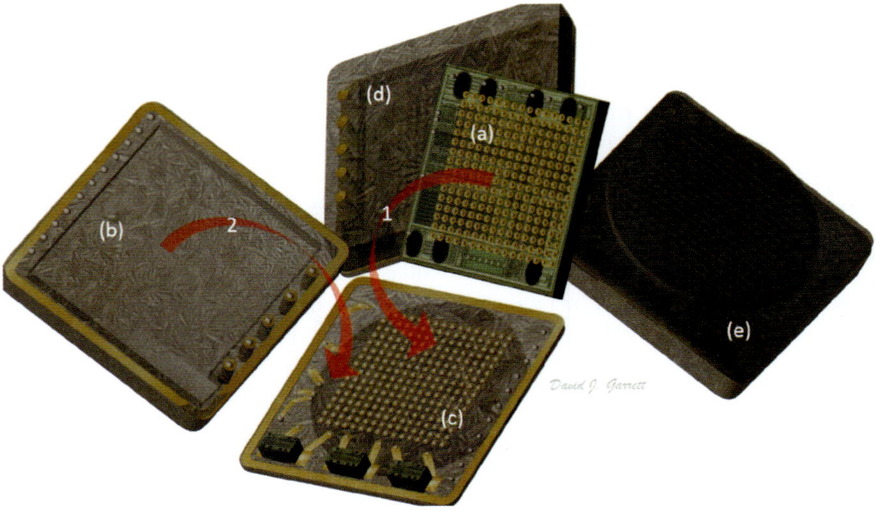

Figure 3.10 Proposed design of an all-diamond retinal prosthesis comprising a diamond capsule including a diamond back (b), a diamond feed-through and electrode array (c) with gold internal wiring. A dedicated stimulator chip (a) is encapsulated in the diamond case. (d) and (e) show rear and front views of the assembled device. The N-UNCD electrode array designed to sit directly on the retina is shown in (e).

method to generate flexible boron-doped diamond electrode arrays, again for retinal stimulation [125]. The arrays caused minimal gliosis implanted chronically under the retina of rats. Stimulation efficacy studies were not described.

3.6 CONCLUSION

There are many current and emerging applications of neural stimulation that will require careful consideration of the principles of safe charge injection from the perspective of both avoiding electrode degradation and avoiding damage to tissue. This is particularly true for applications employing microelectrodes to achieve selective and patterned stimulation. The high charge densities required with microelectrodes are challenging to deliver without inducing irreversible reactions at an electrode. While the search for new electrode materials with enhanced charge injection capabilities and stability will continue, future work will also need to focus on understanding how much electrochemical irreversibility can be tolerated and, importantly, how the properties of these microelectrodes are affected by chronic implantation.

REFERENCES

1. Shannon, R.V. (1992) A model of safe levels for electrical-stimulation. *IEEE Trans. Biomed. Eng.*, **39**: 424–426.

2. McCreery, D.B., Agnew, W.F., Yuen, T.G.H. and Bullara, L.A. (1990) Charge-density and charge per phase as cofactors in neural injury induced by electrical-stimulation. *IEEE Trans. Biomed. Eng.*, **37**: 996–1001.
3. McCreery, D.B., Agnew, W.F., Yuen, T.G.H. and Bullara, L.A. (1988) Comparison of neural damage induced by electrical-stimulation with faradaic and capacitor electrodes. *Ann. Biomed. Eng.*, **16**: 463–481.
4. Agnew, W.F., Yuen, T.G.H. and McCreery, D.B. (1983) Morphologic changes after prolonged electrical-stimulation of the cats cortex at defined charge-densities. *Exp. Neurol.*, **79**: 397–411.
5. Brown, W.J., Babb, T.L., Soper, H.V., Lieb, J.P., Ottino, C.A. and Crandall, P.H. (1977) Tissue-reactions to long-term electrical-stimulation of cerebellum in monkeys. *J. Neurosurg.*, **47**: 366–379.
6. Pudenz, R.H., Bullara, L.A., Jacques, S. and Hambrecht, F.T. (1975) Electrical stimulation of the brain. *III: The neural damage model. Surg. Neurol.*, **4**: 389–400.
7. Yuen, T.G.H., Agnew, W.F., Bullara, L.A., Jacques, S. and McCreery, D.B. (1981) Histological-evaluation of neural damage from electrical-stimulation – considerations for the selection of parameters for clinical-application. *Neurosurgery*, **9**: 292–299.
8. Yuen, T.G.H., Agnew, W.F. and Bullara, L.A. (1984) Histopathological evaluation of dog sacral nerve after chronic electrical-stimulation for micturition. *Neurosurgery*, **4**: 449–455.
9. Gilman, S., Dauth, G.W., Tennyson, V.M., Kremzner, L.T. and Defendini, R. (1975) Morphological and biochemical effects of chronic cerebellar stimulation in monkey. *Trans. Am. Neurol. Assn.*, **100**: 9–14.
10. Agnew, W.F., McCreery, D.B., Yuen, T.G.H. and Bullara, L.A. (1993) MK-801 protects against neuronal injury induced by electrical-stimulation. *Neuroscience*, **52**: 45–53.
11. Schrader, L.M., Stern, J.M., Wilson, C.L., Fields, T.A., Salamon, N. *et al.* (2006) Low frequency electrical stimulation through subdural electrodes in a case of refractory status epilepticus. *Clin. Neurophysiol.*, **117**: 781–788.
12. Abejon, D. and Feler, C.A. (2007) Is impedance a parameter to be taken into account in spinal cord stimulation? *Pain Physician*, **10**: 533–640.
13. Anon (1995) A randomized controlled trial of chronic vagus nerve stimulation for treatment of medically intractable seizures. The Vagus Nerve Stimulation Study Group. *Neurology*, **45**: 224–230.
14. Kinoshita, M., Ikeda, A., Matsumoto, R., Begum, T., Usui, K.. *et al.* (2004) Electric stimulation on human cortex suppresses fast cortical activity and epileptic spikes. *Epilepsia*, **45**: 787–791.
15. Peyron, R., Faillenot, I., Mertens, P., Laurent, B. and Garcia-Larrea, L. (2007) Motor cortex stimulation in neuropathic pain. Correlations between analgesic effect and hemodynamic changes in the brain. A PET study. *Neuroimage*, **34**: 310–321.
16. Medtronic (2010) Activa neurostimulation systems for deep brain stimulation, N'Vision Clinician Programmer with Software, 8840, 8870, 2010, and related product literature.
17. Herzog, J., Volkmann, J., Krack, P., Kopper, F., Potter, M. *et al.* (2003) Two-year follow-up of subthalamic deep brain stimulation in Parkinson's disease. *Mov. Disord.*, **18**: 1332–1337.
18. Haberler, C., Alesch, F., Mazal, P.R., Pilz, P., Jellinger, K. *et al.* (2000) No tissue damage by chronic deep brain stimulation in Parkinson's disease. *Ann. Neurol.*, **48**: 372–376.

19. Burbaud, P., Vital, A., Rougier, A., Bouillot, S., Guehl, D. *et al.* (2002) Minimal tissue damage after stimulation of the motor thalamus in a case of chorea-acanthocytosis. *Neurology*, **59**: 1982–1984.

20. de Balthasar, C., Patel, S., Roy, A., Freda, R., Greenwald, S. *et al.* (2008) Factors affecting perceptual thresholds in epiretinal prostheses. *Invest. Ophthalmol. Vis. Sci.*, **49**: 2303–2314.

21. Mahadevappa, M., Weiland, J.D., Greenberg, R.J. and Humayun, M.S. (2005) Perceptual thresholds and electrode impedance in three retinal prosthesis subjects. *IEEE Trans. Neural Syst. Rehab. Eng.*, **3**: 201–206.

22. Veraart, C., Raftopoulos, C., Mortimer, J.T., Delbeke, J., Pins, D. *et al.* (1998) Visual sensations produced by optic nerve stimulation using an implanted self-sizing spiral cuff electrode. *Brain Res.*, **813**: 181–186.

23. Shepherd, R.K. and McCreery, D.B. (2006) Basis of electrical stimulation of the cochlea and the cochlear nucleus. *Adv. Oto-Rhino-Laryngol.,* **64**: 186–205.

24. Huang, C.Q., Carter, P.M. and Shepherd, R.K. (2001) Stimulus induced pH changes in cochlear implants: An *in vitro* and *in vivo* study. *Ann. Biomed. Eng.*, **29**: 791–802.

25. Fujikado, T., Kamei, M., Sakaguchi, H., Kanda, H., Morimoto, T. *et al.* (2011) Testing of semichronically implanted retinal prosthesis by suprachoroidal-transretinal stimulation in patients with retinitis pigmentosa. *Invest. Ophthalmol. Vis. Sci.*, **52**: 4726–4733.

26. Kuncel, A.M. and Grill, W.M. (2004) Selection of stimulus parameters for deep brain stimulation. *Clin. Neurophysiol.*, **115**: 2431–2441.

27. Bard, A.J. and Faulkner, L.R. (2007) *Electrochemical Methods: Fundamentals and Applications*. Wiley and Sons Inc., London.

28. Weiland, J.D., Anderson, D.J. and Humayun, M.S. (2002) *In vitro* electrical properties for iridium oxide versus titanium nitride stimulating electrodes. *IEEE Trans. Biomed. Eng.*, **49**: 1574–1579.

29. Robblee, L.S., Kelliher, E.M., Langmuir, M.E., Vartanian, H. and McHardy, J. (1983) Preparation of etched tantalum semimicro capacitor stimulation electrodes. *J. Biomed. Mater. Res.*, **17**: 327–343.

30. Rose, T.L., Kelliher, E.M. and Robblee, L.S. (1985) Assessment of capacitor electrodes for intracortical neural stimulation. *J. Neurosci. Meth.*, **12**: 181–193.

31. McCreery, D.B., Pikov, V.and Troyk, P.R. (2010) Neuronal loss due to prolonged controlled-current stimulation with chronically implanted microelectrodes in the cat cerebral cortex. *J. Neural. Eng.*, **7**: 9.

32. Shepherd, R.K., Linahan, N., Xu, J., Clark, G.M. and Araki, S. (1999) Chronic electrical stimulation of the auditory nerve using non-charge-balanced stimuli. *Acta Oto-Laryngol.*, **119**: 674–684.

33. Loeb, G.E., Richmond, F.J.R. and Baker, L.L. (2006) The BION devices: injectable interfaces with peripheral nerves and muscles. *Neurosurgical Focus*, **20**: E2.

34. Brummer, S.B. and Turner, M.J. (1977) Electrical-stimulation with pt electrodes. 2: Estimation of maximum surface redox (theoretical non-gassing) limits. *IEEE Trans. Biomed. Eng.*, **24**: 440–443.

35. Rose, T.L. and Robblee, L.S. (1990) Electrical-stimulation with pt electrodes. 8: Electrochemically safe charge injection limits with 0.2 ms pulses. *IEEE Trans. Biomed. Eng.*, **37**: 1118–1120.

36. Cogan, S.F., Troyk, P.R., Ehrlich, J. and Plante, T.D. (2005) *In vitro* comparison of the charge-injection limits of activated iridium oxide (AIROF) and platinum-iridium microelectrodes. *IEEE Trans. Biomed. Eng.*, **2**: 1612–1614.

37. Leung, R.T., Shivdasani, M.N., Nayagam, D.A.X. and Shepherd, R.K. (2014) *In vivo* and *in vitro* comparison of the charge injection capacity of platinum macro electrodes. *IEEE Trans. Biomed. Eng.*, **62**(30): 849–857.

38. Robblee, L.S., McHardy, J., Marston, J.M and Brummer, S.B. (1980) Electrical-stimulation with pt electrodes. 5: The effect of protein on pt dissolution. *Biomaterials*, **1**: 135–139.

39. Robblee, L.S., McHardy, J., Agnew, W.F. and Bullara, L.A. (1983) Electrical-stimulation with pt electrodes. 7: Dissolution of pt electrodes during electrical-stimulation of the cat cerebral-cortex. *J. Neurosci. Meth.*, **9**: 301–308.

40. Shepherd, R.K., Murray, M.T., Houghton, M.E. and Clark, G.M. (1985) Scanning electron-microscopy of chronically stimulated platinum intracochlear electrodes. *Biomaterials*, **6**: 237–242.

41. Morton, S.L., Daroux, M.L. and Mortimer, J.T. (1994) The role of oxygen reduction in electrical-stimulation of neural tissue. *J. Electrochem. Soc.*, **141**: 122–130.

42. Cogan, S.F., Ehrlich, J., Plante, T.D., Gingerich, M.D. and Shire, D.B. (2010) Contribution of oxygen reduction to charge injection on platinum and sputtered iridium oxide neural stimulation electrodes. *IEEE Trans. Biomed.Eng.*, **57**: 2313–2321.

43. McCreery, D.B., Lossinskym A. and Pikov, V. (2007) Performance of multisite silicon microprobes implanted chronically, in the ventral cochlear nucleus of the cat. *IEEE Trans. Biomed. Eng.*, **54**: 1042–1052.

44. McCreery, D.B. (2008) Cochlear nucleus auditory prostheses. *Hear. Res.*, **242**: 64–73.

45. Terasawa, Y., Tashiro, H., Nakano, Y., Osawa, K. and Ozawa, M. (2013) Safety assessment of semichronic suprachoroidal electrical stimulation to rabbit retina. *Annual International Conference of the IEEE Engineering in Medicine and Biology Society: IEEE Engineering in Medicine and Biology Society Conference*, October 11, ed2013, pp 3567–3570.

46. Brummer, S.B., Robblee, L.S. and Hambrecht, F.T. (1983) Criteria for selecting electrodes for electrical-stimulation – theoretical and practical considerations. *Ann. NY Acad. Sci.*, **405**: 159–171.

47. Robblee, L.S., Lefko, J.L. and Brummer, S.B. (1983) Activated Ir – an electrode suitable for reversible charge injection in saline solution. *J. Electrochem. Soc.*, **130**: 731–733.

48. Cogan, S.F., Troyk, P.R., Ehrlich, J., Gasbarro, C.M. and Plante, T.D. (2007) The influence of electrolyte composition on the *in vitro* charge-injection limits of activated iridium oxide (AIROF) stimulation electrodes. *J. Neural. Eng.*, **4**: 79–86.

49. Robblee, L.S., Mangaudis, M.J., Lasinsky, E.D., Kimball, A.G. and Brummer, S.B. (1985) Charge injection properties of thermally-prepared iridium oxide films. *MRS Online Proceedings Library.* **1985**:55.

50. Meyer, R.D., Cogan, S.E., Nguyen, T.H, and Rauh, R.D. (2001) Electrodeposited iridium oxide for neural stimulation and recording electrodes. *IEEE Trans. Neur. Syst. Rehab. Eng.*, **9**: 2–11.

51. Cogan, S.F., Ehrlich, J., Plante, T.D., Smirnov, A., Shire, D.B. and Gingerich, M., *et al.* (2009) Sputtered iridium oxide films for neural stimulation electrodes. *J. Biomed. Mater. Res. Part B.*, **89B**: 353–361.

52. Cogan, S.F., Guzelian, A.A., Agnew, W.F., Yuen, T.G.H. and McCreery, D.B. (2004) Over-pulsing degrades activated iridium oxide films used for intracortical neural stimulation. *J. Neurosci. Meth.*, **137**: 141–150.

53. Kane, S.R., Cogan, S.F., Ehrlich, J., Plante, T.D., McCreery, D.B. and Troyk, P.R. (2013) Electrical performance of penetrating microelectrodes chronically implanted in cat cortex. *IEEE Trans. Biomed. Eng.*, **60**: 2153–2160.

54. Huang, C.Q., Shepherd, R.K., Carter, P.M., Seligman, P.M. and Tabor, B. (1999) Electrical stimulation of the auditory nerve: Direct current measurement *in vivo*. *IEEE Trans. Biomed. Eng.*, **46**: 461–470.

55. Merrill, D.R., Bikson, M. and Jefferys, J.G.R. (2005) Electrical stimulation of excitable tissue: design of efficacious and safe protocols. *J. Neurosci. Meth.*, **141**: 171–198.

56. Cogan, S.F. (2008) Neural stimulation and recording electrodes. *Ann. Rev. Biomed. Eng.*, Palo Alto: Annual Reviews, pp 275–309.

57. Beebe, X. and Rose, T.L. (1988) Charge injection limits of activated iridium oxide electrodes with 0.2 ms pulses in bicarbonate buffered saline. *IEEE Trans. Biomed. Eng.*, **35**: 494–495.

58. Cogan, S.F. (2006) *In vivo* and *in vitro* differences in the charge-injection and electrochemical properties of iridium oxide electrodes. *28th Annual International Conference of the IEEE Engineering in Medicine and Biology Society*, vols **1–15**. IEEE, New York, pp 6388–6391.

59. Garrett, D.J., Ganesan, K., Stacey, A., Fox, K., Meffin, H. and Prawer. S. (2012) Ultra-nanocrystalline diamond electrodes: optimization towards neural stimulation applications. *J. Neural. Eng.*, **9**: 10.

60. Cogan, S.F. *et al.* (2007) Polyethylenedioxythiophene (PEDOT) coatings for neural stimulation and recording electrodes. *Materials Research Society Meeting*, Boston, p. Abstract QQ2.7.

61. McHardy, J., Geller, D. and Brummer, S.B. (1977) Approach to corrosion control during electrical-stimulation. *Ann. Biomed. Eng.*, **5**: 144–149.

62. Troyk, P.R., Detlefsen, D.E., Cogan, S.F., Ehrlich, J., Bak, M. *et al.* (2004) "Safe" charge-injection waveforms for iridium oxide (AIROF) microelectrodes. *Proceedings of the 26th Annual International Conference of the IEEE Engineering in Medicine and Biology Society*, vols 1–7. IEEE, New York, pp 4141–4144.

63. Peterson, D.K., Nochomovitz, M., Dimarco, A.F. and Mortimer, J.T. (1986) Intramuscular electrical activation of the phrenic-nerve. *IEEE Trans. Biomed. Eng.*, **33**: 342–351.

64. Strange, K.D. and Hoffer, J.A. (1999) Restoration of use of paralyzed limb muscles using sensory nerve signals for state control of FES-assisted walking. *IEEE Transactions on Rehabilitation Engineering. IEEE Eng. Med. Biol.Soc.*, **7**: 289–300.

65. Cogan, S.F. (2007) Corrosion resistance of stainless steel nerve cuff electrodes in the neurostep. *12th Annual Conference of the International FES Society Philadelphia*: IFESS, 2007.

66. Cogan, S.F., Jones, G.S., Hills, D.V., Walter, J.S. and Riedy, L.W. (1994) Comparison of 316lVM and MP35N alloys as charge injection electrodes. *J. Biomed. Mater. Res.*, **28**: 233–240.

67. Heinze, J. (1993) Ultramicroelectrodes in electrochemistry. *Angew Chem-Int. Edit Engl.*, **32**: 1268–1288.

68. Cogan, S.F., Troyk, P.R., Ehrlich, J., Plante, T.D. and Detlefsen, D.E. (2006) Potential-biased, asymmetric waveforms for charge-injection with activated iridium oxide (AIROF) neural stimulation electrodes. *IEEE Trans. Biomed. Eng.*, **53**: 327–332.

69. Hu, Z., Troyk, P.R., Brawn, T.P., Margoliash, D. and Cogan, S.F. (2006) *In vitro* and *in vivo* charge capacity of AIROF microelectrodes. *28th Annual International Conference of the IEEE Engineering in Medicine and Biology Society*, vols 1–15. IEEE, New York, pp 5372–5375.

70. Butson, C.R and McIntyre, C.C. (2005) Tissue and electrode capacitance reduce neural activation volumes during deep brain stimulation. *Clin. Neurophysiol.*, **116**: 2490–2500.

71. Hibbert, D.B., Weitzner, K., Tabor, B. and Carter, P. (2000) Mass changes and dissolution of platinum during electrical stimulation in artificial perilymph solution. *Biomaterials*, **21**: 2177–2182.

72. Cui, X.Y., Wiler, J., Dzaman, M., Altschuler, R.A. and Martin, D.C. (2003) *In vivo* studies of polypyrrole/peptide coated neural probes. *Biomaterials*, **24**: 777–787.

73. Green, R.A., Hassarati, R.T., Bouchinet, L., Lee, C.S., Cheong, G.L.M. *et al.* (2012) Substrate dependent stability of conducting polymer coatings on medical electrodes. *Biomaterials*, **3**: 5875–5886.

74. Richardson, R.T., Wise, A.K., Thompson, B.C., Flynn, B.O., Atkinson, P.J. *et al.* (2009) Polypyrrole-coated electrodes for the delivery of charge and neurotrophins to cochlear neurons. *Biomaterials*, **30**: 2614–2624.

75. Kim, D.H., Richardson-Burns, S.M., Povlich, L., Abidan M.R., Spanninga, S. *et al.* (2007) Soft, Fuzzy and Bioactive Conducting Polymers for Improving the Chronic Performance of Neural Prosthetic Devices. CRC Press, USA.

76. Wilks, S.J., Koivuniemi, A.S., Thongpang, S., Williams, J.C. and Otto, K.J. (2009) Evaluation of micro-electrocorticographic electrodes for electrostimulation. *Annual International Conference of the IEEE, Eng. Med. Biol. Soc.*, vols **1–20**. IEEE, New York, pp 5510–5513.

77. Venkatraman, S., Hendricks, J., King, Z.A., Sereno, A.J., Richardson-Burns, S. *et al.* (2011) in In Vitro *and* In Vivo *Evaluation of PEDOT Microelectrodes for Neural Stimulation and Recording. IEEE Trans. Neur. Syst. Reh. Eng.*, **19**: 307–316.

78. Green, R.A., Matteucci, P.B., Hassarati, R.T., Giraud, B., Dodds, C.W.D. *et al.* (2013) Performance of conducting polymer electrodes for stimulating neuroprosthetics. *J. Neural. Eng.*, **10**: 11.

79. Yamato, H., Ohwa, M. and Wernet, W. (1995) Stability of polypyrrole and poly(3,4-ethylenedioxythiophene) for biosensor application. *J. Electroanal. Chem.*, **97**: 163–170.

80. Green, R.A., Lovell, N.H., Wallace, G.G. and Poole-Warren, L.A. (2008) Conducting polymers for neural interfaces: challenges in developing an effective long-term implant. *Biomaterials*, **29**: 3393–3399.

81. Pratt, C. (1966) *Conducting Polymers*. Kingston University, London.

82. Green, R.A., Baek, S., Lovell, N.H. and Poole-Warren, L.A (2010) Nanostructured Conductive Polymers as Biomaterials. *Nanostructured Conductive Polymers*. John Wiley & Sons, Ltd, Chichester UK, pp 707–736.

83. Lovell, N.H., Hallum, L.E., Chen, S.C., Dokos, S., Byrnes-Preston, P. *et al.* (2006) Advances in Retinal Neuroprosthetics. *Handbook of Neural Engineering*. IEEE/Wiley Press, New York, pp 337–356.

84. Baek, S., Green, R.A. and Poole-Warren, L.A. (2014) The biological and electrical trade-offs related to the thickness of conducting polymers for neural applications. *Acta Biomater.*, **10**(7): 3048–3058.

85. Baek, S., Green, R.A., Granville, A., Martens, P.J. and Poole-Warren, L.A. (2013) Thin film hydrophilic electroactive polymer coatings for bioelectrodes. *J. Mat. Chem. B.*, **1**: 3803–3810.

86. Ludwig, K.A., Uram, J.D., Yang, J.Y., Martin, D.C. and Kipke, D.R. (2006) Chronic neural recordings using silicon microelectrode arrays electrochemically deposited with a poly(3,4-ethylenedioxythiophene) (PEDOT) film. *J. Neural. Eng.*, **3**: 59–70.

87. Hassarati, R.T., Goding, J.A., Baek, S., Patton, A.J., Poole-Warren, L.A. and Green, R.A. (2014) Stiffness quantification of conductive polymers for bioelectrodes. *J. Poly. Sci. Part BL Polymer Physics*, **52**: 666–675.

88. Zhou, K., Thouas, G.A., Bernard, C.C., Nisbet, D.R., Finkelstein, D.I. *et al.* (2012) Method to impart electro- and biofunctionality to neural scaffolds using graphene-polyelectrolyte multilayers. *ACS Appl. Mater. Interfaces*, **4**: 4524–4531.

89. Cui, X.Y., Lee, V.A., Raphael, Y., Wiler, J.A., Hetke, J.F. *et al.* (2001) Surface modification of neural recording electrodes with conducting polymer/biomolecule blends. *J. Biomed. Mater. Res.*, **56**: 261–272.

90. Wadhwa, R,, Lagenaur, C.F. and Cui, X.T. (2006) Electrochemically controlled release of dexamethasone from conducting polymer polypyrrole coated electrode. *J. Controlled Release*, **110**: 531–541.

91. Green, R.A., Lovell, N.H. and Poole-Warren, L.A. (2010) Impact of co-incorporating laminin peptide dopants and neurotrophic growth factors on conducting polymer properties. *Acta Biomater.*, **6**: 63–71.

92. Green, R.A., Lovell, N.H. and Poole-Warren, L.A. (2009) Cell attachment functionality of bioactive conducting polymers for neural interfaces. *Biomaterials*, **30**: 3637–3644.

93. Collier, J.H., Camp, J.P., Hudson, T.W. and Schmidt, C.E. (2000) Synthesis and characterization of polypyrrole-hyaluronic acid composite biomaterials for tissue engineering applications. *J. Biomed. Mater. Res.*, **50**: 574–584.

94. Green, R.A., Hassarati, R.T., Goding, J.A., Baek, S., Lovell, N.H. *et al.* (2012) Conductive hydrogels: mechanically robust hybrids for use as biomaterials. *Macromol. Biosci.*, **12**: 494–501.

95. Lim, K.S., Alves, M.H., Poole-Warren, L.A. and Martens, P.J. (2013) Covalent incorporation of non-chemically modified gelatin into degradable PVA-tyramine hydrogels. *Biomaterials*, **34**: 7097–7105.

96. DeLong, S.A., Moon, J.J. and West, J.L. (2005) Covalently immobilized gradients of bFGF on hydrogel scaffolds for directed cell migration. *Biomaterials*, **26**: 3227–3234.

97. Cheong, G.L.M., Lim, K.S,, Jakubowicz, A., Martens, P.J., Poole-Warren, L.A. and Green, R.A. (2014) Conductive hydrogels with tailored bioactivity for implantable electrode coatings. *Acta Biomater.*, **10**: 1216–1226.

98. Hassarati, R.T., Dueck, W.F., Tasche, C., Carter, P.M., Poole-Warren, L.A. and Green, R.A. (2014) Improving cochlear implant properties through conductive hydrogel coatings. *IEEE Trans. Neur. Syst. Rehab. Eng.*, **22**: 411–418.

99. Kim, D.H., Abidian, M. and Martin, D.C. (2004) Conducting polymers grown in hydrogel scaffolds coated on neural prosthetic devices. *J. Biomed. Mat. Res. Part A*, **71A**: 577–585.

100. Kim, D.H., Wiler, J.A., Anderson, D.J., Kipke, D.R and Martin, D.C. (2010) Conducting polymers on hydrogel-coated neural electrode provide sensitive neural recordings in auditory cortex. *Acta Biomater.*, **6**: 57–62.

101. Iijima, S. (1991) Helical microtubules of graphitic carbon. *Nature*, **354**: 56–58.

102. Fabbro, A., Prato, M. and Ballerini, L. (2013) Carbon nanotubes in neuroregeneration and repair. *Adv. Drug Del. Rev.*, **65**: 2034–2044.

103. Garrett, D.J., Flavel, B.S., Shapter, J.G., Baronian, K.H.R. and Downard, A.J. (2010) Robust forests of vertically aligned carbon nanotubes chemically assembled on carbon substrates. *Langmuir,* **26**: 1848–1854.

104. Garrett, D.J., Brooksby, P.A., Rawson, F.J., Baronian, K.H.R. and Downard, A.J. (2011) Reproducible fabrication of robust, renewable vertically aligned multiwalled carbon nanotube/epoxy composite electrodes. *Analyt. Chem.*, **83**: 8347–8351.

105. Bareket-Keren, L. and Hanein, Y. (2013) Carbon nanotube-based multielectrode arrays for neuronal interfacing: progress and prospects. *Frontiers in Neural Circuits*, **6**: 16.

106. Keefer, E.W., Botterman, B.R., Romero, M.I., Rossi, A.F. and Gross, G.W. (2008) Carbon nanotube coating improves neuronal recordings. *Nat. Nanotechnol.*, **3**: 434–439.

107. Saxena, T., Karumbaiah, L., Gaupp, E.A., Patkar, R., Patil, K. *et al.* (2013) The impact of chronic blood-brain barrier breach on intracortical electrode function. *Biomaterials*, **34**: 4703–4713.

108. Karumbaiah, L., Saxena, T., Carlson, D., Patil, K., Patkar, R. *et al.* (2013) Relationship between intracortical electrode design and chronic recording function. *Biomaterials*, **34**: 8061–8074.

109. David-Pur, M., Bareket-Keren, L., Beit-Yaakov, G., Raz-Prag, D. and Hanein, Y. (2014) All-carbon-nanotube flexible multi-electrode array for neuronal recording and stimulation. *Biomed. Microdevices*, **16**: 43–53.

110. de Asis, E.D., Nguyen-Vu, T.D.B., Arumugam, P.U., Chen, H., Cassell, A.M. *et al.* (2009) High efficient electrical stimulation of hippocampal slices with vertically aligned carbon nanofiber microbrush array. *Biomed. Microdevices*, **11**: 801–808.

111. Sorkin, R., Greenbaum, A., David-Pur, M., Anava, S., Ayali, A. *et al.* (2009) Process entanglement as a neuronal anchorage mechanism to rough surfaces. *Nanotechnol.*, **20**: 8.

112. Kolarcik, C.L., Catt, K., Rost, E., Albrecht, I.N., Bourbeau, D. *et al.* (2015) Evaluation of poly(3,4-ethylenedioxythiophene)/carbon nanotube neural electrode coatings for stimulation in the dorsal root ganglion. *J. Neural. Eng.*, **12**: 15.

113. Samba, R., Herrmann, T. and Zeck, G. (2015) PEDOT-CNT coated electrodes stimulate retinal neurons at low voltage amplitudes and low charge densities. *J. Neural. Eng.*, **12**: 11.

114. Donaldson, K., Poland, C.A., Murphy, F.A., MacFarlane, M., Chernova, T. and Schinwald, A. (2013) Pulmonary toxicity of carbon nanotubes and asbestos – similarities and differences. *Adv. Drug Del. Revs.*, **65**: 2078–2086.

115. Nayaga, D.A.X., Williams, R.A., Chen, J., Magee, K.A., Irwin, J. *et al.* (2011) Biocompatibility of Immobilized aligned carbon nanotubes. *Small*, **7**: 1035–1042.

116. Ganesan, K., Garrett, D.,J, Ahnood, A., Shivdasani, M.N., Tong, W. *et al.* (2014) An all-diamond, hermetic electrical feed-through array for a retinal prosthesis. *Biomaterials*, **35**: 908–915.

117. Thalhammer, A., Edgington, R.J., Cingolani, L.A., Schoepfer, R. and Jackman, R.B. (2010) The use of nanodiamond monolayer coatings to promote the formation of functional neuronal networks. *Biomaterials*, **31**: 2097–2104.

118. Xiao, X.C., Wang, J., Liu, C., Carlisle, J.A. and Mech, B. *et al.* (2006) *In vitro* and *in vivo* evaluation of ultrananocrystalline diamond for coating of implantable retinal microchips. *J. Biomed. Mater. Res. Part B*, **77B**: 273–281.

119. Halpern. J.M., Cullins, M.J., Chiel, H.J. and Martin, H.B. (2010) Chronic *in vivo* nerve electrical recordings of *Aplysia californica* using a boron-doped polycrystalline diamond electrode. *Diamond Rel. Mat.*, **19**: 178–181.

120. Birrell, J., Carlisle, J.A., Auciello, O., Gruen, D.M. and Gibson, J.M. (2002) Morphology and electronic structure in nitrogen-doped ultrananocrystalline diamond. *Appl. Phys. Lett.*, **81**: 2235–2237.

121. Kim, Y.T., Ito, Y., Tadai, K., Mitani, T., Kim, U.S. *et al.* (2005) Drastic change of electric double layer capacitance by surface functionalization of carbon nanotubes. *Appl. Phys. Lett.*, **87**: 3.

122. Hadjinicolaou, A.E., Leung, R.T., Garrett, D.J., Ganesan, K., Fox, K. *et al.* (2012) Electrical stimulation of retinal ganglion cells with diamond and the development of an all diamond retinal prosthesis. *Biomaterials*, **33**: 5812–5820.

123. Garrett, D.J., Saunders, A.L., McGowan, C., Specks, J., Ganesan, K. *et al.* (2016) *In vivo* biocompatibility of boron doped and nitrogen included conductive-diamond for use in medical implants. *J. Biomed. Mat. Res. Part B: Appl. Biomat.*, **104**(1): 19–26.

124. Shivdasani, M.N., Villalobos, J. and Garrett, D.J. Unpublished work.

125. Bergonzo, P., Bongrain, A., Scorsone, E., Bendali, A., Rousseau, L. *et al.* (2011) 3D shaped mechanically flexible diamond microelectrode arrays for eye implant applications. *The MEDINAS Project. Irbm.*, **32**: 91–94.

4

PRINCIPLES OF RECORDING FROM AND ELECTRICAL STIMULATION OF NEURAL TISSUE

JAMES B. FALLON

Bionics Institute and Medical Bionics Department, University of Melbourne, East Melbourne, Victoria, Australia

PAUL M. CARTER

Cochlear Ltd, Macquarie Park, NSW, Australia

4.1 INTRODUCTION

Great advances have occurred in lower power, miniaturised computers in recent years. Impressive as they are, for these machines to be useful in the field of neurobionics they must be able to interface well with the body's nervous system. The hard, electronic world of microcomputing is inherently very different from the soft, ionic world of the body and the interface between these two systems remains one of the major challenges in the field of neurobionics. This chapter explores various aspects of that interface by examining the principles and methods used in both recording from and stimulation of neural tissue. It is divided into four major sections: the first provides an overview of the anatomy and physiology of neural tissue; the next section describes the principles of recording from neural tissue and gives an overview of the current methods in this area; the third section describes the principles behind the electrical stimulation of neural tissue, and the final section describes the main principles behind the safe stimulation of neural tissue.

Neurobionics: The Biomedical Engineering of Neural Prostheses, First Edition.
Edited by Robert K. Shepherd.

4.2 ANATOMY AND PHYSIOLOGY OF NEURAL TISSUE

The human body contains around 100 billion neurons and as many support cells (glia in the central nervous system (CNS) and Schwann cells in the peripheral nervous system (PNS)), most of which are located within complex networks within the CNS (composed of the brain and the spinal cord) (Moffett *et al.* 1993). In contrast, the neurons within the PNS (composed of the cranial and spinal nerves) can be grouped into afferent and efferent neurons that transmit information to or from the CNS respectively. The target cells of neural prostheses are the neurons, the "active" cells of the neural tissue, and will be the focus of this chapter. The general configuration of a neuron is illustrated in Figure 4.1, and is comprised of a nucleus within the cell body, which in turn is surrounded by a number of dendrites. Typically a single, although sometimes two, axons emerge from the cell body, and after traversing some distance terminates in a number of synaptic connections to other nerves, muscles or glands. The precise configuration of each neuron is dependent on its role. For example, the length of the axon can vary from less than one millimetre to more than a metre and can be either wrapped in a myelin sheath or unmyelinated. In the PNS, axons are myelinated by individual Schwann cells, while within the CNS, a single oligodendrocyte will myelinate several axons.

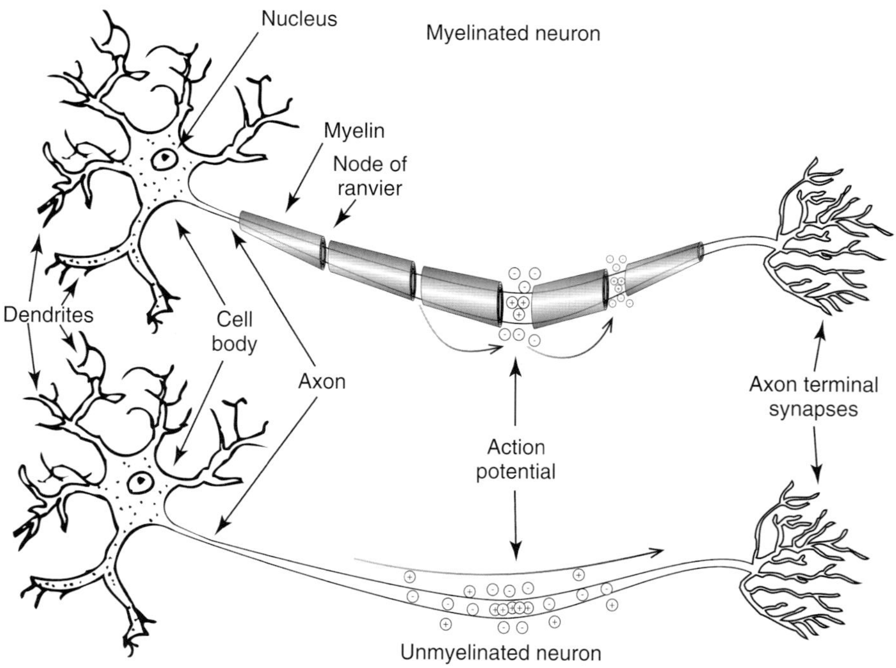

Figure 4.1 Schematic of a myelinated and unmyelinated peripheral neuron demonstrating the basic structure and anatomy. Also demonstrated is the propagation of an action potential, and resulting accumulation of ions, along the axon of each of the neurons.

In the PNS and spinal cord, individual axons serving similar purposes are grouped together into fascicles, which are in turn grouped into large nerve fibres or bundles. Note that the term "nerve" normally refers to one of these fibres or bundles of axons, whereas the term "neuron" represents an individual nerve cell. In contrast, in the brain, neurons involved in similar tasks are grouped into nuclei or cortical regions, as identified by the anatomist Korbinian Brodmann at the start of 20th century (Brodmann 1909). These regions make up the so-called grey matter of the brain. They are typically connected together by groups of axons which make up the white matter of the brain. The basic wiring diagram of much of the PNS is quite well understood. However, the connections within the CNS, sometimes referred to as the connectom, are still an area of current active research (Sporns 2013). Independent of their role, the same general physiological principles apply to all nerves.

4.2.1 Active neurons

A distinguishing feature of neurons is that they are "excitable" cells, that is, they are capable of initiating and conducting action potentials. Action potentials are travelling waves of voltage change along a neuron's axon, and are the primary method by which the nervous system transmits information. Action potentials arise from voltage and time-dependent changes in the selective permeability of the neural membrane to ions (Figure 4.2), particularly potassium (K^+) and sodium (Na^+) ions, and were first described by Hodgkin and Huxley (1952) using the squid giant axon. The permeability to K^+ and Na^+ is predominately controlled by voltage gated ion channels. At rest, the internal potential of a neuron with respect to its surround tissues (i.e. its

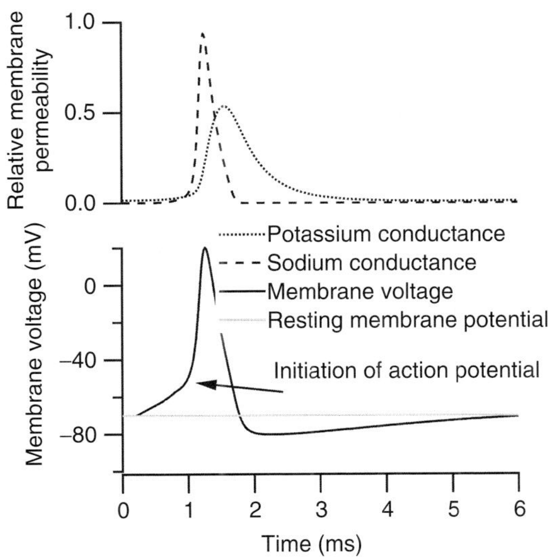

Figure 4.2 Example of changes in potassium and sodium conductances during an action potential as derived from the Hodgkin-Huxley equations (Hodgkin and Huxley 1952).

resting membrane potential) is typically around –70 mV. This potential is driven by the difference in intracellular and extracellular concentrations of K^+ and Na^+ and is described by the Goldman equation (Equation 4.1) (Goldman 1943). The typical extracellular concentrations of these ions are 4 mM and 145 mM respectively, while the typically intracellular concentrations are 140 mM and 10 mM respectively (Moffett *et al.* 1993). These concentration gradients are predominately established and maintained by sodium-potassium adenosine triphosphatase, more commonly known as the sodium-potassium pump, located in the cell membrane. For each molecule of adenosine triphosphate converted to adenosine diphosphate, the sodium-potassium pump moves two potassium ions in and three sodium ions out of the cell. In a typical neuron, the sodium-potassium pump is responsible for over half of the cell's energy needs.

Goldman Equation

$$E_m = E_{K^+}P_{K^+}/P_{tot} + E_{Na^+}P_{Na^+}/P_{tot^+}E_{Cl^-}P_{Cl^-}/P_{tot} \quad (4.1)$$

where

E_{m} is the membrane potential, measured in volts
E_{X} is the equilibrium potential for ion X, also in volts
P_{X} is the relative permeability of ion X in arbitrary units (e.g. Siemens for electrical conductance)
P_{tot} is the total permeability of all permeant ions, in this case $P_{\mathrm{K}}{}^+ + P_{\mathrm{Na}}{}^+ + P_{\mathrm{Cl}}{}^-$

At rest, the membrane permeability to K^+ is much greater than to Na^+, and therefore the resting membrane potential is near the equilibrium potential for K^+. To initiate an action potential, the membrane must first be depolarised by either a synaptic input or an external stimulus (Section 4.4.5). As the membrane is depolarised from around –70 mV, voltage sensitive Na^+ channels rapidly begin to open, allowing a pulse of Na^+ to enter the cell and resulting in further depolarisation (Figure 4.2). The opening of these channels is well described by the Hodgkin-Huxley equations (Hodgkin and Huxley 1952) and is evident as a rapid increase in the membrane permeability to Na^+ ions. The self-reinforcing activation of the voltage sensitive Na^+ channels causes the membrane potential to increase towards the Nernst potential of Na^+ of approximately 40 mV. However, the depolarisation also causes a slower activation of voltage sensitive K^+ channels and an inactivation of the Na^+ channels. The combined effect of the reducing Na^+ permeability and increasing K^+ permeability is to slow and then reverse the depolarisation of the membrane, with a peak voltage of around +15 mV. As the membrane begins to repolarise, the Na^+ channels remain inactivated and the K^+ channels slowly begin to inactivate. The still increased K^+ permeability helps to repolarise the cell membrane, often with an overshoot to a level closer to the K^+ Nernst potential (–80 mV) evident as a hyperpolarisation.

While the Na^+ channels remain inactivated, the neuron is incapable of initiating another action potential and is said to be in its absolute refractory period. As the Na^+

channels de-inactivate and the K^+ permeability slowly returns to normal, the neuron is capable of initiating another action potential, but requires additional outside depolarisation and is said to be in its relative refractory period. In the absence of any external stimuli, for most neurons the voltage sensitive channels will slowly return to their resting states and the cell membrane will return to around the resting membrane potential. Refractory periods vary significantly (Enderle and Bronzino 2012), with mammalian auditory neurons for example having an absolute refractory period of around 1 ms and a relative refractory period of around 5 ms; while for cortical neurons, the absolute and relative refractory periods can be of the order of hundreds of milliseconds.

In an unmyelinated axon, the depolarisation of one part of the membrane will be coupled to adjacent regions of the membrane, also causing a depolarisation (Figure 4.1). The strength of the coupling is determined by basic cable theory (Rall 2011), and is well approximated by an exponential decay function. For unmyelinated axons an action potential, once initiated, will sweep down the length of the axon in a continuous wave, decaying in amplitude as it goes. From cable theory, it is clear that the length constant of the exponentially decaying amplitude is proportional to the square root of the diameter, and therefore action potentials propagate more rapidly along larger fibres.

The purpose of myelination is to effectively extend the length constant for action potential propagation by increasing the membrane resistance in the areas where myelin surrounds the cell, therefore directing more of the current flow along the length of the axon. Using cable theory, it can be shown that the optimum ratio of fibre diameter to myelin diameter to maximise the length constant while maintaining overall fibre diameter is 60% axon, 40% myelin, that is 1.6:1. Not surprisingly, these proportions are closely followed in nature. The addition of more myelin above this ratio would result in a decrease in the area of the axon lumen, resulting in an increase in the effective resistance to current flow along the axon, decreasing propagation. The myelin sheath increases the membrane resistance enough to prevent the depolarisation of the neuronal membrane underlying the myelin, and therefore gaps in the myelin sheath are required to allow the regeneration of the action potential. These approximately 1 μm long gaps in the myelin are known as the nodes of Ranvier (Figure 4.1). In myelinated axons, current flowing from one node must produce a threshold depolarisation at the next, and to aid this, the density of voltage sensitive Na^+ channels is high at the nodes of Ranvier. This jumping from node to node, or saltatory conduction, results in a significant increase in the conduction velocity of the action potential for a given axon diameter. While increasing the spacing between nodes increases the conduction velocity, it also increases the chances of the action potential failing to propagate; therefore the spacing of nodes is approximately proportional to the diameter of the axon for most neurons. A ratio of inter-node distance to fibre diameter of approximately 50:1 to 100:1 is typical.

The axon of a nerve cell essentially acts as a transmission line, faithfully transmitting the action potential from the site of initiation to its terminal. At the axonal terminal, while some synapses activate effector organs such as muscles and glands, the majority of neurons then make synaptic connections to dendrites on other neurons. It is at the synapse between neurons that information processing occurs; both in the form combining input from multiple neurons and in changing synaptic strength.

As the action potential enters the axon terminal or synapse, the depolarisation of the membrane causes a release of neurotransmitter in the gap between the neurons, the synaptic cleft. The neurotransmitter causes receptors in the post-synaptic cell to alter the membrane conductances that results in either an excitatory post-synaptic potential (EPSP) or an inhibitory post-synaptic potentials (IPSP). Typically the EPSPs and IPSPs from many pre-synaptic neurons on many synapses then summate, and potentially cause the post-synaptic neuron to generate an action potential. Clearly the relative strengths and numbers of the EPSPs and IPSPs will affect the response of the post-synaptic cell. The strength of individual connections can be altered over relatively short time periods (minutes to hours) by a process known as spike-timing dependent plasticity (Caporale and Dan 2008), during which the strength of a synaptic connection is increased, if the pre-synaptic neuron is active (via activity from another synapse) immediately before the post-synaptic neuron activates, but is decreased if the post-synaptic neuron is active immediately before the pre-synaptic neuron activates. In this way, inputs that are involved in the activation of the post-synaptic neuron are strengthened, while those that are not are weakened. This modification of synaptic strength occurs in both the adult as well as the developing nervous system. The strength of synaptic connections between neurons can also be changed over longer time periods (days to weeks) by changes in the number of connections, typically by the growth or retraction of synaptic spines (small protuberances from the target neuron), which can result in dramatic changes of function (Keck *et al.* 2008). Synaptogenesis tends to peak in early development and has led to the concept of critical or sensitive periods early in life during which the brain is more sensitive to modification by external stimuli; although even the adult brain is capable of significant plasticity.

4.3 PHYSIOLOGICAL PRINCIPLES OF RECORDING FROM NEURAL TISSUE

In the late 18th century, Luigi Galvani was the first to report evidence of electrical activity in neural tissue, but it was not until the 20th century that the first recordings of this electrical activity were made (Adrian 1954). Since then there has been rapid progress in our ability to record the activity of neural tissue and we can now record activity over a large range of scales, from a single ion channel in a cell membrane to the massed activity of large brain regions. This section explores the principles and techniques required to record neural activity, with examples of a number of common techniques to record the electrical signals generated by neural tissue. The section will conclude with a brief discussion of functional imaging techniques, the current trend in examining the activity within neural tissue of the CNS.

4.3.1 Theory of recording

The measurement of biological voltages and current involves conversion of ionic currents in tissue to electron currents in wires. This conversion is done at the electrode–electrolyte interface via a combination of transfer of charge across the interface and changes in the charge distribution at the interface. The charge is typically distributed in a thin double layer on the electrode, a thin charge layer in the electrolyte, known at the Helmhotz layer, and a more diffuse layer in the electrolyte, sometime referred to as the Gouy layer. Electrodes are said to be perfectly polarisable if no charge transfer occurs between this double layer in the electrolyte and the metal electrode. Real electrodes fall between perfectly polarisable and perfectly non-polarizable electrodes, depending on the degree of charge transfer.

Perfectly polarisable electrodes are essentially purely capacitive, with no transfer of charge and noble metals such as platinum often approach this. Polarisable electrodes have the advantage that because there is no charge transfer, there is no chemical reactions that occur at the interface, just a dense build-up of charge. This dense build-up of charge is also their major disadvantage, as any disturbance of the interface, such as physical movement, results in large artefacts due to the disturbance of this mobile layer of charge. Perfectly non-polarizable electrodes have an unhindered exchange of charge between the electrodes and the electrolyte and silver-silver chloride electrodes often approach this. As there is no build-up of charge at the electrode-electrolyte interface, this type of electrode is less susceptible to movement artefacts, but always involves some form of chemical reaction. Therefore, the use of these electrodes must be limited to situations in which any products of the chemical reaction will not have any adverse impacts, particularly in terms of toxicity to the underlying tissue and limits their use to acute applications.

The action potential is arguably the fundamental signal of interest when recording from neural tissue, and for a single neuron this signal is typically sufficient to fully describe its behaviour. As described previously, action potentials result from changes in membrane conductance due to the opening and closing of ion channels. At the other extreme of scale, often it is not the activity of a single neuron but rather a large number of neurons that are functionally connected that is of interest. These changes in scale are reflected in the different amplitudes and bandwidth of different neural signals of interest (Figure 4.3). Individual action potentials are typically tens of mV in amplitude with a relatively wide bandwidth. In contrast, recordings of massed brain activity from the scalp (electroencephalography, EEG) are typically 10–100 μV in amplitude and limited in frequency to between 0.1 and 70 Hz. The large differences in amplitude and bandwidth for different signals require the use of electrodes with different electrical characteristics. Electrode choice is also important in determining the selectivity of the recording, with smaller, higher impedance electrodes required for recordings from physically smaller areas.

4.3.2 Recording electrodes

For recording from single neurons, or small sections of the membrane of a single neuron, the glass micropipette is the electrode of choice. These electrodes are made from capillaries of quartz glass that are heated and then drawn to produce tip diameters of 5 μm or less. The micropipette is filled with an ionic fluid and a silver-silver chloride electrode placed inside. These electrodes can be used to record from single or multiple ion channels in a section of membrane using the patch clamp technique.

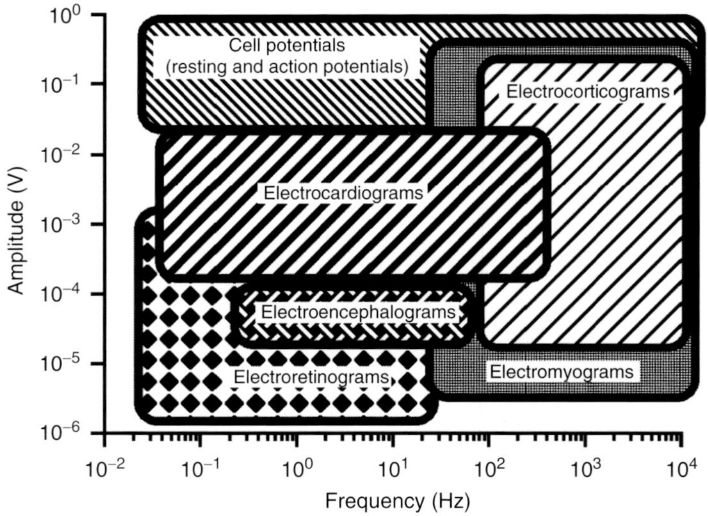

Figure 4.3 Amplitudes and bandwidth of several common biological signals.

Typically these recordings are done in isolated cell cultures, although they can also be performed *in vivo*. There are a number of variations on the technique (Figure 4.4A), but all involve making a high resistance seal between the electrode and a patch of cell membrane, often referred to as the gigaseal. The "cell-attached" configuration allows the recording of current through the ion channels within the patch of membrane with control of the immediate extracellular environment. Both the "inside-out" and "outside-out" configurations allow for control of both the extracellular and intracellular environment of the patch, but obviously without the normal intracellular responses due to the removal of the remainder of the cell. The "whole-cell" configuration records the current through multiple channels and is similar to "sharp" electrode recordings described later.

Glass micropipettes (B) or insulated metal microelectrodes (typically tungsten, (C) with impedances of tens of M Ω (~1 μm diameter), or so-called sharp electrodes, can be used for recording potentials from within or more typically adjacent to individual neurons. Intracellular electrodes can be used to record the resting membrane potential, postsynaptic potentials and action potentials. In contrast, extracellular electrodes can only record action potentials, but because they do not penetrate into the cell, can be used for long periods of recording without damage to the cell. Extracellular recordings using metal microelectrodes of a single neuron can also be performed in awake, behaving animals, including in people (Vallbo 1989).

Often the activity of multiple neurons needs to be recorded simultaneously, particularly when attempting to determine the functional connections between them. Obviously, multiple single glass or metal microelectrodes can be used, but simply adding more electrodes becomes impractical. Therefore, multichannel electrode arrays have been developed based on a range of different technologies for both *in vitro* and *in vivo*

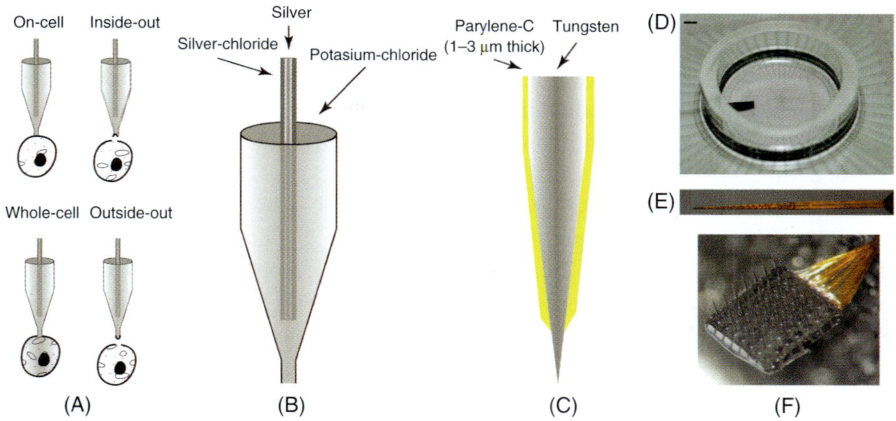

Figure 4.4 Example of recording electrodes from patch-clamp to multi-electrode arrays. There are a number of variations of "patch-clamp" recordings that vary, depending on which part of the cell membrane is attached to the glass micropipette and the remainder of the cell. (A and B) Glass microelectrodes are typically made from quartz glass that has been pulled to a sharp tip and then filled with high molarity potassium chloride solution, while metal microelectrodes (C) are typically insulated with a thin covering of parylene-C. (D–F) Examples of multi-electrodes arrays from (D) Multichannel System, (E) NeuroNexus and (F) Blackrock Microsystems.

recordings. The electrodes for *in vitro* arrays are typically made of titanium nitrite with up to a few hundred, approximately 20 μm diameter electrodes in a couple of square millimetres (Figure 4.4D). These arrays are designed to record from tissue that has been cultured on the array, or from acute brain slices. The arrays have become increasingly thin and transparent and can now be coupled with imaging techniques to provide information about both intra- and extracellular events.

There are two main types of *in vivo* arrays, linear and planar, which are designed for different target tissues. The linear arrays, typified by the arrays produced by NeuroNexus (www.neuronexus.com) (Figure 4.4E), are made on silicon substrates and are designed to penetrate into neural structures (e.g. a cortical column). The arrays have electrodes distributed along their length, and therefore allow signals to be recorded throughout the insertion depth. In contrast, the planar arrays, silicon substrates typified by the Blackrock Microsystems (www.blackrockmicro.com) (Figure 4.4 F), or microwires typified by MicroProbes (www.microprobes.com) are designed to record across a plane within a neural structure, due to the electrodes being located on the tips of the arrays. The *in vivo* arrays all have electrode impedances in the hundreds kΩ range and therefore these extracellular recordings are typically from a combination of a single neuron and/or a small group of neurons. The recordings can be performed in awake, behaving animals, and have also been used clinically as part of a brain–machine interface (Hochberg *et al.* 2012) (Chapter 9). However, an issue with the use of these arrays is the degradation in neural signal often observed over long durations of use related to an ongoing foreign body response to the array. Much of this response is thought to relate to the mechanical mismatch between arrays and the neural tissue (several orders of magnitude difference in Young's modulus), and to micro-motion of the arrays (Karumbaiah *et al.* 2013). These issues have led to the development of thin flexible arrays based on polymers such as polyimide and parylene that offer a better mechanical match to the neural tissue (Lind *et al.* 2010; Norlin *et al.* 2002; Tooker *et al.* 2012) (Chapter 2).

Even the most flexible penetrating electrode will cause some damage to the underlying tissue, which is particularly undesirable in clinical applications such as sub-dural recording when localising the source of focal epilepsy (Cook *et al.* 2013). In such situations, a planar electrode array of approximately 5 mm diameter platinum electrodes in a grid can be placed directly on the surface of the brain. These electrodes are not capable of recording responses from individual neurons, or even small groups of neurons, but rather record the activity of a small region of neurons. However, by placing the electrodes in close contact with the underlying neural tissue and with modern signal processing techniques, they can provide a spatial resolution of around 10 mm (Cook *et al.* 2013).

Electrodes attached to the skin, used to record electromyograms (EMGs), electrocardiograms (ECGs) and EEGs, provide even lower spatial resolution, but have the advantage of being non-invasive. These electrodes are typically silver-silver chloride electrodes of more than 10 mm diameter to reduce their impedance. Even though silver-silver chloride electrodes are relatively insensitive to movement artefacts, to further minimise this artefact the electrode is often kept off the skin by a sponge containing a potassium-chloride paste, therefore limiting the physical movement at the electrode itself. The signals recorded from these electrodes are typically generated by large regions of neural tissue or whole muscles.

Due to the poor spatial selectivity of surface electrodes, the temporal patterns of the responses are often used to identify the neural generators. For example, the auditory brainstem response (ABR) and cortical auditory evoked potential (CAEP, with signals in a similar range to EMGs, ECGs and EEGs) can be used to assess the function of the auditory system (Figure 4.5). The ABR consists of five waves that are each generated by a different sub-thalamic nucleus (Figure 4.5A) and can be used to objectively assess the performance of these lower auditory nuclei, while the latency of the P1 wave in the CAEP (Figure 4.5B) has been suggested as an objective measure of normal development of the auditory pathway (Sharma *et al.* 2005).

4.3.3 Amplification

Once the neural signals have been converted from ionic currents in the tissues to electron currents in wires by appropriate electrodes, there is typically a need to amplify the milli- to micro-volt signals before further processing. Ideally, the amplification should:

- not disturb the signal being measured (i.e. have high input impedance compared to the effective electrode impedance);
- be low distortion;

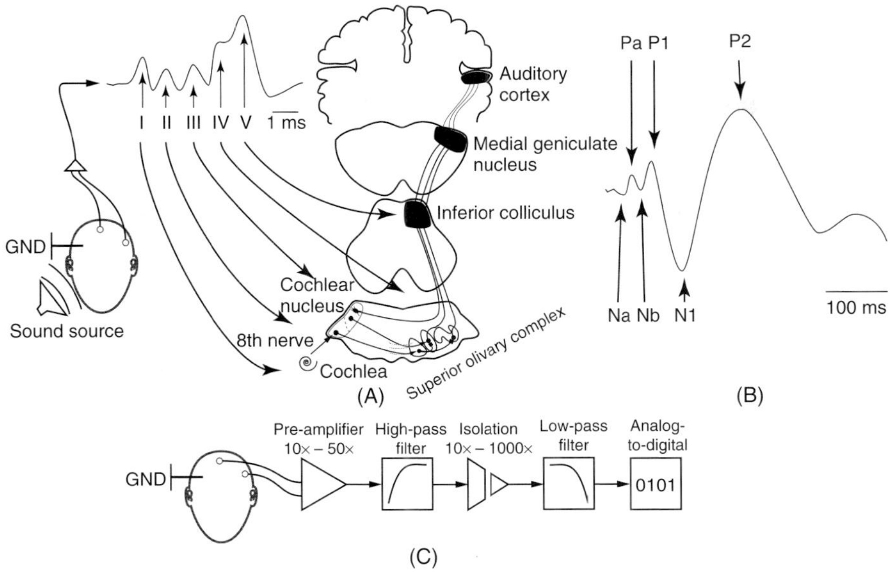

Figure 4.5 Examples of far-field recording techniques. Stylised ABR (A) and CAEP (B) recordings demonstrating the standard recording configuration of differential recording from scalp electrodes relative to a ground electrode (GND). The individual waves (I – V) in the ABR are generated by different sub-thalamic nuclei, as illustrated by the arrows in A. The waves in the CAEP (Na, Pa, Nb, P1, N1 and P2) represent different aspects of cortical processing. (C) Typical stages of amplification and signal conditioning.

- have a high rejection ratio of unwanted signals, in particular a high common mode rejection ratio (i.e. does not amplify signals common to both electrodes so that signals common to recording electrodes such as a heartbeat are eliminated);
- have minimal risk of electric shock from breakdown or from other sources (i.e. provide electrical isolation);
- have good overvoltage protection; and fast recovery following saturation.

In reality, no one amplifier can provide all this, so amplification and signal processing is divided into several stages (Figure 4.5C).

The first stage of amplification has high input impedance, good common mode rejection, and as much gain as possible to minimise the effects of noise introduced by the other stages. However, the amplification is limited by the need to ensure the amplifier is not saturated due to low frequency signals or noise. Ideally this amplification is directly connected to the electrode. The high-pass filter removes any slowly varying voltages such as electrode drift, motion artefacts, etc., while the isolation stage provides extra gain and protection against supply or ground currents. Isolation is typically in the form of either optical isolation (utilising light emitting diodes and optocouplers) or transformer-based isolation (utilising the coupling of high frequency signals through the transformer windings). Finally, the signal is low-pass filtered to prevent aliasing before typically being sampled by an analogue-to-digital converter.

Due to the very small amplitude of many of the biological signals, it is very easy for them to be swamped by unwanted noise. Sources of noise include: noise from the amplifier; power line interference; movement artefacts; other biological signals; and artefacts from electrical stimulation if present. Some noise sources, such as power liner interference and movement artefacts, can be minimised by careful physical positioning of electrodes (e.g. the potassium-chloride paste in surface electrodes) and wires (e.g. twisting wires to reduce magnetic coupling, removing ground loops) and effectively removed by filtering (e.g. at 50/60 Hz notch filter for mains). Band-pass filtering is often very effective at removing unwanted biological signals due to the different bandwidths of the neural signals (Figure 4.3), although this is not always possibly (e.g. separating EEGs and electroretinograms (ERGs)). Finally, for bionic devices that involve both stimulating and recording from neural tissue, electrical artefact from the stimulation can be several orders of magnitude larger than the neural signals. Special techniques (e.g. double pulses, alternating stimulus polarity, third stimulus phase cancelation and artefact subtraction) are required for this purpose (Heffer and Fallon 2008). One of the simplest techniques is to use a sample-and-hold circuit at the input of the amplifier that is triggered by the stimulation to ensure the amplifier does not saturate, but this technique does not work if the signal and artefact overlap in time.

4.3.4 Imaging

Directly measuring the electrical signals produced by neural tissue is not the only way to record its activity. Functional imagining techniques are playing an increasingly important role, with the techniques falling into two broad categories. The first technique images changes in intracellular concentrations of ions (typically calcium) related to the depolarization of the cell membrane and is typically used for *in vitro* or small animal work. The second technique images the increased metabolic demands of active neural tissue (typically changes in haemodynamics). Once limited to low temporal resolution, these techniques are rapidly improving and increasingly taking over from more traditional electrophysiological techniques.

Calcium imaging requires the introduction of calcium indicators that change fluorescence when bound by calcium ions. These were historically chemical indicators, such as Fura-2, an aminopolycarboxylic acid, that were bulk loaded into tissue for *in vitro* experiments. While effective, the chemical indicators are not specific and can be damaging to tissue. Therefore, more recently, genetically-encoded indicators, often based on a fusion of a calcium-modulated protein and a fluorescent protein, have become more prevalent (Tian *et al.* 2012). The use of calcium imaging has progressed to be able to perform *in vivo* imaging of networks of neurons with millisecond precision in awake, behaving animals (Helmchen *et al.* 2013); although it is still not suitable for a clinical setting.

One of the earliest functional imaging techniques was positron emission tomography that used a radioactive form of fludeoxyglucose, a glucose analogue, to produce a static three-dimensional image of relative glucose uptake.

Blood-oxygen-level dependent (BOLD) contrast functional magnetic resonance imaging (fMRI) currently dominates functional imaging BOLD fMRI measures the difference in magnetization between oxygen-rich and oxygen-poor blood that results from the haemodynamic changes with neural activity. The effective spatial and temporal resolution of BOLD fMRI are limited by the haemodynamics, which are localised to within a few millimetres of the neural activity, but occur with a lag of a second or two and occur over several seconds. Therefore, while BOLD fMRI can have excellent spatial resolution, the temporal resolution will also be limited by the haemodynamics of the neural tissue. There are also issues with using these techniques with bionic devices due to the use of permanent magnets and ferrous metals. Therefore, new techniques that take advantage of the different absorption spectra of oxygen-rich and oxygen-poor blood have been used for intrinsic optical imaging *in vivo* (Hillman 2007), which allows the imaging of activity over large cortical regions. Functional near infrared spectroscopy (fNIRS) (Scholkmann *et al.* 2014) also takes advantage of changes in absorption spectra and can be used non-invasively in humans to provide similar resolution to BOLD fMRI without the need for the expensive high magnetic fields, but is limited to a maximum of around 4 cm depth.

4.4 PRINCIPLES OF STIMULATION OF NEURAL TISSUE

4.4.1 Introduction

The first therapeutic electrical stimulation of neural tissue is thought to be by the Romans, who applied the electrically active Torpedo fish to the head in an attempt to relieve migraines (see Chapter 1 for a comprehensive history of neural prostheses). However, it was not until the beginning of the 20th century that electricity could be controlled sufficiently well to allow experimentation and characterisation of its effects on neural tissue. Then, in the 1950s, the invention of the silicon transistor allowed fully implantable, controllable neural stimulators to become a reality. Later, the work of McNeal (1976) greatly contributed to our understanding of electrical stimulation of nerve tissue. McNeal described the myelinated axon of a neuron as a discrete electrical cable where each node of the cable is represented by a node of Ranvier of the axon. This allowed electrical cable theory, well understood since the days of transatlantic telegraph cables, to be applied to the conduction properties of a neuron. The McNeal model, although relatively simple in concept, has proven extremely useful in predicting neural behaviour in response to electrical stimulation.

4.4.2 Principles of neural stimulator design

This section looks at the main features of neural stimulators related to how they stimulate neural tissue.

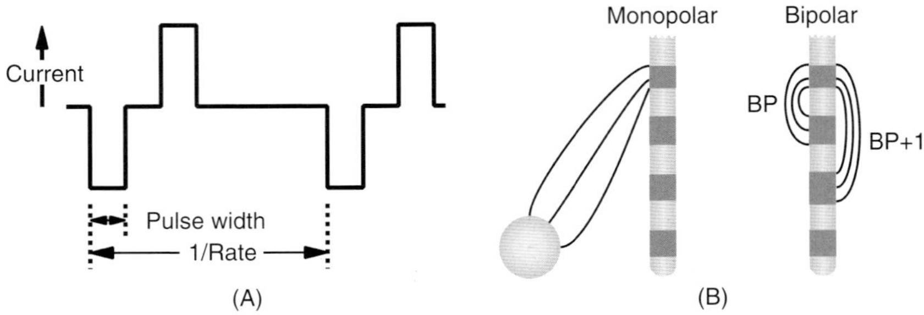

Figure 4.6 (A) Schematic of a constant current, biphasic output pulse train for a typical neural stimulator. (B) Representation of monopolar and bipolar stimulation modes typically used by neural stimulators.

4.4.2.1 Stimulus waveform A typical neural stimulator output current waveform is shown in Figure 4.6A.

The main features of the waveform are that it is:

- *Constant current*: This makes it easy to control the total charge delivered in each phase of the biphasic pulse. Fine control of the delivered charge can be achieved by accurately controlling the current and/or the duration of each pulse. Some stimulators (e.g. pacemakers) opt for the application of voltage rather than current controlled pulses. Such stimulators are typically easier and cheaper to design and implement. However, they suffer from the problem that the delivered charge – an important parameter in controlling neural excitation – varies with impedance, so they are unsuitable for situations when fine control of the neural response is required.
- *Biphasic and charge balanced*: The waveform has two phases of equal duration and opposite current, leading to the delivery of nominally zero net charge to tissue. This is required to prevent the flow of DC through tissue, which can be harmful if above a certain limit (see later in this chapter). Biphasic pulsing as a means of minimising the harmful effects of tissue stimulation was first proposed by Lilly *et al.* (1955). The safety aspects of neural stimulation are explored later this chapter.
- *Cathodic first*: The electrode closest to neural tissue is usually a cathode (current sink) in phase one of the pulse. In many applications cathodic pulses require less charge than anodic pulses to obtain a given neural response so this polarity is commonly used. Note that in bipolar mode stimulation, shown in Figure 4.6B and described in more detail later, both electrodes are close to neural tissue so the polarity of the pulse is arbitrary.
- *Repeated at a certain rate*: Most stimulators deliver repetitive stimulation of differing rates according to their application. Natural neural signals are generally repetitive in nature and a single, isolated stimulation pulse will rarely evoke a noticeable physiological response. So stimulators apply repeated pulse trains to mimic the natural activation rates of the neurons they are targeting. The rates used vary widely, depending on the application. Deep brain and spinal

cord stimulators use rates of a few tens of Hertz, whereas cochlear implants use rates of many thousands of Hertz. The perceived intensity of the stimulation is usually higher at increased stimulation rates, as is found in retinal stimulation for example (Shivdasani *et al.* 2014). In more established fields such as cochlear implants, sophisticated models of perceptual intensity have emerged (Francart *et al.* 2014).

4.4.2.2 Electrode configuration or mode Neural stimulators employ different electrode configurations (sometimes called modes) in order to excite neural tissue. Figure 4.6B shows the two most popular clinical stimulation modes (monopolar and bipolar stimulation) employed by a range of bionic devices.

In monopolar stimulation, a current is passed between an electrode in close proximity to the tissue being stimulated and a large remote "ground", "reference" or "return" electrode. The electrode near the targeted neural tissue is often known as the active electrode. An advantage of monopolar stimulation is that thresholds are often several times lower than other stimulation modes; however, the spatial spread of neural excitation may be greater than with other stimulation modes – something undesirable for a stimulator wishing to target small, specific groups of nerves.

In bipolar stimulation, a current is passed between two electrodes, both located near the neural tissue being targeted. Fine control of the region of stimulation can often be obtained as the current is mostly confined to a small region close to the active and return electrodes. Cuff electrodes, such as used with vagal nerve stimulation (Groves and Brown 2005), usually employ bipolar mode stimulation, since targeted stimulation of small groups of neurons within the cylindrical nerve bundle is required. The region of excitation can be varied by choosing different active and return electrodes (e.g. labelled as BP and BP + 1 in Figure 4.6B). Biploar mode is effective at containing the spread of current to the target region of neural tissue, for example to avoid unwanted stimulation of other, nearby nerves. Thresholds are typically significantly higher than for monopolar stimulation however, leading to increased power consumption and reduced time between battery changes for portable stimulator systems.

Many other stimulation modes have been proposed. Tripolar stimulation for example uses current of one polarity on a central active electrode and usually half the current of the opposite polarity on two adjacent electrodes. It has been shown to provide more highly focused (also known as spatially selective) stimulation than bipolar stimulation but, however, at the cost of efficiency. It can also provide stimulation which is less highly focused and has lower power consumption if it is partially combined with monopolar stimulation (i.e. return current is shared between the two adjacent electrodes and a distant ground electrode). This potentially allows a "sweet spot" between degree of focusing and the power consumption to be chosen (Litvak *et al.* 2007).

An example of a more advanced stimulation mode is known as Multipolar or "Phased Array" mode (van den Honert and Kelsall 2007). In this mode, shown to minimise the spatial spread of excitation in cochlear implants, the current is sourced from multiple electrodes simultaneously in weightings derived from previously measured electrode impedances. It potentially provides the highest degree of focusing of all modes, although the weights can be chosen to reduce the degree of spatial selectivity and power if desired to optimise these two parameters as required. Despite the plethora of available modes, monopolar and bipolar modes remain by far the most commonly used modes in commercial neural stimulators today.

4.4.3 Modelling nerve stimulation

The goal of most models of nerve behaviour under electrical stimulation is to predict whether and when an action potential is initiated. By doing this for all the neurons in a nerve bundle, the response of the whole nerve to electrical stimulation can be predicted. As discussed earlier in this chapter, an action potential occurs when the internal potential of a neuron with respect to its surrounding tissue rises by around 15 mV or more from its resting potential of around –70 mV. A two-step process can be used to determine the potential inside each node of Ranvier as a function of time and therefore determine whether an action potential has been initiated. It turns out that to a good approximation both steps can be treated independently, much simplifying the process.

4.4.3.1 Step one The first step is to calculate the potentials at the nodes of Ranvier in the tissue surrounding the axon. These can vary with time, although for a constant current stimulation source, as is often the case for modern neural stimulators, these potentials are usually assumed to be constant while the current is flowing. Note that most tissue is quite reactive as well as resistive in the frequency region of interest to neural stimulators, so to assume purely resistive tissue and hence constant tissue potentials for constant current stimulation is sometimes inaccurate (Butson and McIntyre 2005). However, the assumption is often made and for the purposes of explaining the principles of neural stimulation it is a useful simplification which we will adopt throughout this section.

Figure 4.7 illustrates how voltages are induced in tissue for the case of a point source of current in the middle of a large, homogeneous volume of tissue. This models the case of a needle electrode used in cortical stimulation for example. The tissue in Figure 4.7 contains a section of the neuron's axon showing several nodes of Ranvier. Assuming the return electrode is at some distant point, the current spreads uniformly through the tissue (which is of uniform conductivity), generating a voltage in the tissue. This voltage, as a function of the distance, r, from the electrode, can be derived from the continuous form of Ohms law, $E = \sigma J$, to give Equation 4.2:

$$V(r) = I_s / 4\pi r \sigma \tag{4.2}$$

where:

$V(r)$ is the voltage at a point in tissue
I_s is the current from the point source
σ is the conductivity of the tissue
r is the distance between the point source and the point in question

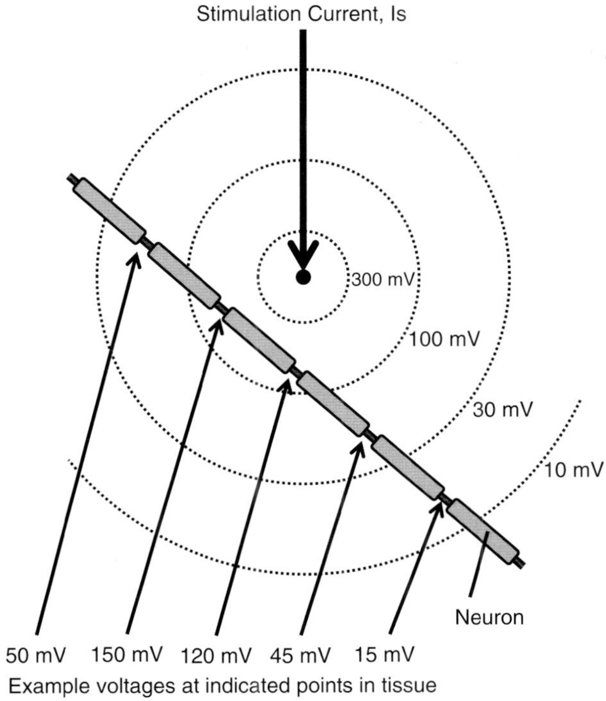

Figure 4.7 A point source of current in uniform tissue containing the axon of a neuron. Dotted circles with marked voltages are the spheres of constant voltage at various radii from a central point source of constant current. The approximate voltages appearing at five nodes of Ranvier for the section of neuron shown are indicated at the bottom of the picture.

4.4.3.2 Step two Once tissue voltages are established at each node of Ranvier, step two is to take these voltages and apply them to the McNeal model of the axon. This allows the voltages inside the axon at each node of Ranvier to be predicted as a function of time. To do this comprehensively requires the Hodgkin-Huxley equations to be solved for each node and for each time step. Specialist software has been written and is available to do this (Reilly and Diamant 2011). However, it is possible to gain an understanding of many of the properties of nerves in response to electrical stimulation without resorting to solving the Hodgkin-Huxley equations. To a reasonable approximation, the nerve membrane at each node of Ranvier can be treated as a capacitor and the intercellular fluid between nodes of Ranvier can be treated as a resistor. Knowledge of the McNeal model and a basic understanding of capacitor and resistor circuits is then sufficient to explain most characteristics of nerves under electrical stimulation.

For more complex geometries than the one shown in Figure 4.7, electro-anatomical models are often defined in software (Tran *et al.* 2015). Such models frequently use

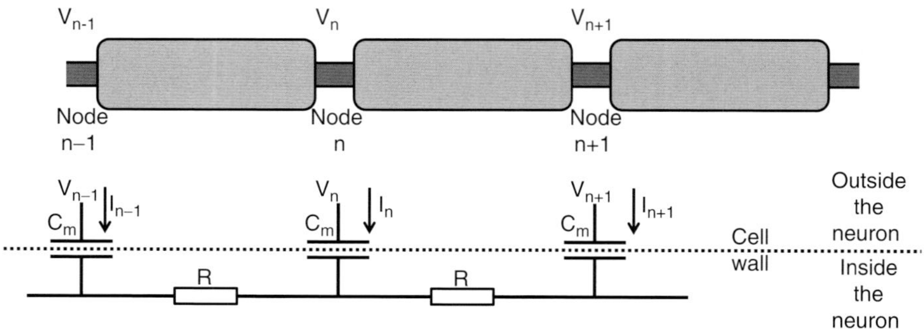

Figure 4.8 A myelinated neuron with an idealised schematic of a the McNeal equivalent circuit shown underneath

the same two-step approach described above: first calculate the voltage at each node using a geometrical anatomical model and then apply those voltages to the nerve model to calculate when and where the nerve is activated.

4.4.4 The activating function

The activating function was first proposed by Rattay (1986). It is derived from the McNeal model and provides the rate of change of the voltage across the nerve membrane at each node of Ranvier, given a set of external voltages imposed by electrical stimulation. If the rate of change of nerve membrane voltage is known, then the time taken (i.e. the pulse width for a constant current pulse) for the membrane voltage to change by the required amount to activate the nerve can be readily calculated. The activating function therefore provides a useful tool for predicting the extent of depolarisation or hyperpolarisation of any node of Ranvier and how easily it is "activated" by a given stimulating current.

Figure 4.8 shows a myelinated neuron with an idealised and simplified schematic of the McNeal equivalent circuit shown underneath.

Even without a detailed understanding of the McNeal circuit, it is clear that the voltages, V_n, etc., imposed by the external stimulation current can induce currents, I_n, etc., to flow across the cell membrane at each node of Ranvier. When this membrane current flows, it changes the voltage across the cell membrane as its capacitance, C_m, becomes charged or discharged. As discussed earlier in this chapter, if the voltage inside the neuron *increases* (with respect to the voltage outside the neuron), the node is said to be *depolarised*. If the voltage inside the neuron *decreases*, the node is said to be *hyperpolarised*. If the node is depolarised by more than about 15 mV, an action potential will be initiated and the nerve is said to be "activated".

Equation 4.3 shows a simplified equation for the activating function derived from Rattay (1986) for the case shown in Figure 4.8, where the membrane capacitance has no leakage path across it:

$$dV_{m,n}/dt = 1/RC_m[V_{n-1} - 2V_n - V_{n+1}] \qquad (4.3)$$

where:

$dV_{m,n}/dt$ is the rate of change of the membrane voltage, $V_{m,n}$, at node n

R is the internal resistance of the axon between successive nodes of Ranvier (assumed constant throughout the axon)

C_m is the capacitance of the membrane at its nodes of Ranvier

V_{n-1}, V_n and V_{n+1} and are the external voltages imposed at nodes n–1, n and n + 1 respectively by the stimulating current

Note from Equation 4.3 that the term $[V_{n-1} - 2V_n - V_{n+1}]$ is equivalent to the discrete second derivative of the node voltage V_n. Therefore the activating function is proportional to the second derivative of node voltage.

4.4.5 Properties of nerves under electrical stimulation

With knowledge of the McNeal model and the activating function, most of the properties of nerves under electrical stimulation can be analysed and understood. We will now list some of the major properties and explain them in the context of these models.

4.4.5.1 Sites of neural activation We have seen in the previous section that the activating function is proportional to the discrete second derivative of voltage along the nodes of the neuron's axon. In practice, this tends to be maximised at:

- the ends of the axon
- sharp bends in the axon
- regions where the surrounding tissue resistivity changes suddenly
- regions where stimulus current density changes (e.g. due to anatomical constraints)

For simple geometries, such as the case of a straight axon passing near a point current source shown in Figure 4.7, the activating function can be derived mathematically. It turns out that in this case the neuron is most easily activated at a site nearest to the current source by a cathodic (current sink) pulse. For such a pulse, at sites a few nodes of Ranvier away in both directions, the activating function becomes negative and an action potential can in principle be blocked from propagating along the axon at these regions. However, this normally happens at much higher levels than the threshold of activation for the nerve, so action potentials can be activated and will propagate over a range of cathodic pulses amplitudes. Anodic pulses applied in the same configuration can also activate the nerve, although they do so at a higher charge level than for cathodic pulses (Reilly and Diamant 2011).

4.4.5.2 The strength-duration curve The strength duration curve can be used in neurology as a way of characterising and diagnosing nerve pathologies. It consists of a plot of the current required to cause a particular level of stimulation (usually threshold) of a nerve as a function of the duration of current pulse used.

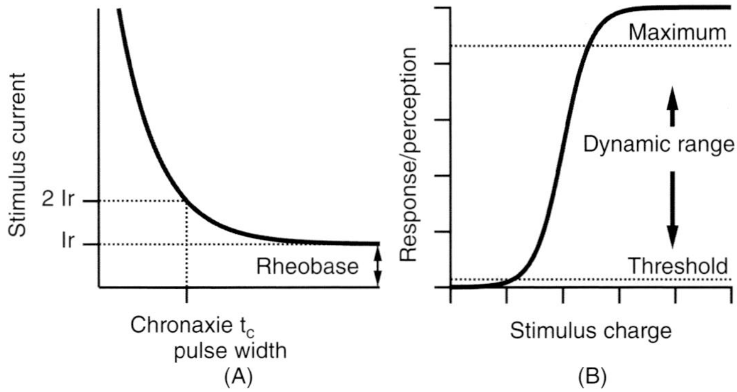

Figure 4.9 (A) Typical strength-duration curve showing the threshold activating pulse width versus the stimulus current. The rheobase current (Ir) and the chronaxie pulse width (t_c) are also illustrated. (B) Typical recruitment curve illustrating the threshold and maximum responses and the dynamic range.

Experiments conducted by Lapique (1905) established the general shape of the curve (Figure 4.9A), which was found to fit well with an equation of the following form:

$$I_s = I_r/[1 - \exp(-\ln 2t/t_c)] \tag{4.4}$$

where:

I_s is the stimulation current that evokes the response (e.g. threshold)

t is the duration of the applied stimulation current pulse that evokes the response

I_r is the rheobase current – the current that evokes the response for very long pulse durations

t_c is the chronaxie – the duration of the pulse at which twice the rheobase current evokes the response

Strength-duration curves can easily be obtained in clinical practice by applying constant current pulses of different durations to the nerve and measuring the current required to achieve nerve activation. Monophasic, biphasic or other wave shapes can be applied and pulses can be applied singularly or repeated in bursts, all of which modify the threshold and other properties of the curve. However, they can all still be well described using Equation 4.4 above, albeit with different values for the chronaxie and rheobase.

From Equation 4.4, note that these two constants, the chronaxie and the rheobase, are sufficient to completely define the strength-duration curve. The ideal model of a non-leaky nerve membrane, as depicted in Figure 4.8, results in a strength-duration curve with a rheobase of zero. In this case the strength-duration curve is a simple hyperbola of the form y = k/x. However, in practice, this is never observed experimentally. Introduction of nerve membrane "leakiness", which can be crudely modelled as a resistor in parallel with the membrane capacitance C_m in Figure 4.8, causes the rheobase to be non-zero. The chronaxie can be thought of as a measure of the time constant of the nerve and the rheobase as a measure of the "leakiness" of the nerve membrane.

4.4.5.3 Dynamic range Most nerves consist of many thousands of individual neurons and these multiple paths are used to convey various aspects of the information being transmitted. For example, the number of afferent neurons simultaneously activated within a certain time window is usually related to the perceived magnitude of the sensory input. In order to mimic natural behaviour, neural stimulators often try and obtain a graduated neural response by stimulating a controlled number of neurons in an individual nerve bundle. This can sometimes be achieved by carefully controlling the magnitude of the delivered stimulus charge.

The number of nerves activated for a given electrode configuration can be plotted as a function of the applied stimulus charge and usually reveals a sigmoid shaped "recruitment curve" similar to that of Figure 4.9B. The sigmoid shape arises as the increasing charge first reaches a threshold when activation (or recruitment) begins; then climbs steeply as an increasing number of individual neurons are activated; and finally plateaus when all the available neurons are activated. Often the maximum limit will be reached before the plateau in the recruitment curve occurs, when the response becomes uncomfortably strong or hazardous. This is labelled as the Maximum level in Figure 4.9B. The difference between the current at the Maximum level and the current at the Threshold level is known as the dynamic range.

The y-axis in Figure 4.9B can be based on a perceptual scale (e.g. the intensity of the response), on an objective measure (e.g. the force applied by a limb under control of neural stimulation) or it can be a directly measured neural response (e.g. the amplitude of the electrically evoked compound action potential as measured in a cochlear implant system).

4.4.5.4 Nerves are activated by total delivered charge Most modern neural stimulators aim to operate at pulse widths shorter than the chronaxie of the strength-duration curve. In this part of the curve, the extent of depolarisation or hyperpolarisation of the nerve membrane is roughly proportional to the product of the stimulus current, I_s, and the duration of the applied current pulse, t. Inspection of Equation 4.3, the activating function, reveals that the rate of change of membrane voltage is proportional to the external voltages that are imposed on the neuron's nodes of Ranvier. For constant current stimulation in resistive tissue, these voltages are proportional to the stimulating current that flows, I_s. It follows therefore, that the extent of depolarisation (or hyperpolarisation) experienced at each node of Ranvier is proportional to both the stimulus current, I_s, and the time for which it flows, t. Since the product of current and time is charge ($Q = I.t$), neural activation can be thought of as a proportional function of the charge delivered from a stimulating pulse. The total charge axiom is a useful rule of thumb and applies for most neural stimulators and for most pulse shapes (biphasic, monophasic, sinusoid, exponential decay, etc.). Note, however, that most modern stimulators apply biphasic, charge balanced pulses (i.e. the net charge applied after a stimulation pulse pair is zero), so the rule must be applied to just one phase or polarity of the biphasic current pulse.

4.4.5.5 Large myelinated neurons are more easily activated than small, unmyelinated neurons As discussed earlier in this chapter, neurons can vary in size and shape considerably. Sensory afferent neurons and motor neurons are typically large and myelinated and have high conduction velocities. By contrast, sensory pain neurons are typically small and unmyelinated and have low conduction velocities. Since sensory and motor neurons are usually the target of neural stimulation, it is fortunate from a neural stimulation point of view that they are usually larger than pain neurons. The reason small neurons are less easily stimulated than large ones is because the distance between nodes of Ranvier is less for small neurons (earlier in the chapter we saw that the ratio of axon diameter to internode distance is roughly the same for most axons). Therefore the driving voltages at the nodes of Ranvier (V_{n-1}, V_n etc. in Equation 4.3) are proportionally smaller for small neurons, meaning they are less likely to be excited.

Unmyelinated neurons can be thought of and modelled as neurons with very little distance between nodes (McNeal 1976). A continuous form of the activating function can be derived rather than the discrete version in Equation 4.3. This predicts similar behaviour of unmyelinated neurons to that of myelinated ones (e.g. relationship of activation threshold with fibre diameter). However, one difference is that the continuous form of the activating function predicts unmyelinated neurons to have a higher activation threshold than myelinated ones and this is found to be true in practice.

4.5 SAFETY OF ELECTRICAL STIMULATION

It is generally accepted that there are two primary mechanisms whereby neural stimulators can damage tissue – metabolic stress and electrochemical stress. Metabolic stress occurs when nerves are stimulated unnaturally (e.g. overly quickly) to a point where their natural mechanisms for maintaining homeostasis break down and damage occurs. Electrochemical stress occurs when the charge density (charge per unit area) delivered through an electrode is high enough to cause harmful electrochemical reactions to occur. Both these mechanisms will be considered in more detail later in this section.

In addition to the above mechanisms, heating of tissue from electrical stimulation can also occur, but is rarely a problem for all but the highest powered devices such as cardiac defibrillators. Also, all neural stimulators can potentially stimulate unwanted nerve populations, sometimes causing hazardous side effects (e.g. stimulation of non-auditory parts of the brainstem by an auditory brainstem implant). However, this is normally well managed through careful fitting procedures. Neither of these mechanisms will be discussed in more detail here.

4.5.1 Safe stimulation limits

The Shannon limit is probably the most widely-used model to define safe stimulation for a neural stimulating electrode (Shannon 1992). Shannon proposed that damage-free stimulation could be predicted with just two variables – charge, Q, and charge density, D – according to Equation 4.5:

$$\log(D) + \log(Q) \leq k \quad (4.5)$$

where:

Q = the charge per phase in μC
D = the charge density at the electrode surface in $\mu C.cm^{-2}$
k = an arbitrary constant found to equal 1.9 to define the region of damage free stimulation

Shannon based his limit on the comprehensive studies of McCreery *et al.* (1990), who stimulated platinum disc electrodes placed in the cortex of anaesthetised cats for a number of hours. He noted that if the charge and charge density as applied in the left-hand side of Equation 4.5 was less than k = 1.9, no damage was observed. A sketch showing the region of damage (shaded) as reported by Shannon is shown in Figure 4.10.

Shannon and McCreery both commented that the damage mechanism observed in the McCreery studies was likely to be metabolic in origin (they used the term "mass action", because neurons are recruited en masse during electrical stimulation).

McCreery's electrodes were all sufficiently large that the charge density in these experiments was much lower than that required to cause electrochemical stress. For small electrodes, such as those used in today's cochlear implants and retinal prostheses for example (~0.1–0.5 mm^2 in area), electrochemical stress is important to

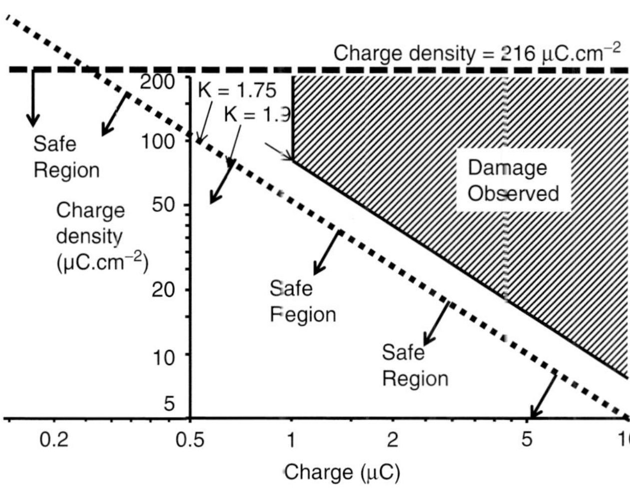

Figure 4.10 Graph of the region where Shannon reported damage was observed (shaded). The region now considered safe in cochlear implants is indicated by dotted lines and arrows.

consider as charge densities can approach the electrochemically active region. These electrodes lie in the top left region of the graph in Figure 4.10, which is an area not examined in the McCreery studies. For these electrodes an electrochemistry limit is required, based on charge density alone, in addition to the Shannon limit (see later in this chapter for more discussion on this). Very little *in vivo* work has been performed on the electrochemically safe limit. However, the *in vitro* electrochemistry literature suggests a safe electrochemistry limit may be in the range from 100 $\mu C.cm^{-2}$ (Rose and Robblee 1990) to 500 $\mu C.cm^{-2}$ (Brummer and Turner 1977a).

The regions below and to the left of the dotted lines shown in Figure 4.10 represent proposed charge and charge density limits in a cochlear implant standard currently under public review (AAMI 2015). The horizontal dotted line at the top of Figure 4.10 lies at the charge density value of 216 $\mu C.cm^{-2}$, which represents the proposed electrochemical limit in that standard. The diagonal dotted line shows a value of $k = 1.75$, which represents the proposed metabolic limit. The proposed permissible safe region for cochlear implant operation is shown by arrows below and to the left of the two dotted lines.

Note that Shannon suggested a value of $k = 1.5$ could be used as a safe limit to allow for a margin of error over the observed limit of $k = 1.9$ in the McCreery studies. Cochlear implants electrodes are different from the electrodes used by McCreery, as they do not lie in direct contact with the tissue they are stimulating. Therefore a less conservative value of $k = 1.75$ is considered appropriate for these electrodes (AAMI 2015). For electrodes lying in direct contact with neural tissue, the value of $k = 1.5$, which lies further from the damage region in Figure 4.10, and perhaps a lower value for the charge density limit may be more appropriate. A good review of the work of Shannon, McCreery and others related to the safety limits for cortical neural stimulation is available in Merrill *et al.* (2005).

The processes of metabolic and electrochemical stress will now be discussed in more detail.

4.5.2 Metabolic stress

Neural stimulators usually attempt to mimic the natural behaviour of the nerve populations they target. If they succeed in producing natural stimulation then presumably the nerve is under no more metabolic stress than it would be in its natural state and the stimulation must therefore be metabolically safe. However, it is often difficult for stimulators to accurately reproduce natural neural firing patterns. Their electrodes are usually large compared to the size of a single neuron and they often recruit a group of neurons with each stimulation pulse in a manner that would not occur naturally. In addition, neural stimulators are sometimes programmed to stimulate at rates higher than those that occur in nature (e.g. cochlear implants stimulate at many thousands of pulses per second, whereas individual auditory neurons can respond with only a few hundred action potentials per second). So neural stimulators can potentially activate nerves beyond their metabolically safe limits, which leave open the possibility of metabolic stress and temporary or permanent nerve damage.

Studies have found a number of factors are important in determining metabolic stress and these are discussed below.

4.5.2.1 ***Charge per phase*** Earlier in this chapter we saw that to a first order the charge delivered in a stimulation pulse is the parameter responsible for depolarising a neuron's membrane and therefore determining how many neurons in a group of nerves are activated. Higher charge levels correspond to greater numbers of activated neurons, which in turn place more metabolic stress on the nerve. It is hardly surprising then that most studies report charge per phase as an important determinant of neural injury (McCreery *et al.* 1990; Tykocinski *et al.* 1995). Note that the electrochemical stress mechanism, described later in this chapter, also increases strongly with the delivered charge, so the dependency of injury on delivered charge applies regardless of whether the primary damage mechanism is metabolic or electrochemical stress.

The exact mechanisms of metabolic injury due to neural stimulation are still under review. Suffice to say that stimulation of a nerve, which causes a response well beyond the subjective comfortable level of the subject for prolonged periods, appears to be damaging (Tykocinski *et al.* 1995).

4.5.2.2 ***Pulse timing*** Pulse timing (rate, duty cycle and duration) have all been shown to affect the metabolic stress experienced by the stimulated nerve. Acute experiments using anaesthetised animals have been used to test the effect of pulse timing on metabolic stress. At stimulation intensities which would have been well above comfortable levels in awake animals, it was found that higher rates, higher duty cycles and longer durations all increased levels of damage in cochlear implants (Tykocinski *et al.* 1995). Similar findings have been observed in other neural stimulation systems.

Conversely, at levels which were subjectively comfortable in awake animals, cochlear stimulation at rates of 2000 pulses per second have been shown to be safe over a period of several months (Xu *et al.* 1997). Although 2000 pulses per second is well in excess of the natural rate of auditory neurons (which typically activate up to a few hundred times per second), it may be that individual neurons are self-limiting with regard to rate due to their refractory period. The activation rate of individual neurons may follow a sub multiple of the stimulation rate at rates above the inverse of their refractory period. There is an argument for cochlear implants and perhaps other forms of sensory prostheses that the overall subjective comfort of an applied pulse train may provide a limit to the metabolic harm it can cause – in other words, if the stimulus feels comfortable for the subject, then metabolic stress is unlikely to occur. This is supported by an implanted population of over 300 000 cochlear implant recipients worldwide, most using stimulation rates well above the natural auditory neuron firing rate, where there is no evidence of metabolic stress. However, the same is not likely to apply for devices such as cortical prostheses, where there is no direct sensory feedback mechanism.

Duty cycle – the proportion of time stimulation a pulse train is actively applied to tissue – appears to have a strong effect on damage. One hundred percent duty cycles were found to cause significantly more damage than 50% duty cycle in some studies (Agnew *et al.* 1989; Tykocinski *et al.* 1995).

4.5.2.3 Electroneural anatomy The anatomy of the stimulation site and the electrode's location within that anatomy play an important part in determining the metabolic stress experienced by the nerve. Individual anatomies can be complex, so electro-anatomical modelling is sometimes used to determine parameters such as charge density at the neural tissue. Much of the current flow in cochlear implant stimulation for example, occurs in tissue other than the auditory nerve where it will not affect the nerve (Tran *et al.* 2015). When considering metabolic stress, clearly only current that flows across the target neural structures is relevant. Hence the total current should be reduced by a factor based on the electro-anatomical modelling to determine the likelihood of metabolic stress in anatomies where the electrode is not in direct contact with neural tissue.

4.5.3 Electrochemical stress

This section presents an overview of the harm that can occur from electrochemical by-products and electrode dissolution during neural stimulation. More details on the mechanisms involved can be found in Chapter 3.

Platinum is the most widely-used neural stimulator electrode material. In the 1970s, seminal work by Brummer and Turner (1977b) demonstrated that excessive charge density (charge per unit area of electrode surface) can give rise to potentially harmful electrochemical reactions at the platinum electrode surface as follows:

At the anode (where current is sourced)

1. Oxidation of water $2H_2O \rightarrow O_2 \uparrow +4H^+ + 4e^-$
2. Oxidation of saline, e.g. $Cl^- + H_2O \rightarrow ClO^- + 2H^+ + 2e^-$
3. Oxidation of platinum, e.g. $Pt + 6Cl^- \rightarrow PtCl_6^{2-} + 4e^-$
4. Oxidation of organics, e.g. $C_6H_{12}O_6 + 6H_2O \rightarrow 6CO_2 + 24H^+ + 24e^-$

At the cathode (where current is sunk)

5. Reduction of water $2H_2O + 2e^- \rightarrow H_2 \uparrow +2OH^-$

Brummer and Turner proposed that these reactions could cause damage to tissue surrounding the electrode by:

- changing its pH (reactions 1, 2, 4 and 5)
- causing gas bubbles (reactions 1 and 5)
- the production of toxic by-products (reactions 2, 3 and 4)
- dissolution of the platinum electrode (reaction 3)

The following factors have been found to be important in determining the level of electrochemical stress.

4.5.3.1 Charge balanced stimulation and direct current (DC) The charge density limit described above is necessary but not sufficient to avoid electrochemical stress. In most neural stimulator applications, a series of repeated biphasic pulses are delivered to the electrodes. If a small charge is left on an electrode following each stimulation pulse, then the net charge delivered to that electrode can quickly build up over time with repeated pulsing. Eventually it will reach a point where the maximum charge density limit is reached and electrochemical reactions will begin to occur (Merrill *et al.* 2005).

This net delivery of charge with repeated pulsing can be defined in electrical engineering terms as direct current or DC. For this reason, DC is potentially very harmful to neural stimulators and measures are invariably put in place to control it. Levels of DC of 2–3 uA have been shown to be harmful in the cochlea (Shepherd *et al.* 1991), where it produced bone growth and reduction of auditory neurones in the cochlea. The safe level of DC is currently defined as less than 100 nA in the implantable device standard related to cochlear implants and auditory brainstem implants (EN45502-2-3, 2010).

For this reason, most neural stimulators use nominally charge balanced stimulation pulses, such as the biphasic ones shown in Figure 4.6A as proposed by Lilly *et al.* (1955). However, even the use of nominally charged balanced stimulation may not be sufficient to maintain DC below the required limit. It is common therefore for neural stimulators to include capacitors in series with their electrodes to block the delivery of DC to tissue and/or to short electrodes after each stimulation pulse to recover any net charge delivered to tissue.

4.5.3.2 Charge density As discussed previously, provided DC is adequately controlled, charge density is the single-most important factor in determining the level of electrochemical stress. In electrical engineering terms the electrode/tissue interface behaves similarly to a capacitor, which becomes charged and discharged as stimulation current flows across it. As the interface becomes charged, the voltage between the metal electrode and the tissue changes. The hazardous electrochemical reactions mentioned above are all driven by the Butler-Volmer equation (Böckris *et al.* 2002), which states that the rate at which the reactions occur is an exponential function of the electrode/tissue voltage. When the electrode is at rest or within a certain negative or positive voltage window of its resting voltage (sometimes referred to as the "water window"), the reaction rates are negligible. Once the water window is exceeded, either in the positive or negative voltage direction, reactions can proceed at sufficient rates to cause tissue damage.

Note that electrochemical damage can occur if *either* the DC limit *or* the electrochemical charge limit is exceeded. Stimulators delivering no DC can cause electrochemical damage if the delivered charge per phase exceeds the safe charge density limit. Equally, stimulators delivering individual pulses of very low charge density can cause electrochemical damage if DC is not adequately controlled.

4.5.3.3 Pulse timing The timing of the applied pulses defines their pulse width (or pulse shape if it is not a square wave), the rate at which they are applied and their duty cycle (the percentage of time for which a particular pulse train is applied). All these parameters can have an effect on the amount of electrochemical reactions that occur. For example, a stimulator delivering 2000 pulses per second will move the electrode voltage towards its electrochemically active region 10 times more frequently than the same stimulator delivering the same pulses 200 times per second. Although charge density is the most relevant factor in determining the level of electrochemical stress pulse, timing also plays a role, although the interaction between charge density and pulse timing is complex and is still under review.

Other pulse timing parameters may also be relevant. For example, long pulse widths may generate less by-products than short ones (for the same amount of injected charge), because the capacitance of the electrode/tissue is greater at low frequencies. This means that the voltage excursion on the electrode is less for longer pulse widths (for the same injected charge) and this in turn may lead to less electrochemical stress. An alternate view is that longer pulse widths generate more electrochemical by-products, because the harmful voltages are applied to the electrodes over a longer period. Full characterisation of the effect of pulse timing on electrochemical stress is yet to be performed.

4.5.3.4 Electrode material Many noble metals which are biologically safe within the body are unsafe when used as a stimulating electrode. For example, gold is considered biocompatible and is well tolerated by the body in its passive state. However, if it is raised to a positive potential of a few hundred millivolts with respect to its resting potential, as can happen during stimulation, it will readily dissolve. Only a very limited number of metals will not dissolve when their voltage is raised significantly from their resting potentials. Platinum, titanium and iridium are such metals and are the main stimulating electrode materials currently used commercially.

Other non-metallic materials such as conducting polymers, which provide a "bridge" between the metal of the stimulator electronics and tissue, have also been studied for use as electrode materials. They allow large charges to be transferred into tissue without significantly raising their potentials with respect to tissue and have shown promise in *in vitro* studies. However, currently, they are not used in significant volumes commercially due to a lack of long-term *in vivo* safety data. Delamination of the polymer coating from the metal substrate is one of the major risks to be eliminated with such materials, although progress is being made to prevent this (Hassarati *et al.* 2014).

4.5.3.5 Electrode shape Current flowing from an electrode into tissue does not do so uniformly across the electrode's surface. Recent modelling work incorporating electrode and tissue geometries suggests that electrochemical reactions tend to occur more readily at the edges and corners of electrodes rather than in their centre (Sue *et al.* 2013). Hence electrode shapes that minimise the number of corners and edges (e.g. spheres or disks) may produce lower levels of electrochemical by-products compared to shapes with more corners and edges (for the same electrode area and delivered charge). However, at present there is limited *in vitro* or *in vivo* data available to substantiate this work.

4.6 CONCLUSION

We have provided a review of the principles and practice relating to perhaps the most critical interface in neurobionics – the one between the electrode and the nerve. Although a lot has been achieved in this area, there is still much to be done. Some areas where future work will likely focus include: miniaturisation of electrodes to allow more targeted neural control; improved matching of mechanical characteristics between the electrode material and the target tissue to reduce tissue damage and improve electrode longevity; and changes to the interface to use other modalities (e.g. optical stimulation). Some of the technologies currently being developed (e.g. microfabrication technology) show great promise for providing new materials and fabrication methods for use in neural electrodes. However, much work is still required to prove the safety and reliability of these technologies to the satisfaction of the regulators. The basic principles and practice of interfacing to neural tissue for recording and stimulation are well understood, but significant developments are required to achieve the ultimate aim of being to stimulate or record from individual nerves over the decades a neurobionics device is expected to operate.

REFERENCES

AAMI (2015) American National Standard AAMI/CI86:2015, Cochlear implant systems – Safety, performance and reliability.

Adrian, E.D. (1954) The basis of sensation; some recent studies of olfaction, *B.M.J.,* **1**: 287–290.

Agnew, W.F., McCreery, D.B., Yuen, T.G. and Bullara, L.A. (1989) Histologic and physiologic evaluation of electrically stimulated peripheral nerve: considerations for the selection of parameters, *Ann. Biomed. Eng.,* **17**: 39–60.

Böckris, J.O.M., Reddy, A.K.N. and Gamboa-Aldeco, M.E. (2002) Fundamentals of electrodics, in *Modern Electrochemistry,* 2nd Edition, vol **2A**: Böckris, J.O.M, Reddy, A.K.N. and Gamboa-Aldeco, M. (eds), Kluwer Academic, New York.

Brodmann, K. (1909) *Vergleichende Lokalisationslehre der Groshirnrinde*. Barth, Leipzig.

Brummer, S.B. and Turner, M.J. (1977a) Electrical stimulation with Pt electrodes. II: Estimation of maximum surface redox (theoretical non-gassing) limits. *IEEE Trans. Biomed. Eng.,* **24**: 440–443.

Brummer, S.B. and Turner, M J. (1977b) Electrochemical considerations for safe electrical stimulation of the nervous system with platinum electrodes. *IEEE Trans. Biomed. Eng.,* **24**: 59–63.

Butson, C.R. and McIntyre, C.C. (2005) Tissue and electrode capacitance reduce neural activation volumes during deep brain stimulation. *Clinical Neurophysiology: official journal of the International Federation of Clinical Neurophysiology,* **116**: 2490–500.

Caporale, N. and Dan, Y. (2008) Spike timing-dependent plasticity: a Hebbian learning rule. *Ann. Rev. Neurosci.,* **31**: 25–46.

Cook, M.J., O'Brien, T.J., Berkovic, S.F., Murphy, M., Morokoff, A. *et al.* (2013) Prediction of seizure likelihood with a long-term, implanted seizure advisory system in patients with drug-resistant epilepsy: a first-in-man study. *The Lancet Neurology,* **12**: 563–571.

EN45502-2-3 (2010) Active implantable medical devices. Particular requirements for cochlear and auditory brainstem implant systems – the European standard for cochlear and auditory brainstem implants.

Enderle, J.D. and Bronzino, J.D. (2012) *Introduction to Biomedical Engineering*. Academic Press, Burlington MA.

Francart, T., Innes-Brown, H., McDermott, H.J. and McKay, C.M. (2014) Loudness of time-varying stimuli with electric stimulation. *J. Acoust. Soc. Am.,* **135**: 3513–3519.

Goldman, D. E. (1943) Potential, impedance, and rectification in membranes. *J. Gen. Physiol.,* **27**: 37–60.

Groves, D.A. and Brown, V.J. (2005) Vagal nerve stimulation: A review of its applications and potential mechanisms that mediate its clinical effects. *Neurosci. Biobehav. Rev.,* **29**: 493–500.

Hassarati, R.T., Dueck, W.F., Tasche, C., Carter, P.M., Poole-Warren, L.A. and Green, R.A. (2014) Improving cochlear implant properties through conductive hydrogel coatings. *IEEE Trans. Biomed. Eng.,* **22**: 411–8.

Heffer, L.F. and Fallon, J.B. (2008) A novel stimulus artifact removal technique for high-rate electrical stimulation,. *J. Neurosci. Meth.,* **170**: 277–284.

Helmchen, F., Denk, W. and Kerr, J.N. (2013) Miniaturization of two-photon microscopy for imaging in freely moving animals. *Cold Spring Harbor Protocols,* pp 904–913.

Hillman, E.M. (2007) Optical brain imaging *in vivo*: techniques and applications from animal to man. *J. Biomed. Optics,* **12**: 051402.

Hochberg, L.R., Bacher, D., Jarosiewicz, B., Masse, N.Y., Simeral, J.D. *et al.* (2012) Reach and grasp by people with tetraplegia using a neurally controlled robotic arm. *Nature,* **485**: 372–375.

Hodgkin, A.L. and Huxley, A.F. (1952) A quantitative description of membrane current and its application to conduction and excitation in nerve. *J. Physiol.,* **117**: 500–544.

Karumbaiah, L., Saxena, T., Carlson, D., Patil, K., Patkar, R., Gaupp, E.A. *et al.* (2013) Relationship between intracortical electrode design and chronic recording function. *Biomaterials,* **34**: 8061–8074.

Keck, T., Mrsic-Flogel, T.D., Vaz Afonso, M., Eysel, U.T., Bonhoeffer, T. and Hubener, M. (2008) Massive restructuring of neuronal circuits during functional reorganization of adult visual cortex. *Nature Neurosci.,* **11**: 1162–1167.

Lapique, L. (1905) Rescherches quantitatives sur l'excitation electrique des nerfs traitee comme une polarization. *J. Physiol.(Paris),* **9**: 622–635.

Lilly, J.C., Hughes, J.R., Alvord, E.C., Jr. and Galkin, T.W. (1955) Brief, noninjurious electric waveform for stimulation of the brain. *Science*, **121**: 468–469.

Lind, G., Linsmeier, C.E., Thelin, J. and Schouenborg, J. (2010) Gelatine-embedded electrodes – a novel biocompatible vehicle allowing implantation of highly flexible microelectrodes. *J. Neural Eng.,* **7**: 046005.

Litvak, L.M., Spahr, A.J. and Emadi, G. (2007) Loudness growth observed under partially tripolar stimulation: model and data from cochlear implant listeners. *J. Acoust. Soc. Am.,* **122**: 967–981.

McCreery, D.B., Agnew, W.F., Yuen, T.G. and Bullara, L. (1990) Charge density and charge per phase as cofactors in neural injury induced by electrical stimulation. *IEEE Trans. Biomed. Eng.,* **37**: 996–1001.

McNeal, D.R. (1976) Analysis of a model for excitation of myelinated nerve. *IEEE Trans. Biomed. Eng.,* **23**: 329–337.

Merrill, D.R., Bikson, M. and Jefferys, J.G. (2005) Electrical stimulation of excitable tissue: Design of efficacious and safe protocols. *J. Neurosci. Meth.,* **141**: 171–198.

Moffett, D.F., Moffett, S.B. and Schauf, C.L. (1993) *Human Physiology: Foundations and Frontiers*. William C Brown Publ., St Louis MO.

Norlin, P., Kindlundh, M., Mouroux, A., Yoshida, K. and Hofmann, U.G. (2002) A 32-site neural recording probe fabricated by DRIE of SOI substrates. *J. Micromech.. Microeng.,* **12**: 414–419.

Rall, W. (2011) Core conductor theory and cable properties of neurons, in *Comprehensive Physiology*, Suppl. 1: *Handbook of Physiology. The Nervous System, Cellular Biology of Neurons*, pp 39–97., doi: 10.1002/cphy.cp010103

Rattay, F. (1986) Analysis of models for external stimulation of axons. *IEEE Trans. Biomed. Eng.,* **33**: 974–977.

Reilly, J.P. and Diamant, A.M. (2011) *Electrostimulation: Theory, Applications and Computer Model*. Artech House, Boston, London.

Rose, T.L. and Robblee, L.S. (1990) Electrical stimulation with Pt electrodes. VIII: Electrochemically safe charge injection limits with 0.2 ms pulses. *IEEE Trans. Biomed. Eng.,* **37**:1118–1120.

Scholkmann, F., Kleiser, S., Metz, A.J., Zimmermann, R., Mata Pavia, J. *et al.* (2014) A review on continuous wave functional near-infrared spectroscopy and imaging instrumentation and methodology. *Neuroimage,* **85**(1): 6–27.

Shannon, R. V. (1992) A model of safe levels for electrical stimulation. *IEEE Trans. Biomed. Eng.,* **39**: 424–426.

Sharma, A., Martin, K., Roland, P., Bauer, P., Sweeney, M.H. *et al.* (2005) P1 latency as a biomarker for central auditory development in children with hearing impairment. *J. Am. Acad. Audiol.,* **16**: 564–573.

Shepherd, R.K., Matsushima, J., Millard, R. and Clark, G.M. (1991) Cochlear pathology following chronic electrical stimulation using non charge balanced stimuli. *Acta Oto-Laryngologica,* **111**: 848–860.

Shivdasani, M.N., Sinclair, N.C., Dimitrov, P.N., Varsamidis, M., Ayton, L.N. *et al.* (2014) Factors affecting perceptual thresholds in a suprachoroidal retinal prosthesis. *Invest. Ophthal. Vis. Sci.,* **55**: 6467–6481.

Sporns, O. (2013) The human connectome: Origins and challenges. *Neuroimage,* **80**: 53–61.

Sue, A., Tran, P., Wong, P., Li, Q. and Carter, P. (2013) Time-domain finite element models of electrochemistry in intracochlear electrodes. *Engineering in Medicine and Biology Society (EMBC), 35th Annual International Conference of the IEEE*, pp 1554–1557.

Tian, L., Hires, S.A. and Looger, L.L. (2012) Imaging neuronal activity with genetically encoded calcium indicators. *Cold Spring Harbor Protocols,* pp 647–656.

Tooker, A., Tolosa, V., Shah, K.G., Sheth, H., Felix, S. *et al.* (2012) Polymer neural interface with dual-sided electrodes for neural stimulation and recording. *Annual International Conference of the IEEE Engineering in Medicine and Biology Society Conference,* pp 5999–6002.

Tran, P., Sue, A., Wong, P., Li, Q. and Carter, P. (2015) Development of HEATHER for cochlear implant stimulation using a new modeling workflow. *IEEE Trans. Biomed. Eng.,* **62**: 728–735.

Tykocinski, M., Shepherd, R.K. and Clark, G.M. (1995) Reduction in excitability of the auditory nerve following electrical stimulation at high stimulus rate. *Hearing Res.,* **88**: 124–142.

Vallbo, Å. (1989), Single fibre microneurography and sensation, in *Hierarchies in Neurology.* Springer, London, pp 93–109.

van den Honert, C. and Kelsall, D.C. (2007) Focused intracochlear electric stimulation with phased array channels. *J. Acoust. Soc, Am.,* **121**: 3703–3716.

Xu, J., Shepherd, R.K., Millard, R.E. and Clark, G.M. (1997) Chronic electrical stimulation of the auditory nerve at high stimulus rates: a physiological and histopathological study, *Hearing Res.,* **105**: 1–29.

PART II

DEVICE DESIGN AND DEVELOPMENT

5

WIRELESS NEUROTECHNOLOGY FOR NEURAL PROSTHESES

Arto Nurmikko and David Borton
Department of Engineering & Physics, Brown University, Providence, RI, USA
Ming Yin
Blackrock Microsystems, Salt Lake City, UT, USA

5.1 INTRODUCTION

Contemporary investment on brain science research by many recent national and international initiatives has sharpened the focus on the development of direct brain-controlled neural prostheses. With an increasing understanding at microscopic level of functional circuits in the brain and the nervous system by experimental mapping tools and theoretical models, neuroengineers are challenged to create truly mobile electronic and related neurotechnology platforms for disabled individuals that enable direct electronic wireless communication between the central nervous system and the external physical world. We use the term "communication" here in a bidirectional electronic sense: on the one hand, command signals from the brain for intended action by the brain or other parts of the nervous system must be adequately recorded ("read out") and decoded into electronically useful formats to actuate assistive technologies (such as operating a computer mouse today and perhaps enabling future control of humanoid robots, semi-autonomous vehicles and so on). On the other hand, providing a proxy for lost sensory function implies the need to advance neuromodulation techniques ("write-in") to inform the brain of

Neurobionics: The Biomedical Engineering of Neural Prostheses, First Edition.
Edited by Robert K. Shepherd.

the relevant physical environment. For example, how might a lost sense of touch by hand or a specific digit be directly conveyed to the brain? The challenge (and opportunity) to invent means of bidirectional "read-out" and "write-in" as the basis of closed-loop communication is a tall order, even for restoring very basic neurological functions. This chapter outlines the exploratory research by reviewing ongoing neuroengineering research on wireless brain interfaces in non-human primates (NHPs) through the critical steps required toward human translation.

For a practical closed-loop functional wireless brain interface, sophisticated engineering hardware systems require integration with neurocomputational models. One example might be implementing data-driven model-based algorithms to optimize sensory cortical write-in for truly biomimetic brain motor control of, for example, a haptic artificial hand. In terms of current capabilities we are still limited by our understanding of the brain and the nervous system at multiple scales, from microcircuits to networks connecting functional brain areas. Yet, from a neural engineering point of view it is encouraging that extracellular electrophysiological recordings from ensembles of neurons (at single neuron resolution) have enabled subsequent decoding models to translate intention by paralyzed patients to operate a robotic hand, a keyboard and related basic communication devices by direct cortical commands (with the patient using intact visual percepts to "close the loop", as described in Chapter 9). By contrast, our understanding of encoding comparably detailed sensory information by write-in directly to the cortex, say at the level of individual digits, remains imperfect and poorly understood at best.

Presently, we are technically limited in our ability to stimulate specific neural circuits at a high level of spatial and temporal resolution by electrical current injection, which may be fundamental in their lack of biophysical specificity (Chapter 4). Other physical modalities such as optical and ultrasound neuromodulation on the microscale are under active research and may open new avenues for sensory input to the brain. Nonetheless, as shown by examples elsewhere in this book, applying electrical stimulation as a physical agent to correct errant brain behavior or to augment specific lost function have been met with remarkable success in several specific cases of neuromodulation-based technology. These include device systems for human use such as deep brain stimulators, cochlear implants, and most recently emergent retinal prostheses (Chapters 7 and 8). Related technologies, such as subdural electrical stimulators for epilepsy, vagus nerve stimulators in attempts to treat a suite of disorders, as well as functional electrical stimulation of peripheral nerves or directly of paralyzed muscles have likewise emerged as valuable tools for clinical use in research.

To map the scope of this chapter for the reader, we focus here on recent research where the intersection of fundamental brain science, clinical translation of brain-machine interfaces and state-of-the-art microelectronics join in synergy for the development of next generations of brain-controlled neural prostheses – especially those for mobile subjects. The specific engineering emphasis is on implants capable of high data rate wireless neural recording with electrical inputs from multi-element microprobes such as microelectrode arrays (MEAs), or their optoelectronic and chemicals sensor counterparts. With current intracortical probes capable of approximately 100–200 nodes of full spectrum electrophysiology, with microfabrication approaches for higher density multichannel probes under way, an ability to continuously stream large quantities of neural data to body-external receivers is required for communicating plans and intent of the brain to issue, for example, motor commands. As of this writing, the status of wireless brain interfaces, whether at systems or component/device levels, is still at a relatively early phase. However, it is encouraging how recent results from animal experiments (mainly NHPs) have laid the groundwork whereby prototype wireless electronic brain interfaces are poised to embark on the path to clinical translation deployment.

Unsurprisingly, and given the present sophistication of wireless telecommunication technologies, there is much activity to adapt these for wearable or implanted biomedical devices. We will not discuss here the vast field of wearable (and typically) external-to-body biomedical devices for personal health diagnostics with future trends towards whole-body area networks, but it is clear that wireless methods will play an essential role in this very rapidly advancing health care technology sector (Zhang *et al.* 2014).

One key thread running through this chapter is that the development of any wearable or implantable active electronic wireless neural interface must be placed within a total *system* perspective. For wireless neural implants in particular, this includes restrictions placed by anatomical and physiological constraints, limitations in size and shape, ease of surgical implantation, and so on. Figure 5.1 shows the system-based approach from top-down and bottom-up points of view, by listing the nested layers of distinct subsystems and their interfaces which should be viewed holistically from the starting point of a technical roadmap.

Accordingly, the chapter is organized as follows: An overview of the basic technical requirements and challenges associated with wireless electronic access to

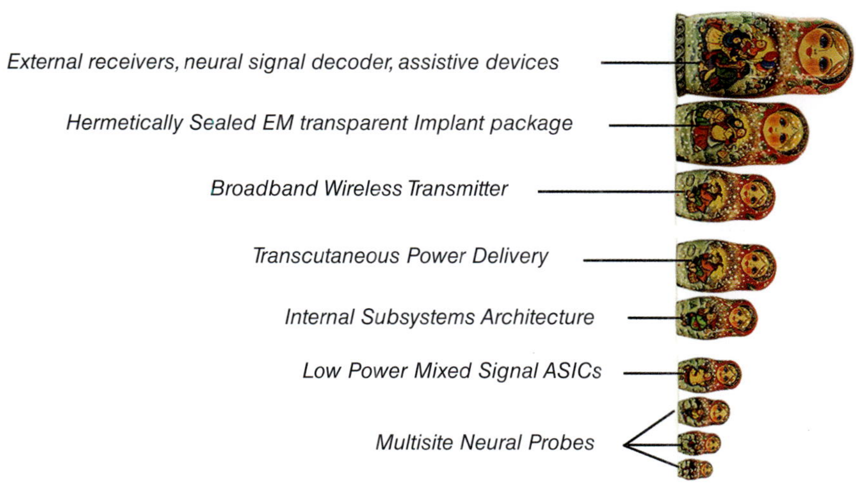

Figure 5.1 "Nested Doll" representation of a wireless neural implant as a system, highlighting multiple subsystems and their interfaces.

the brain from broader neuroengineering perspective is outlined Section 5.2. This is followed in Section 5.3 by description and needs for specialized electronics, from integrated circuit designs to subsystem blocks, which constitute the electronic heart of a high-data rate, wireless implant. Concrete examples of building wireless brain interfaces for NHPs are discussed in Section 5.4, including methods of telemetry. Delivering power and power management is the subject of Section 5.5, followed by the critically important issues for hermetic sealing in Section 5.6. In Section 5.7, selected examples are shown of today's wireless cortical interfaces applied to freely-moving NHPs. Concluding remarks of the overall prospects, opportunities and challenges for successful translation of wireless brain interface technologies are discussed in Section 5.8, with a few comments on the emerging compact neurocomputing platforms as body wearable modules while communicating both with wireless implants and external devices.

With apologies to many accomplished research teams and only for the sake of continuity, we mainly use the work from the authors' laboratories as the template in this chapter, where emphasis is placed on the goal of developing implants for recording broadband neural signals from the cortex in free-moving animal (NHP) models. There are superb efforts under way in many laboratories across the world, ranging from development of highly advanced multichannel and multifunctional microprobes to systems-level approaches to implement sophisticated closed-loop neural interfaces. We have made an attempt to cite as many of these as possible within the space constraints of this chapter.

5.2 RATIONALE AND OVERVIEW OF TECHNICAL CHALLENGES ASSOCIATED WITH WIRELESS NEUROELECTRONIC INTERFACES

Modern neurotechnology requires synergy of neuroscience, electrical engineering, materials science, physics, chemistry and microbiology – all against the underlying clinical sciences of neurology and neurosurgery. The synergy intersects two broad goals:

1. unraveling the neural code at increasing levels of circuit resolution, to decipher brain and overall nervous system functions such as sensing, planning and specific commands, and
2. the strategic application of such learning to create new medical device technologies for clinical, prosthetic and rehabilitative use.

The central theme in this book is to innovate technologies for people with neurological disabilities where restoration of or substitution for lost neural function is sought. Accordingly, the principal aim of the sections below is to offer to the reader snapshots of recent progress with implantable wireless brain interfaces, particularly in the development of techniques for reading out (i.e. recording) motor intentions from the brain of primates. The readout techniques considered here are based on extracellular high-resolution electrophysiology, but we note the very important parallel which all-optical techniques are beginning to yield for such detailed circuit mapping (Barretto *et. al.* 2011; Ziv *et al.* 2013).

In this chapter we will skirt the technically very challenging question to improve the write-in (neuromodulation or neurostimulation) functions of a future electronic brain interface by accessing a large number of target sites. Briefly, whether by electrical or other biophysical means (optical, ultrasound, magnetic, etc.), non-pharmacological stimulation of the brain and the nervous system has a long history in both research and clinical therapies. Today, the availability of real-time information about targeted neural microcircuits at high spatial and temporal resolution is one cornerstone to advance "smart" neurostimulation techniques in the future, which reach well beyond the present largely empirical approaches, such as deep brain stimulation to manage Parkinson's disease or essential tremor (Chapter 8). It should be noted that there is considerable asymmetry in state-of-the-art between recording and stimulation of neural circuits. Due to the complexity of the neural code, recording information, for example from the motor cortex about the intent to move an upper limb, can require vast amounts of multichannel data for a high performance neural prosthesis. By contrast, the use of electrical stimulation has been typically limited to a few channels only so far (whether in the cortex, subcortical structures, spinal cord or elsewhere), in part because of technical, physical and physiological limitations – as well as the many unknowns related to finding an optimal stimulation "code" (multisite stimulus applied with a specific space-and-time pattern by technically sophisticated microarrays).

As with any active electronic biomedical devices, the engineering of direct electronic pathways to the brain for medical bionics applications requires considerable reliance on customized low-power microelectronics, both of which are challenged and constrained by the specialized performance requirements that apply especially to chronic implants. Fortunately the vast resources of both analog and digital silicon-based complementary metal-oxide semiconductor (CMOS) microelectronics provide a vast design space for a neuroengineer. Still, rational design for application specific integrated circuits (ASICs) must be viewed in the full system context, which in turn requires teaming and intensive collaboration between neuroengineers, neuroscientists, clinicians from neurology and neurosurgery, and when possible, with the end users (Chapter 11).

Here we focus on wireless read-out of multichannel neural information from the primate brain, and more specifically the neocortex. Figure 5.2 shows a simplified schematic of the basic electronic architecture of such an open-loop wireless electronic brain-interface system. Ideally, the wireless link should be bidirectional in its signal, data and command communication such as in cochlear implants. Insofar as the brain, one example is a recently developed "closed-loop" deep brain stimulation system. Here the electrical stimulation parameters are informed by real-time recording by dual use of the implant electrodes in subcortical structures such as the subthalamic nucleus and the thalamus (Afshar *et al.* 2013). Likewise, a recent US Food and Drug Administration (FDA) approved neuromodulation system developed by Neuropace (Morrell 2011) for epilepsy diagnosis offers an example where in each case relatively low data rates are sufficient given the specific targeted clinical applications using

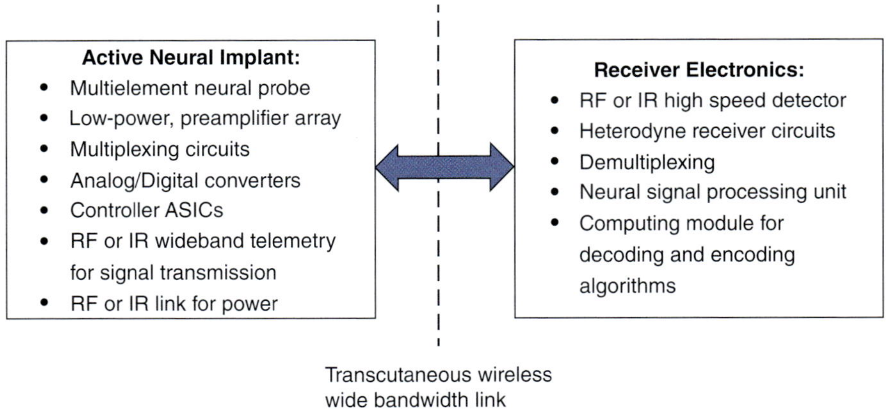

Figure 5.2 Schematic of electronic subsystems in a transcutaneous wireless link between an active neural implant, external peripherals and further downstream links.

the vagus nerve as the input/output channel in this sense and stimulate electronic interface. A third example, is a recent development by the group of Benabid and colleagues of a 64-channel wireless electrocorticography (ECoG) system, which uses transcutaneous inductive coupling as a means to power a system, likewise targeting disorder monitoring such as epilepsy (Mestais *et al.* 2015).

By contrast to the above low- to moderate data rate systems, the types of wireless neural interfaces envisioned here aim at providing the highest possible data rates for future brain–machine interface applications, especially for multifunctional neural prostheses. Acquiring as much neural information from the brain as possible implies an ability to capture action potentials, field potentials, high frequency oscillations and various brain rhythms from brain microcircuits. True, the ambition of retrieving (and delivering) the technologically maximum amount of information to/from the brain from external electronic devices is not necessary for specific clinical, and assistive applications and will depend a great deal on the specific device configurations, for example of intracortical MEAs as well as physiological and anatomic factors. Nonetheless, we suggest that in the long run, planting the seeds for truly wideband bidirectional brain interfaces is important, even if this requires pushing the microelectronic and related neurotechnologies beyond today's limits. By wideband, we refer in this chapter to transcutaneous neural data rates starting at the digital equivalent of hundreds of Mbits/sec for chronic wireless implants.

From an electronic systems point of view, the block diagram of Figure 5.2 breaks down some of the key technical components in a high-performance brain interface. Examples of how these components can be integrated will be considered in the sections below. Briefly, multichannel neural signals from an electrode array are first directed to low power, low noise ASICs, which typically provide multichannel analog amplification, conditioning and multiplexing, followed by analog-to-digital (A/D) conversion. A set of digital controller circuits handles communication protocols including facilitation of protocols for possible multichannel electrical stimulation by dedicated separate ASICs. One key component is a low-power radio frequency (RF) transmitter ASIC whose speed (and efficiency) dictates the maximum data rates at a distance to external receivers, decoders/encoders and assistive devices. At the full system level, an engineering goal is to create a mobile implantable electronic brain interface which communicates with external assistive devices via a wearable neural signal processor and decoders, as we shall see below.

5.3 WIRELESS BRAIN INTERFACES REQUIRE SPECIALIZED MICROELECTRONICS

5.3.1 Lessons learned from cabled neural interfaces

Techniques employing wired (cabled) broadband neural recording have culminated in the remarkable demonstrations of the utility of multichannel extracellular electrophysiological recording of neural population dynamics in first-in-human clinical trials (Hochberg *et al.* 2006, 2012) and Collinger *et al.* (2013) (Chapter 9). These ongoing trials have mainly focused on the motor cortex, with intracortical MEAs of up to 200 electrodes connected via a percutaneous cable/connector assembly to external electronics. The detected neural signals, conducted by transcranial and percutaneous fine wire bundles, are typically first amplified outside the skull/skin, followed by A/D conversion. The digitized data is in turn cabled to dedicated computers for subsequent decoding and data storage. While focusing on system electronics in this chapter, we underscore the major advances which have been made in decoding of neural population dynamics recorded by MEA electrodes as well as by subdural ECoG electrodes. These advances have made it possible to achieve increasingly sophisticated prosthetic control (Gilja *et al*, 2015: Simeral *et al.* 2011). Increasing the number of MEAs (across primary and premotor cortices) has provided further enrichment through innovative decoding for brain control of robotic arm/hand movement with multiple degrees of freedom by a patient (Wodlinger *et al.* 2015).

The cabled tethering of the subject to external electronics in ongoing human trials as well as experiments with NHPs imposes severe restrictions on mobility. Nonetheless, the generic neuroelectronic principles for the cabled systems serve as a useful template for engineering miniaturized wireless approaches. As a rough benchmark for the electronics required to stream out broadband neural signals recorded by intracortical MEAs from approximately 100–200 individual channels, they must be capable of continuous wireless digital transmission at data rates on the order of 100 Mbits/sec. Collecting, managing and storage of the cumulative large data pose significant added engineering challenges which we do not address here.

In case of a typical planar intracortical MEA designed to access the cortical layers (through layers I–VI) in a monkey or a human subject, the adjacent electrode spacing is a few hundred microns, laid out as a square lattice in part to optimize recording from individual neurons, though finite “interference” from adjacent neurons (multiunit signals) is a practical challenge for signal analysis (Chapter 4). Given the cortical anatomy (density of neurons), the arrays may, in a successful case, record useful signals at a vast majority (up to 90%) of the electrode sites, even as the underlying neuron organization is spatially rather random. A single action potential spike might generate recorded amplitudes from a few tens of μVs up to more than 100 μV by an electrode of a few tens of micrometers in equivalent diameter and a complex impedance of about Z ~ 500 kΩ at 1 kHz at the metal electrode/brain tissue (most commonly a platinum/electrolyte) interface (Robinson 1968). The non-biological root mean square (RMS) noise background at physiological temperatures is on the order of several μVs over the broadband neural signal spectrum from approximately 1–10 kHz. In addition to these fundamental sources of noise that are common to all electrical systems (Johnson noise, etc.), the brain operates against its own inherently "noisy" background (summation from the more distal distributed networks of brain microcircuits), thereby adding to the challenges of interpreting variable neural signals by signal processing approaches which seek to correlate specific action and behavior. Another critical requirement for a chronic electronic interface with the brain is the electrical and material stability of the microelectrode recording performance. This is still an unresolved question including how reaction by the body tissue on implantable electrodes and related device structures affects their truly long-term chronic performance (say beyond 10 years) and the related issue of device safety (12,13) (Chapter 2).

In their cabled implementations, the electronic ecochain of a broadband neural recording system runs from multichannel preamplifiers to A/D converters (given that the noise levels 12 bit resolution should be more than adequate, though 16-bit A/D converters are also used) to dedicated neural signal processors and beyond. There are currently several rack-mounted commercially available versions, including specific FDA-approved hardware/software, which in turn connect to adjacent computing modules forming an aggregate workstation incompatible with a subject's mobility.

5.3.2 Special demands for compact wireless neural interfaces

The development of any wearable or implantable active electronic wireless neural interface must be placed within a total *system* perspective. For wireless neural implants in particular, this includes restrictions placed by anatomical and physiological constraints, limitations in size and shape, ease of surgical implantation and minimizing heat generation. Here we outline some of the custom microelectronics which in the aggregate amplify, digitize and transmit the recorded neural data as a continuous digital low-error rate stream to external receivers. The overarching circuit design and implementation strategy, especially for implanted wireless neural interfaces, is constrained by an acute challenge – how to achieve maximal performance with a minimum overall power budget and minimal thermal side effects.

Complete sets of neural signals contain a range of complex time-varying waveforms composed of action potentials of single neurons, multi-unit waveforms and lower frequency field potentials, as well as various brain rhythms and oscillations, altogether spanning roughly the frequency range from 1–10 kHz. Thus a question arises whether some initial on-chip preprocessing might be usefully incorporated as part of the ASIC designs within the front end analog amplifier circuits. For example, on-board high pass filtering and threshold detection might be useful if the action potential (spike) firing rates and waveforms are the main target of neural information in a multichannel system. Similarly, incorporating other data conditioning/processing at the ASIC level and some degree of volatile or non-volatile memory is possible, though adding a penalty to the power budget.

A significant body of neural interface specific ASIC-level work exists which informs approaches to advancing neural electronics from their current cabled implementation to a wearable or portable implementation. Among others, Harrison and Charles (2003) laid the basis for the analog front end preamplifier multichannel ASICs, whereby the incoming neural signal on each channel is conditioned and amplified by a low-noise low-power preamplifier. The design principles typically use a capacitive-feedback, folded cascade-type operational transconductance amplifier (OTA) configuration with a source follower output buffer. Noise equivalent performance nearing the 1 μV RMS limit has been implemented by several groups in such CMOS circuits with transistors (FETs) operating in their normal regimes. Wattanapanitch *et al.* (2007) and others have suggested that in comparison to the saturation region, operating a metal-oxide semiconductor field-effect transistor (MOSFET) in the sub-threshold region can reduce the noise by a significant amount (1/4), while maintaining the same power consumption due to the reduction in total charge in the MOSFET. There have also been full system-on-chip (SoC) bench top initial demonstrations where the preamplifier, digital controller and A/D converter circuits are integrated on the same silicon-die by using a single CMOS fabrication process (Gao *et al.* 2012) (Figure 5.3).

At the aggregate system level, a number of wireless microelectronic systems have been built for recording neural activities across multiple species of awake animals (see references listed in Yin *et al.* (2014)). Specific to NHPs, Miranda *et al.* (2010) developed their "HermesD" system mounted externally atop the animal's head, and cabled to the intracortical electrode array that can record and wirelessly transmit 96 channels of broadband neural data through a 3.7–4.1 GHz RF link using frequency shift keying (FSK) modulation. The device was packaged in a 38 × 38 × 51 mm aluminum enclosure for head-mounted use. It consumed 142 mW from a 3.6 V/1200 mAh Li-SOCl battery. The payload consisted of an amplifier/ADC ASIC, a low voltage field programmable gate array (FPGA), and a low power ultra-wideband transmitter consuming approximately 30 mW of power. Rouse *et al.* (2011, 2012) developed a four-channel ECoG-based bi-directional brain-machine interface for motor prosthesis and neurological disease state monitoring and detection. This low data rate device featured four channels of ECoG and local field potential amplification, incorporated spectral analysis, a three-axis accelerometer, algorithm processing, event-based data logging and wireless telemetry for data uploads and algorithm/configuration updates.

The device was packaged for chronically implanted use and demonstrated in an NHP model for brain control of a computer cursor. Harrison *et al.* (2009) reported a SoC 100-channel wireless recording system that provides 100 threshold-based spike detecting data and a single channel broadband neural data. The system was wirelessly powered over a 2.765 MHz inductive link and transmitted data in the 902–928 MHz industrial, scientific, and medical radio (ISM) band by FSK. The system has been validated *in vivo* in the motor cortex of a cat. Rizk *et al.* (2009) presented a 96-channel fully implantable neural data acquisition system. Their system consumed 2 W and

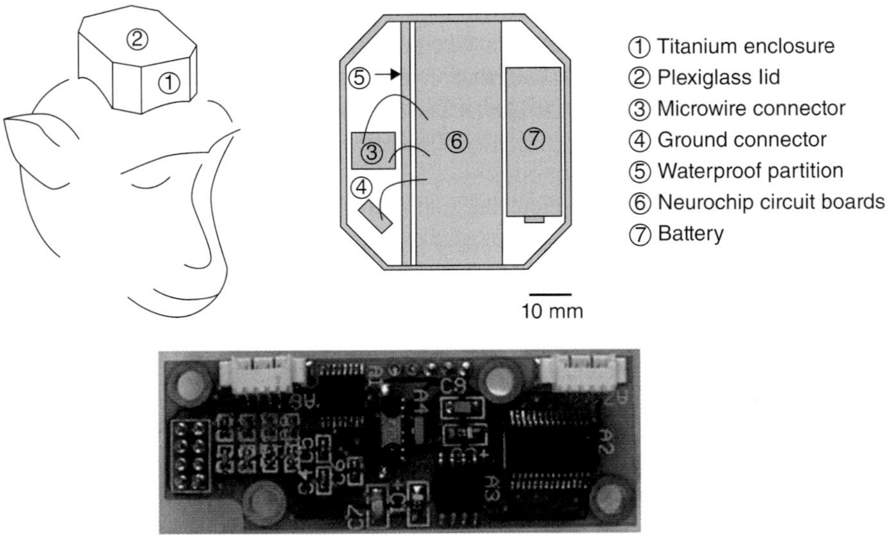

Figure 5.3 One of the first headmounted wireless 32-channel recording devices (Miranda *et al.* 2010) for recording from electrode arrays implanted in rhesus macaque monkeys freely moving in their cages The active electronics of this system exploit selected features of a monolithic integrated neural interface (INI) microchip design which amplifies, digitizes and, in one implementation, transmits neural data across a ~900 MHz wireless channel (Harrison 2010). On a single 2 A-hr battery pack, this system runs contiguously for several days. The total weight of this system including the batteries is 114 g, considerably heavier than human-wearable devices such as earpiece-mounted personal communication packages.

used continuous inductive coupling for powering. The system has been tested both externally in a rodent model and implanted in a sheep model for a short period of 3 hours. Yin and Ghjovanloo (2011) demonstrated a head-mounted 32-channel neural recording device for rodent models. This device was battery powered and consumed 5.6 mW while broadcasting the pulse-width-modulated neural data through an 898 MHz to 926 MHz FSK wireless link. Greenwald *et al.* (2011) reported a 16-channel recording system with a somewhat large form factor, aimed at wireless monitoring of neural activity from awake animals and with the focus on long-term and high fidelity multichannel recording from untethered subjects. Chae *et al.* (2009) in turn described a 128-channel neural recording integrated circuit with an on-the-fly ultra-wideband transmitter capable of transmitting data of up to 90 Mbps.

This device was validated on excised snail neural tissue and included on-board spike sorting. Shahrokhi *et al.* (2010) developed a 128-channel integrated circuit as a candidate for neural interface use, combining eight 16-channel low-power low-noise recording and stimulation ASICs but without wireless telemetry. Among other relevant work we note Szuts *et al.* (2011), who reported on a 64-channel system designed to record and transmit neural spikes and local field potentials for a range of 60 m while consuming 645 mW of power. The device was used to record neural activity from a rat in an outdoor environment. Sodagar *et al.* (2009) introduced a

64-channel miniaturized system for wirelessly transmitting spike occurrences to an external receiver. Their bench top level system had dimensions on the scale of 1.5 cm and consumed 14.4 mW with a 70 MHz to 200 MHz on-off-keying (OOK) wireless link used for transmitting up to 2 Mbps of digital neural data.

The approaches summarized above, as well as others not cited here, have provided a baseline for further improving approaches to broadband wireless neural recording devices, particularly for future human neuroprosthesis use. However, invariably, a number of tradeoffs were necessary to meet specific research needs. In the next section we describe system level approaches to wireless brain interfaces, where specialized microelectronics are configured and integrated for use by mobile primates.

5.4 ILLUSTRATIVE MICROSYSTEMS FOR HIGH DATA RATE WIRELESS BRAIN INTERFACES IN PRIMATES

In this section we use work from the authors' laboratories to illustrate selected system level approaches to wireless neural interfaces. These share specialized microelectronic circuits such as in Section 5.3, and are integrated as implanted or head-mounted external compact modules as part of a full functional microsystem. This research and development has recently culminated in the first (at time of writing) multichannel, broadband, chronically implanted wireless brain interfaces for free-moving NHPs (Borton *et al.* 2013; Yin *et al.* 2013b). More broadly, the engineering approaches and rationale in the design of the types of microsystems described next, including their peripherals, are here viewed from a biomedical device perspective with the ultimate goal to create new neural prostheses. We have built on the work in an NHP context by the groups of Shenoy, Harrison and colleagues, highlighted above (Afshar *et al.* 2011; Barretto *et al.* 2011; Chestek 2009; Miranda *et al.* 2010). Likewise, some early important work for (low channel count) wireless recording from untethered monkeys was demonstrated by Fetz and colleagues (Jackson *et al.* 2007) and Wolf and colleagues (Rizk *et al.* 2009).

Given the success in using the specific types of electrode arrays (MEAs) for cabled recording of neural population dynamics at high data rates for motor prostheses, including their approval by the FDA for human clinical pilot trials (Hochberg 2006), we focus here on such probes as the neural signal source for wide band electrophysiology. The MEAs require multichannel interconnections and some means of close connection or integration to adjacent microelectronics within a fully implantable device. The design "grand challenge" thus involves a specific architecture of the dedicated ASICs and decisions about modalities of telemetry as well as powering of the active circuits. Designs and integration of the multiple heterogeneous elements are significantly impacted, depending on whether the active electronics are fully implanted within the skin envelope or externally head-mounted on the subject (such as for subacute clinical use, or free-moving NHP experiments). Device configurations must account for multiple anatomical, physiological and surgical considerations, all of which impose engineering constraints when making decisions on device form factors, their physical flexibility, fabrication processes and biocompatibility of the multiple material constituents involved. The final layers for the complete ecosystem of a wireless neural prosthesis include the peripheral electronics, such as external receivers of the wireless neural signals and engineering their compatibility with downstream neural signal processing tools, while providing for further communication firmware to access a suite of assistive device technologies aids.

Below, and in the following sections, we compare and contrast three examples of high data rate wireless configurations which differ in their implementation and purpose. While sharing most of the common mixed-signal front end microelectronics, Figure 5.4 shows images of the device configurations which differ as to the spatial layouts and distribution of their functional microelectronic payloads, choice and physical means of telemetry modality, provisioning and managing on-board electrical power for the active electronics. Furthermore, their manifestly different packaging approaches and materials as the wireless modules are tailored according to relevancy as candidates for either basic research in free-moving NHPs or human clinical use, or both. We next use this troika of devices to first address issues relevant to high data rate wireless neural telemetry. Comparison of approaches to power supply and management is discussed in Section 5.5, followed by description of challenges in hermetic packaging in Section 5.6, respectively.

As noted in the Introduction, state-of-the-art wireless interfaces which communicate broadband neural dynamics beyond 100 channels (microelectrode sites, or equivalent recording nodes) demand digital transmission rates on the order of 100 Mbit/sec and above. Corresponding bandwidths are thus considerably larger than for most existing wireless biomedical devices, yet we envision that future electronic brain interfaces will require even larger bandwidths. Given the sophistication, particularly of both RF (microwave) and infrared optical telecommunication (IR) technologies

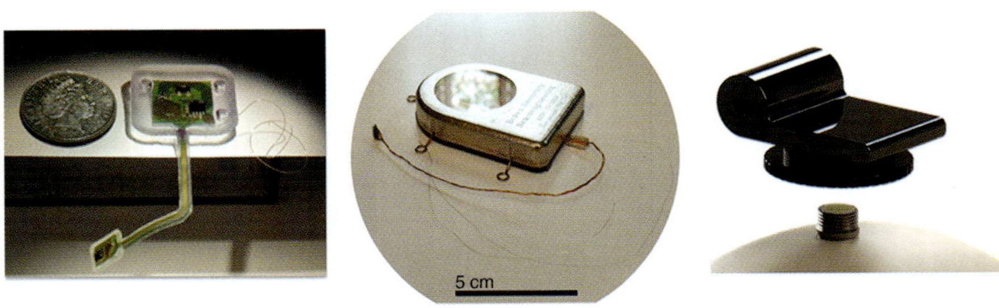

(a) (b) (c)

Figure 5.4 Three case examples of wireless high-data rate brain interface recording devices. (a) An implantable, polymer packaged device (Song *et al.* 2009) with an inductive coupled transcutaneous link for powering and near infrared signal transmitting microlaser. An analog preamplifier/multiplexer ASIC is flip-chip integrated with the intracortical microelectrode array; (b) an implantable 100-channel wireless neurosensor with RF telemetry (3–4 GHz link), housing all active microelectronics and a rechargeable Li-ion battery in a hermetically sealed Ti-enclosure weighing <50 g (Borton *et al.* 2013); (c) a 100-channel externally headmounted, battery-powered wireless neurosensor with RF telemetry up to 200 Mbits/sec (likewise with telemetry in the 3–4 GHz range and below 50 g in weight (Yin *et al.* 2014).

specifically developed for high data rate information transfer (including a pivotal role in the Internet), multiple opportunities exist to adapt elements from these technologies to the types of wireless neuroprosthetic devices considered here. The two primary modalities of RF and IR telemetry are deployed within the device arrangements of Figure 5.4. The device in Figure 5.4a features a transcutaneous digital infrared link at 850 nm wavelength, whereas the devices in Figures 5.4b and c integrate microwave transmitters in the 3–4 GHz frequency regime, with capability to extension to higher frequencies.

Electromagnetic transmission losses through the skin in the selected microwave and IR spectral windows considered here have been chosen to be acceptably low for useful transcutaneous wireless transmission. Practical free space external source-to-receiver distances for neural prostheses are sufficient for short ranges (up to a few meters). Biophysically, in the 1–5 GHz microwave region, losses in the skin in primates are mainly due to absorption by water (polar molecules have permanent dipoles), while infrared losses in the tissue "transparent" window around 850 nm are dominated by light scattering due the optical heterogeneity and microstructure of skin/tissue on the optical wavelength spatial scale. In principle, lower frequency RF radiation can penetrate bone (and the skull), but we focus here on those scenarios where a wireless module is placed subcutaneously, such as between the skin and the skull if implanted within the head.

Consider first the IR telemetry scheme implemented in the device of Figure 5.4a, which integrates a particular solution also to wireless power management microelectronic ASIC payload distribution, and packaging in a specific way. Using a flexible and durable polymer substrate (Kapton or liquid crystal polymer), the device was constructed by assembly of the constituent components in electrically interconnected two-island geometry (Nurmikko *et al.* 2010; Song *et al.* 2005). Embedded in the flexible Kapton substrate were five 25-micron-wide planar wires (for signal and power) connecting the "cortical front end" module to the "cranial back end" module across a narrow bridge. The front end is placed below the skull, while the backend is designed to reside between the skull and the skin of a primate so that the flexible tether threads through a craniotomy in the subject's skull. The corresponding division of the active microelectronic components is shown as a block diagram in Figure 5.5 and their placement relative to the intracortical MEA, skull and skin as a schematic in Figure 5.6. We note that integrating the active mixed-signal ASICs immediately atop the intracortical electrode array is technically demanding, as it requires specialty flip-chip bonding techniques (Song *et al.* 2005). The further idea of placing an entire microsystem including the transmitter onto the MEA below the skull is conceptually attractive, but integrated concept was pursued by packaging approaches using inorganic thin films at the University of Utah by the Solzbacher group (Kim *et al.* 2009), resulting in limited bench top demonstrations. One serious constraint related to integration of ASICs directly on MEAs are limits imposed to their power consumption to prevent heating of the adjacent cortex beyond a 0.1°C clinical boundary condition.

The transcutaneous IR telemetry for the device of Figure 5.4a is achieved using a microcrystal semiconductor laser (vertical cavity surface emitting laser) within the "back end" cranial unit, together with a dedicated A/D ASIC command/control chip

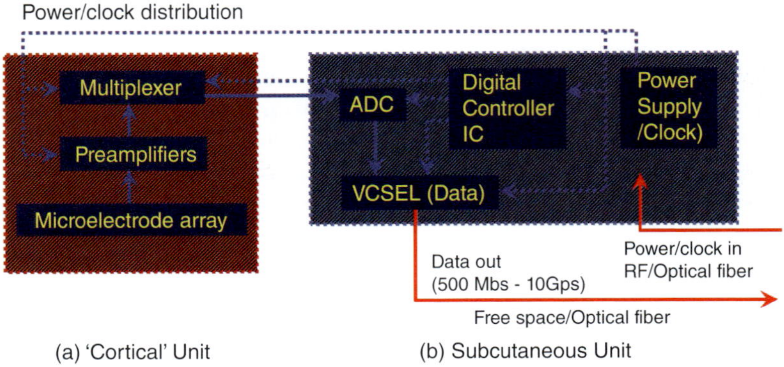

Figure 5.5 Distribution of the electronic and optoelectronic payloads of the implant of Figure 5.4a, with near infrared transcutaneous signal transmission. Power is supplied transcutaneously, either by continuous inductive coupling or by optically powered high efficiency custom multijunction tandem photocell (Song *et al.* 2005; *Park et al.* 2014).

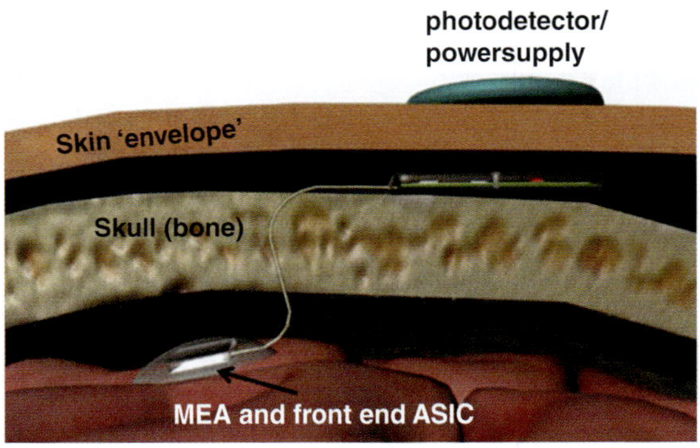

Figure 5.6 Implant placement into a NHP with the cortical front end module linked electrically by short flexible polymer ribbon cable to subcutaneous the back end module.

on the same substrate plane. These low power GaAs-based lasers (~1 mW) have a high electrical-to-optical conversion efficiency (>50%) and are suitable for transmission of the digitized broadband neural data through a subjects skin near 850 nm in the IR to immediately adjacent photoreceivers. Among the generic advantages of an IR wireless transmission modality is the very large bandwidth which modern optical transceiver systems possess (>Gb/sec and beyond). Transmission through a primate's skin (a rhesus macaque) involves light scattering, however, so that the exiting radiation spreads over a large solid angle. Yet a conventional photodiode with an area of approximately 1 cm^2 is readily able to pick up the digital stream when placed within approximately 2 mm from the skin surface, well above the fundamental noise floor (with photon counting optoelectronics as the receiver, the estimated transmission range can readily exceed 1 m). On the opposite side of the polymer substrate is a planar RF receiving coil for enabling inductively coupled (transcutaneous), receiving power and clock to the microsystem (see below). We note that by using optical fibers as output channels for the signal on the one hand, as well as feeding in useful power as well as timing signals, wideband neural data has been extracted from an all-optical device by a percutaneous optical fiber in a rat model. In this case the power was delivered optically via the fiber using a specially designed high efficiency photovoltaic energy converter (Park *et al.* 2013). In terms of safety, the associated laser power levels need to remain within the Class IIIa levels to ensure safety as a medical device is maintained (including eye safety).

Considering the next approaches to RF telemetry, there are many opportunities to adapt technologies from the ubiquitous presence of say Wi-Fi and Bluetooth applications that operate in specific US Federal Communications Commission (FCC) licensed telecommunication bands, broadly in the 1–5 GHz range. The neural interface devices of Figures 4b and c both deploy dedicated, custom-designed low-power RF ASICs in their transmission subsystems. The key functional blocks in their overall RF telemetry scheme are identified in Figure 5.7 (note, however, that the devices in Figures 5.4b and c use different approaches to on-board batteries and overall power management (Section 5.5)).

The RF transmitter/antenna circuit featuring the custom RF ASIC can achieve low bit error rates at ultralow (<20 mW) power dissipation (Yin *et al.* 2013a). The chip

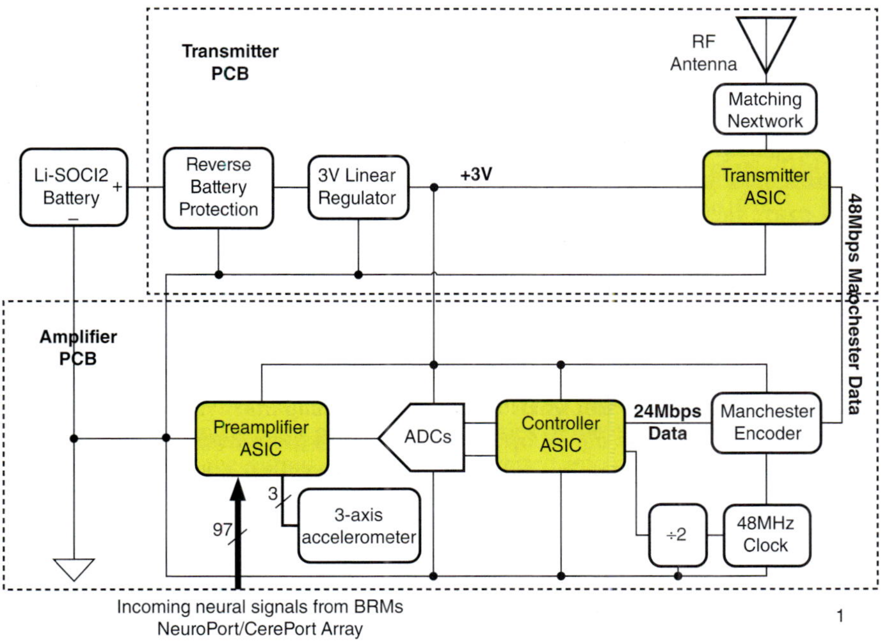

Figure 5.7 Block diagram of the RF telemetry scheme (and battery powering) of the wireless devices of Figures 5.4b and c; both using the same high efficiency, low-power custom RF transmitter ASIC.

was implemented in a state-of-the-art bipolar 180 nm BiCMOS process, achieving a 35% overall power efficiency when 4 mW output power is delivered. The transmitter chip, including a wide tuning-range voltage controlled oscillator (VCO), a wide band OOK up-conversion mixer and a uniquely low power high-efficiency power amplifier (PA), operated on a 3 V battery power supply. Importantly, the frequency tuning range of the VCO is adjustable over one GHz range, enabling the flexibility to choose a carrier frequency according to the wireless environment in the experimental or clinical site. The PA output power is variable from 4–8 mW through an off-chip matching network, to ensure reliable wireless connection to the customized receiver located 1–2 metres away over various ambient conditions and to satisfy regulatory requirements for human use.

A standard approach for detection of RF signals in telecommunication is based on superheterodyne detection. As a concrete example, we show here a receiver configuration implemented to receive the neural data streaming from the implanted or head-mounted devices in Figures 5 4b and c (Y. Min *et al.* 2014). Compact, portable units were developed for receiving the RF modulated neural data, with a built-in capability to route the detected demodulated signals to subsequent neural process electronics for decoding. The system block diagram of the receiver system is shown in Figure 5.8. The signal from the receiving antenna was first amplified by two low noise amplifiers and then down-converted to an intermediate frequency (IF) signal by a microwave local oscillator. After amplitude stabilization by an automatic-gain-control block, the IF signal was fed into an envelope detector that has two envelope filters with cutoff frequencies set by external capacitors. The digital neural data and clock were extracted using a commercial data/clock recovery chip from the restored baseband signal. Precise clock extraction is important in synchronizing with downstream neural decoders and external devices. The recovered

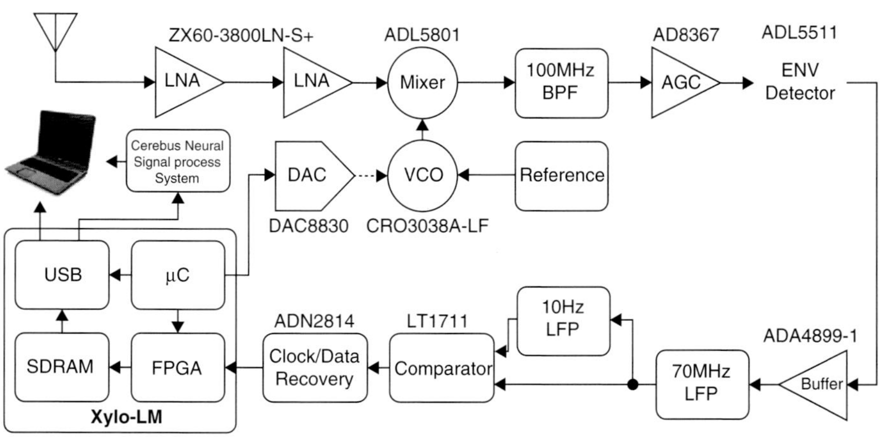

Figure 5.8 Block diagram of a RF superheterodyne receiver, with neural data extracted by demodulation and passed through a dedicated neural signal processor to a laptop computer for decoding and storage.

data was then forwarded to an FPGA and repackaged into 16-bit samples, which is buffered by dedicated ASICs and sent to a neural signal processor and to a PC for further processing and visualization.

For any wireless biomedical device, there are safety considerations with telemetry. From a regulatory point of view, the key metric is the specific absorption rate (SAR) (in the full human body model) of non-ionizing radiation. Both RF (microwave) and infrared frequencies are non-ionizing. The RF spectrum is heavily regulated (by the FCC in the US) by licensing of specific windows of the spectrum (such as reflected in the IEEE 802 standards). Operating in an "unlicensed" band is, however, possible when the RF power and range and more specifically the associated electric field amplitudes fall below approximately 1 mV/meter, which is the case with the devices in Figures 5.4b and c. As for the infrared, and free space propagation, the safety concerns are listed by the Class notation, of which the Class III deployed in the device of Figure 5.4a, for example, it lies well beyond any eye safety limit. Note that in the case of the electromagnetic delivery of power (see below), SAR-related concerns apply equally. Similarly, the use of RF backscattering for "reading out" neural signals must follow SAR guidelines. RF backscattering as a means of recording neural signals is a potentially powerful alternative approach compared to the use of an on-board transmitter chip in an implant, so far applied, for example to insect models (Thomas *et al.* 2012) but outside the scope of this review.

The SAR is a measure of the rate at which energy is absorbed by the human body when exposed to an electromagnetic field. At frequencies between 100 kHz and 10 GHz, the safety limits developed by the International Commission on Non-Ionizing Radiation Protection (ICNIRP) requires localized SAR in the head and trunk area for general public exposure of less than 2 W/kg. The FCC limit for public exposure from cellular telephones is at an SAR level of 1.6 W/kg. The SAR can be calculated using the following equation:

$$SAR = \frac{\sigma E_{tissue}^2}{\rho} \tag{5.1}$$

where σ is the conductivity of the tissue at the frequency of interest in S/m, E_{tissue} is the r.m.s. value of the electric field strength in the tissue in V/m, and ρ is the density of the tissue in kg/m^3. Assuming that the E field is nearly perpendicular to the skin surface, according to the interface conditions for electromagnetic fields, the incident electric field in the skin tissue is:

$$E_{tissue} = \frac{\sigma_{air} + j\omega\varepsilon_{air}}{\sigma_{tissue} + j\omega\varepsilon_{tissue}} \tag{5.2}$$

where σ_{air}, ε_{air}, σ_{tissue}, and ε_{tissue} are the DC conductivity and dielectric constants of air and tissue respectively. Since $\sigma_{air}/\sigma_{tissue}$ is vanishingly small, at 3.5 GHz (for the example considered here), it follows that:

$$E_{tissue} \approx \frac{\varepsilon_{air}}{\varepsilon_{tissue}} E_0 \approx \frac{1}{\varepsilon r_{tissue}} E_0 \tag{5.3}$$

As a numerical example, for our implant device in Figure 5.4b in the low power transmitter mode with the ultra-wide band antenna having 80% efficiency at 3.5 GHz, the total RF output power is on the order of $P_{tx} = 5$ mW and the maximum peak antenna gain is 3.5 dBi. If we assume a minimum distance between the transmitter antenna and the skin tissue as 12 mm, we have $E_0 \approx 60$ V/m from which $E_{tissue} \approx 1.5$ V/m. This corresponds to SAR = 0.004 W/kg, a value well below the current regulatory requirements for safe animal and human use.

5.5 POWER SUPPLY AND MANAGEMENT FOR WIRELESS NEURAL INTERFACES

A pivotal challenge and constraint for electronically active biomedical devices is the choice and approach to supplying power and the corollaries of energy harvesting and storage while ensuring maximum safety for the user. Particularly demanding circumstances apply to high data rate implantable wireless neural interfaces, such as the Ti-enclosed device in Figure 5.4b. There are many examples of advanced technical solutions in the biomedical device industry for lower data rate implants, such as pacemakers and deep brain stimulators to power delivery, storage and energy management. Virtually without exception, these take for granted that the onboard electronic circuits have been configured to minimize power consumption and that the implant is hermetically sealed for chronic use (Section 5.6).

Powering active implants can be broadly classified depending on whether a device carries an on-board implanted battery, or whether the energy is delivered and managed transcutaneously from an external source, either through the same wireless link as the signal or through a separate transmission channel to avoid electromagnetic interference (e.g. a cochlear implant). Some hybrid combination of the two modalities is also possible. The examples in Figure 5.4 reflect this classification, as the devices in Figures 5.4b and c are battery powered, while the implant in Figure 5.4a relies on a continuous transcutaneous power relay scheme. At the same time, the implants in Figures 5.4a and b share the common attribute that transcutaneous power is delivered by inductive coupling, though only during battery recharging episodes for the latter. Inductive coupling, in turn, is but one tier among generalized RF energy delivery approaches which are usually separated into near field (transformer regime) and far field (propagating EM waves) regimes. More recently, a "mid-field" coupling regime (source-to-receiver distance ≈ wavelength) has been studied for wireless powering of small (mm size) active probes (Ho *et al.* 2014). Here we focus on the inductive coupling regime while referring to excellent texts by Van Schuylenbergh and Puers (2009) and Ghovanloo and Kiani in Nikita (2014), as well as more neuroengineering specific research articles (Harrison 2007).

For high-performance neural prostheses, such as the contemporary brain interfaces considered here, today's choice for an onboard battery is a medical grade rechargeable Li-ion battery. Li-ion batteries have developed rapidly in the past 15 years and have reached a level of maturity satisfactory for implantable medical devices. Driven by multiple consumer applications, intense research is under way (van Noorden 2014)

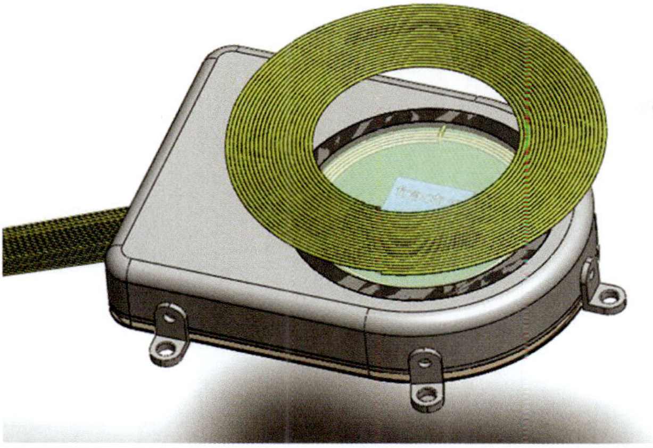

Figure 5.9 Generic schematic of inductive coupling, here in the context of transcutaneous charging of a rechargeable battery through the transparent sapphire window of the Ti-enclosure.

to further improve the two critical figures of merit for compact batteries: (a) energy density (affecting both the size and energy storage) and (b) battery lifetime, measured in number of charge recycles now approaching a value of 5000. Prospects for future improvements (say, by 30%) will directly benefit further advancement of neural prostheses.

To illustrate inductive charging, the hermetically sealed module of Figure 5.4b houses an inductively coupled wireless charger (Figure 5.9) for a medical grade 200 mAh Li-ion rechargeable battery. The wireless charger scheme is shown schematically in Figure 5.10. It deploys a capacitor-inductor resonant circuit to harvest RF energy (choice varying from 100– 1 MHz) from an external primary coil driven by a Class-AB PA. The internal coil of about 1" in diameter and electromagnetically accessible through the sapphire window is custom wound from Litz-type wire to reduce the skin effect for minimum conductive losses. The wireless charger has three safety features: a self-timer that terminates the charging when full, a battery monitor that indicates the status of the battery, and a protection circuit module that protects the battery from over-charging and over-discharging. A low-dropout linear regulator produces a 3 V supply for the rest of the system from the battery. For human use, an additional temperature sensor can be readily added. In NHP use of the prototype system, an inductive power efficiency of 32% was achieved at a 5 mm transcutaneous distance (with an appropriately designed external power coil including associated thermal management).

One important and limiting factor in implementing inductively coupled powering/charging in active medical implants is the induced collateral heating, fundamental collateral in all RF energy transfer schemes. In the frequency range considered here, eddy currents are induced by the time-varying magnetic field component of the total electromagnetic exposure. Since the induced currents are local and mostly confined to the surface of the conductor, there will be competing currents in opposite directions,

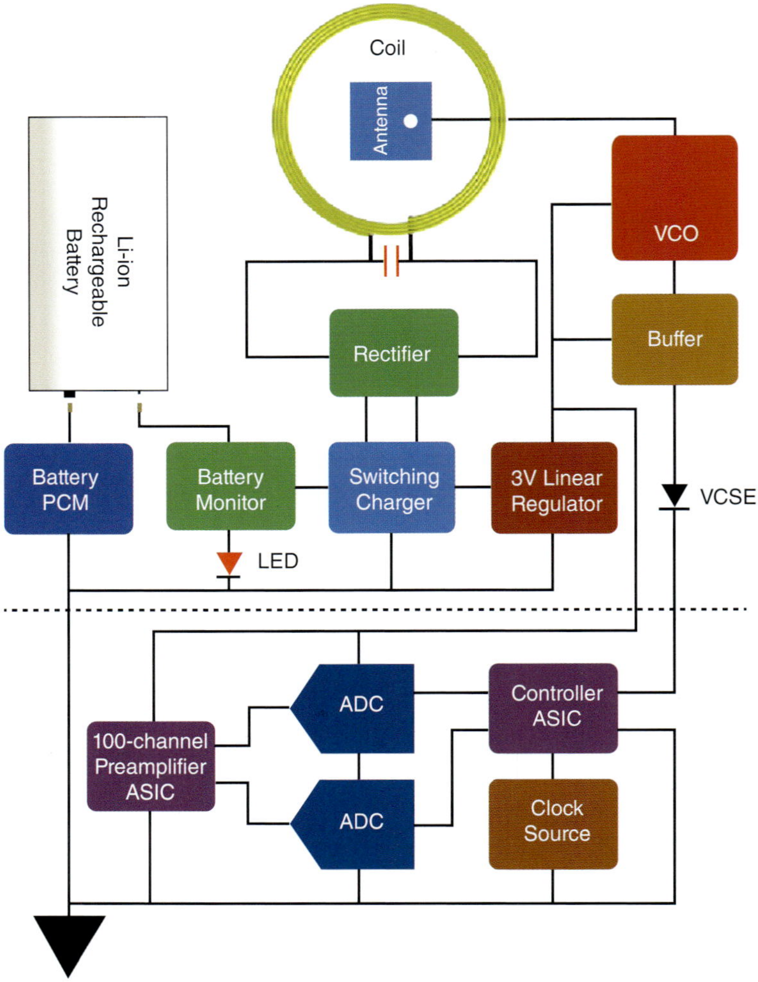

Figure 5.10 Schematic of the circuits within a Ti-enclosure deployed in controlled charging of the on-board Li-ion battery (Yin *et al.* 2013a).

yielding an increase in effective resistance of the material via the skin effect. As an example, for the wireless implant illustrated in Figure 5.4b and the material properties of grade 2 titanium used for the hermetic enclosure, we estimate a skin depth of 0.26 mm at a charging frequency of 2 MHz. Although most of the magnetic flux flows through the sapphire window, a finite fraction of energy spills over to the 1 mm thick titanium shell. Computational models have been used (Borton *et al.* 2013) to optimize the charging geometry and frequency and allow *a priori* quantitative estimates of the temperature rise of the surrounding tissue, including dermis, skull and blood, to balance between acceptable skin heating (<2°C) and maximal rates of power transfer and efficiency (note the contrast with more stringent limits to heating of brain tissue).

We also remark on rapid recent expansion of inductively coupled wireless battery charging technologies for consumer electronics (cell phones, etc.). These have been standardized ("Qi-standard") by the Wireless Power Consortium. Though unproven so far, adapting such a widely used standard for wireless medical implants might become useful. For instance, the use of magnetic flux concentrating ferrite thin sheet materials, the approach can help to eliminate stray power sinks in conducting parts of a hermetically sealed enclosure. The Qi standard is capable of providing up to 5 W of power and its ubiquitous nature has allowed for a multitude of small, lightweight transmitter designs. What could be especially attractive is a design which allows an implantable device to receive power for battery charging while maintaining continuous operation (i.e. neural recording capability), by power management protocols which are consistent with safety considerations.

5.6 PACKAGING AND CHALLENGES IN HERMETIC SEALING

While chronic, electronically active medical devices must satisfy the usual multi-tiered requirements of biocompatibility, the presence of on-board "live" microcircuits adds a further burden for comprehensive protection and safety. In addition to chemical impermeability, electrical impermeability requires packaging of the device against any unintended ionic and electronic flow. The subject is a broad one and plentifully reviewed in the literature (Joung 2013). Here we comment briefly on the principal material approaches to neural implants, while distinguishing "hard" and "soft" packaging classes of enclosure materials, with soft materials referring to polymer-based compounds and derivatives (Guangqiang and Zu 2010).

In the neurotechnology research and development device space, hermetic packaging has been dominated by titanium enclosures, including their requirement for a number of electrical feed-throughs for electrophysiological sensing and stimulation (Chapter 1). While glass enclosures have also met with some success in selected instances, of renewed interests are the possibilities of ceramic materials for chronic neural implants. At this writing, research progress has been made with various approaches to ceramic enclosures for neural implants (Kohler *et al.* 2014), though the critically important validation and regulatory approvals await these for translational use. Fully ceramic packaging is attractive, primarily for reasons of electromagnetic transparency into microwave frequency regime for telemetry and power transmission. Ceramic materials are, however, intrinsically mechanically fragile, hence advanced materials research is continuing in this area.

For titanium-based enclosures, joining dissimilar materials such as alumina and zirconia are frequently employed. Metals such as gold and platinum are typically part of the specialized (proprietary) processes used in the manufacture of ceramic-metal feed-throughs. Taking the specific case example of the high data rate customized implant of Figure 5.4b, the standard Ti-packaging technology was subject to additional challenges at two levels. First, the number of electrical feed-through was unusually high (total of 104 pins), as was the corresponding feed-through density. This placed an extra set of process demands for the specific brazing approach to

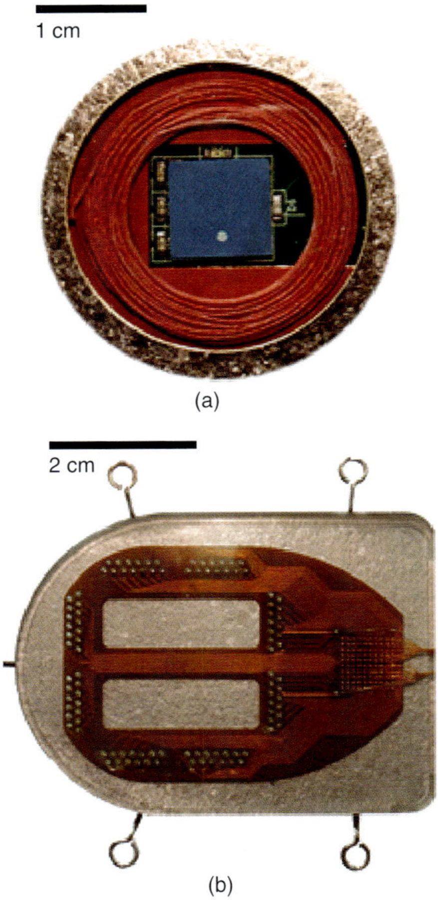

Figure 5.11 Two specific challenges in hermetic sealing. (a) sapphire window brazed to the Ti-enclosure; (b) arrays of metal-ceramic feed-throughs at the bottom of the Ti-can, to which wiring from a Kapton thin film fanout of the inputs from 100 wire bundle from neural sensor are wirebonded. The Kapton film is packaged by a PDMS silicone mold.

ensure reliability and hermeticity in an actual manufacturing setting (Figure 5.11a). Second, the device design called for a relatively large single crystal sapphire window to ensure electromagnetic transparency for both signal transmission and inductive battery recharging (Figure 5.11b). The brazing approach to achieve reliable sealing of the sapphire also required special process attention given the proximity of the sapphire perimeter to the edges of the Ti-enclosure and the differences in thermal expansion coefficients of the two materials. Nonetheless, a high yield outcome was achieved, with bench top He-leak testing validating the hermeticity after manufacture, and perhaps more important, testing the fully functional device after chronic implant to NHPs for more than 2 years without any sign of compromising the integrity of the packaging.

Turning to the soft, polymer-based materials for packaging of neural implants for chronic use, we note the long history of research and development in this field (Haessler *et al.* 2011). The most common material choices involve medical grade silicone (PDMS), polyimide, parylene, liquid crystal polymers, and others used in microelectronic device processing such as SU-8 and curable (photo) resist materials. As of this writing, we are unaware of any active electronic neural or brain implant which is exclusively polymer-packaged and which is on a path to regulatory approvals for human translation. Thus a major materials opportunity awaits bold researchers.

Soft packaging for neural/brain implants is very attractive in principle, due to mechanical flexibility and capability to accommodate complex device form factors. Likewise biocompatibility validation has been verified for many non-electronic polymers. As an example of a challenge and opportunity, consider, for example, the case of the PDMS encapsulated implant of Figure 5.4a, shown here from the packaging perspective of Figure 5.12 (Nurmikko *et al.* 2010; Song *et al.* 2005, 2007). The images suggest the difficulties of using PDMS (including additional parylene sealant layers) as a conformal encapsulant material. In this instance, the cortical front end of the device is formed by integration (by precision flip-chip bonding) of an MEA with a preamplifier/multiplexing ASIC. The back end of the device houses an A/D converter, controller ASICs and (near infrared) telemetry, as well as a planar RF energy harvesting coil for inductive, transcutaneous powering. The back and the front ends were connected by a Kapton ribbon cable (with 10 um gold tracks), with Kapton providing the common material substrate for the entire device. The polymer-based final packaging steps were thus challenged by a geometrically complex and mostly non-planar construct. Furthermore, surgical

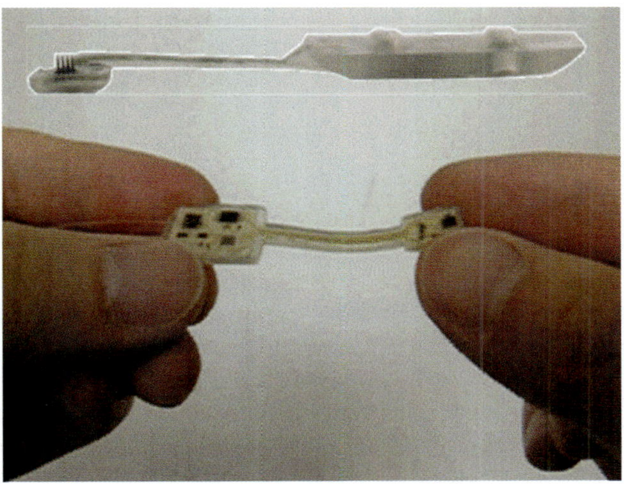

Figure 5.12 PDMS-based packaging of a neural implant. Upper image: profile of the implant with preamplifier/multiplexer ASIC is integrated with the neural sensor. Lower image illustrates the flexibility of a 7-wire ribbon cable connecting the cortical front end to the telemetry back end (Song *et al.* 2007).

implant considerations require careful control of PDMS thickness to maintain flexibility in the tether and to prevent build-up over the electrode array. Nonetheless, we have succeeded in operating the implant up to 6 months in an NHP. In so doing, special attention was placed on a range of packaging process steps to establish clean molding practices. Such practices are critical to ensure good adhesion of the PDMS (or equivalent) to the device, with adhesion being well-known as a main culprit in the failure of soft encapsulation (Donaldson, 1996; Edell, 2002) (Chapter 1).

A challenge remains to advance polymers or other soft materials for packaging electronically active neural implants for potential human use. Beyond the discovery of new material candidates and their physic-chemical application, it is important to embark on an extensive test and assessment program. For example, in our laboratories we developed a testbed (in 50°C saline) to investigate in-plane and cross-PDMS residual electrical conductivity as a means to track the effects of finite permeability, mainly due to unwanted ionic flow and requiring equipment to measure leakages on pA scale. In addition, we have designed and implemented an encapsulation test unit, which simulates the topographical, thermal and electrical stresses put on the encapsulant, including adhesion issues to test leakage current and component functionality over extended periods of soaking, presently up to 6 months in accelerated testing in saline at 52°C. Additional details can be found in Aceros *et al.* (2013) and Nurmikko *et al.* (2010). It is clear that much additional work is needed before polymer or other soft packaging approaches for the types of high-performance neural interfaces described here reach the levels of reliability required for regulatory approval. Such efforts are likely to require new material science approaches, such as exploration of combinations of soft organic polymers with inorganic thin film multilayer barriers or heterogeneous mixtures, solid solutions of inorganic molecules in polymers. Future research might take advantage of all-polymer and/or flexible material systems, where the electrodes and the on-board electronics are based on soft materials to create flexible integrated circuits with a more natural match with the mechanical and mechanically and chemical properties of the encapsulant materials (Vivendi *et al.* 2011; Wenger *et al.* 2015). Entirely new material classes, such as graphene and thin films of dense arrays of carbon nanotubes, are also coming of age for such potential applications (Chapter 2).

5.7 DEPLOYMENT OF HIGH DATA RATE WIRELESS RECORDING IN FREELY MOVING LARGE ANIMALS

With the emphasis of engineering a wireless high-data rate neural interface in the preceding sections, we next show illustrative examples of the use of the particular microsystems in freely moving large animals.

We focus specifically on the battery powered implant and its external head-mounted counterpart sharing the same microelectronics (Figures 5.4b and c) to describe their deployment for wireless recording of broadband neural data in free-moving swine and NHPs (the NHP assessment of performance of the "soft"-packaged implant has been described elsewhere (Nurmikko *et al.* 2010; Song *et al.* 2009).

5.7.1 Sample Case A: Implant in freely moving minipigs in home cage

For validating the wireless implant in a Ti-package, the swine model was chosen as a chronic animal model for reasons which included the immunological similarity to humans. This animal model is cost-effective to test the reliability and safety of the implant. Furthermore, the thickness of the pig's skin (∼1 cm) poses a challenge for both the RF data transmission and the transcutaneous inductive battery recharging. Minipigs move in home cages in idiosyncratic ways (such as rolling around on the floor), which provides a test for mechanical stability of the implant as well as the robustness of the RF data link (Figure 5.13). After thorough bench-top testing, the devices were sterilized and implanted into Yucatan mini-pigs (Agha *et al.* 2013; Borton *et al.* 2013). The 100-element MEA (Blackrock Microsystems) was implanted into the somatosensory association cortex (SSA), which has been implicated for integration neural signals from the somatosensory cortex (sight, sound, touch, etc.). This gave a corollary scientific opportunity to study neural correlates to specific behavior in the animal. From a practical standpoint, the SSA is the most convenient sensory cortex structure to access from the dorsal side of the animal. After implantation, the animal was introduced to a feeding device designed specifically to promote foraging and rooting behavior in pigs (Agha *et al.* 2013). A multiwall foraging device was fabricated where food treats in the wells were masked by set of laterally moveable slats which the animal could slide by using her snout to access the food underneath. By synchronizing the foraging unit and the neural recording instrumentation, wireless neural population data was acquired with the

Figure 5.13 A Yucatan minipig, implanted with the subcutaneous wireless neural sensor implant, engaged in foraging with a fully instrumented wireless feeder device, which synchronizes the animals intent and action to find food (from wells under sets of movable slats) with simultaneously recorded multichannel broadband neural data (Agha *et al.* 2013).

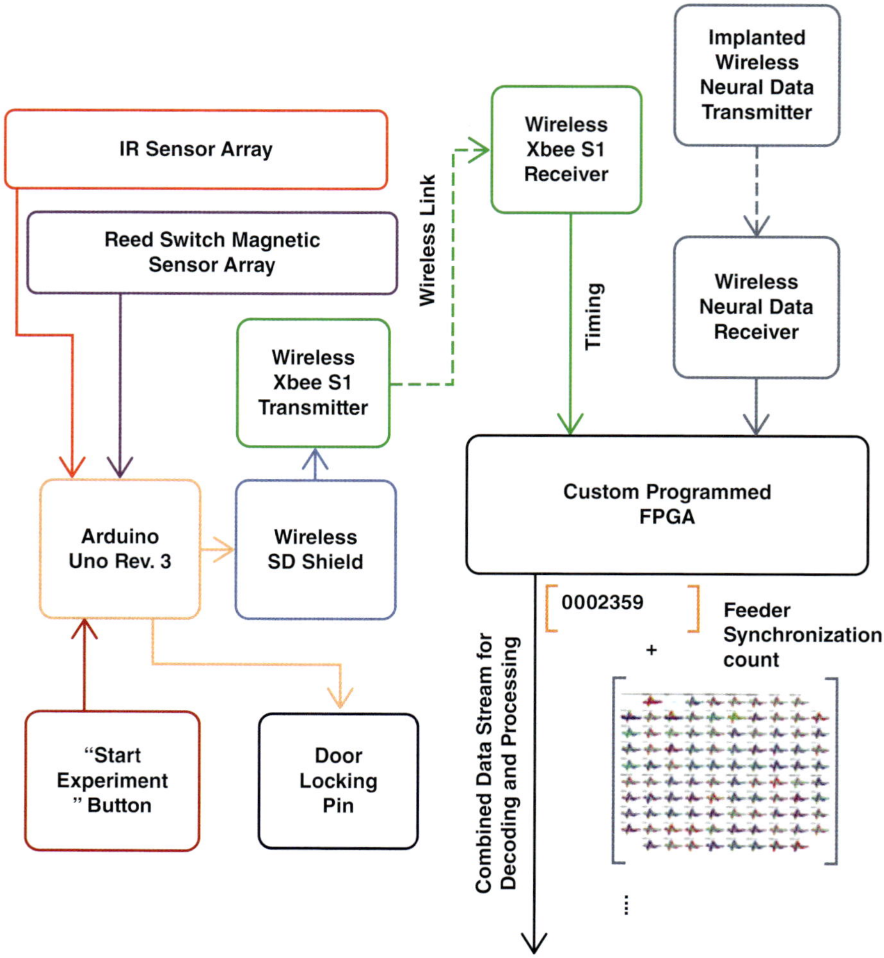

Figure 5.14 Process flow of the design for use of the Arduino microprocessor board and the application of the FPGA (Komar *et al*. 2014).

animal having the freedom to self-initiate an experimental episode. A simple custom programmed FPGA was utilized to develop a method of receiving all neural and behavioral data, to package and synchronize the information for further use. The process flow of the programming of the FPGA is exemplified alongside the process flow of the commercial Arduino microprocessor in Figure 5.14.

5.7.2 Sample Case B: Implant in freely moving non-human primate in home cage

For the NHP model, specific surgical procedures were established and carefully planned to ensure device safety and implant compatibility. The 100-element intracortical MEAs were implanted in the hand/arm area of the left primary motor cortex, while the Ti-enclosed module anchored to the skull was placed in the subcutaneous space atop the head (aided by prior mapping of the anatomic details by MRI and X-ray imaging). Following implant surgery, the wirelessly recorded single unit activity was detected as early as 2 hours into the post-operative period as the animal was returned to its home cage for surgical recovery. Details of the experimental arrangements can be found in the literature (Borton *et al.* 2013; Yin *et al.* 2014); here we highlight salient features and outcomes acquired for a total of four chronically implanted macaque monkeys, with this technology. Typical experimental sessions captured episodes of the monkeys moving freely in their home cages, while spontaneously reaching and grasping for food on their own volition, accessible immediately outside the cage. Prior to a recording session and as needed, the Li-ion battery in the hermetically sealed implant was transcutaneously charged by a compact external device (Section 5.5) Over extended periods of time (>20 months), we also monitored the Li-ion battery status and observed a consistent charge-time to run-time ratio over the complete study of over one year. Figure 5.15 shows a summary of typical wireless neural data and its analysis, captured episodically within time frames of 1–2 hours. The outcomes in the figure include decoding of the population dynamics during movement of the hand/arm of the animal by using neural state-space based analysis (Afshar *et al.* 2011; Truccolo *et al.* 2008; Yu *et al.* 2009). The computational approaches are based on statistical analysis of the multidimensional data via principal component analysis (PCA) similar to those applied to a wide variety of "pattern recognition" problems in science and engineering, where the aim is to discover any common underlying structure among multiple channels of data. Here the analysis aimed to correlate the wireless neural data with particular stereotypes of intended arm/hand movement.

Importantly, the wireless implant has proved reliable in recording both single- and multiunit activity as well as field potentials, initially on the majority of the MEA probe sites (>70). While the channel counts did drop in the time course over one year, due to a decrease in the recording performance of the Blackrock array by complex tissue-material interactions (Chapter 2), the electronic performance of the wireless implant compares well with the percutaneous electronics outside-subject system deployed in recent human clinical trials (Hochberg *et al.* 2006, 2012; Collinger *et al.* 2013) (Chapter 9) and is consistent in quality with the many reports on (chair constrained) NHP publications, where tethered cabling to the animal has been the norm (Achtman *et al.* 2007; Saleh *et al.* 2012; Velliste *et al.* 2008).

5.7.3 Case C: External head mounted wireless neurosensor in freely moving non-human primates

Engineering high-performance, compact lightweight, low-power, external head-mounted wireless devices (Figure 5.4b and description in Section 5.4) has been a very useful parallel outcome in acquiring high-fidelity broadband neural data in spatially unconstrained environments from free-behaving NHPs (Yin *et al.* 2014). While this approach requires a percutaneous connection to the MEA to

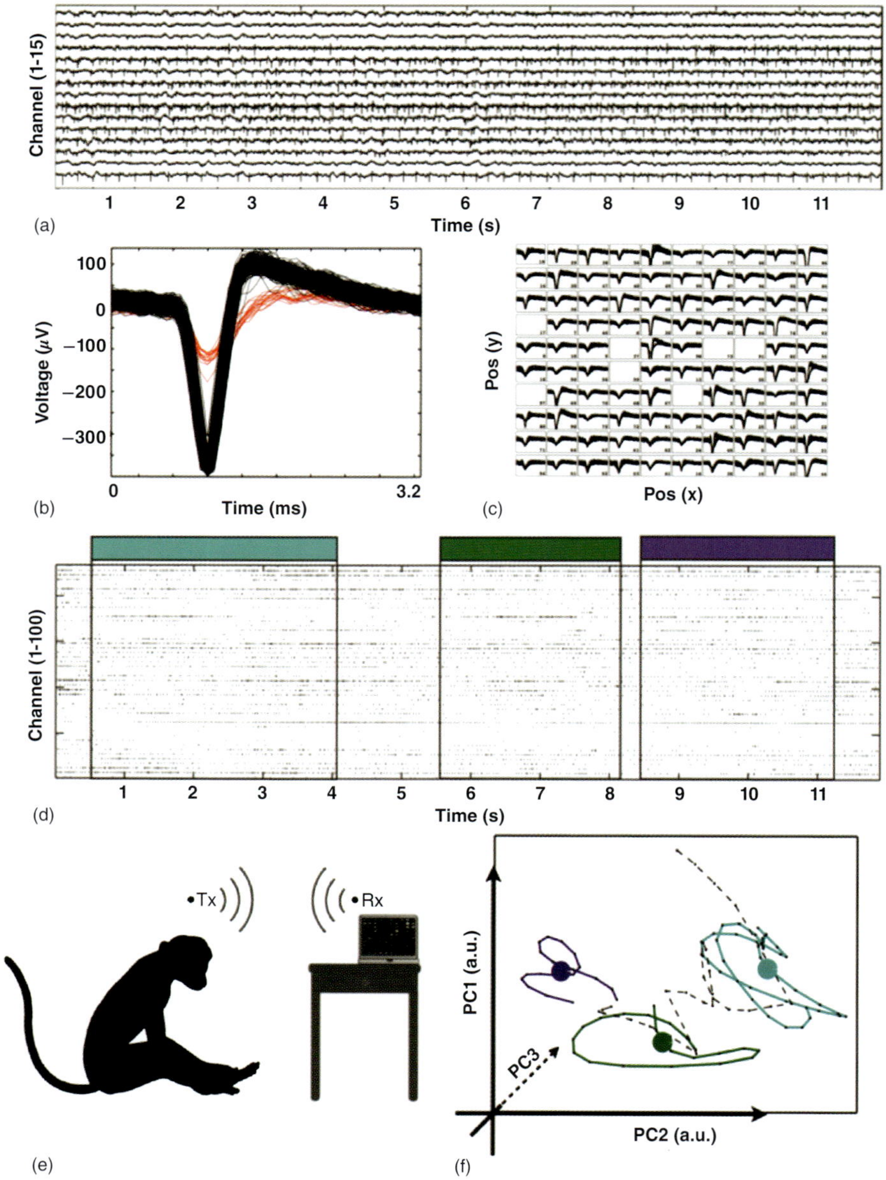

Figure 5.15 Wirelessly recorded data from an implanted neural interface in NHPs. (a) A selection of 15 of the 100 broadband recording channels richness of high-sample rate (20 ksps) data collection. Spikes were extracted (b) on all channels (c), showing single neuron activity on input channels. Raster plot (12 s) marking spike timestamps for all input channels is shown with behavior indicated by color overlay. In (e), the monkey reaches for food while neural data is recorded wirelessly. Spikes across all channels are reduced to a low-dimensional state space (f) through principal component analysis, where such neural trajectories produced during free movement of monkey scratching eye (blue), touching an apple (green) and turning head (purple) (Borton *et al.* 2013).

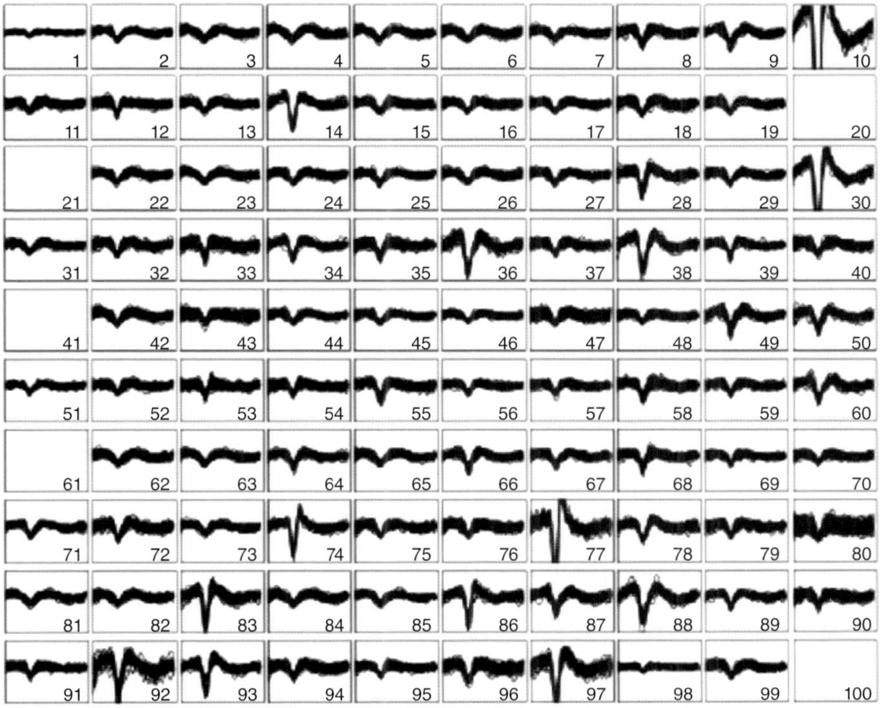

Figure 5.16 Sampling of *in vivo* recorded neural action potential from a rhesus macaque monkey sitting in chair.

access single neuron level population recordings, it provides a cost-effective means of exploiting high data rate wireless brain interfaces to pursue fundamental brain science (e.g. sample data in Figure 5.16). Also, the device configuration with its peripheral electronics support was designed as a platform concept to explore and enable further development upward scalability, such as for acquiring wireless broadband neural data perhaps up to hundreds of channels. Furthermore, the headmounted platform as well as parallels pursued in other laboratories could be readily expanded to include additional functionalities such as multisite, patterned neural stimulation.

Figure 5.17 shows an example of the performance of the head-mounted wireless neural interface, while streaming neural data continuously to monitor cortical states as reflected through neuron population dynamics of an NHP in its home cage (Yin *et al.* 2014). In this instance, the objective was to study sleep-wake transitions during night-time recordings, but the system has also been applied to NHPs during natural locomotion and is being expanded to studies in multiple ecologically equivalent settings. The head-mounted device and its internal components were protected from mechanical and electrical impact by a static-dissipative, carbon fiber reinforced PolyEtherEtherKetone enclosure (Figure 5.4c). The device utilizes a "screw-on" quick interconnect to the skull-mounted pedestal. Small cavities were embedded in

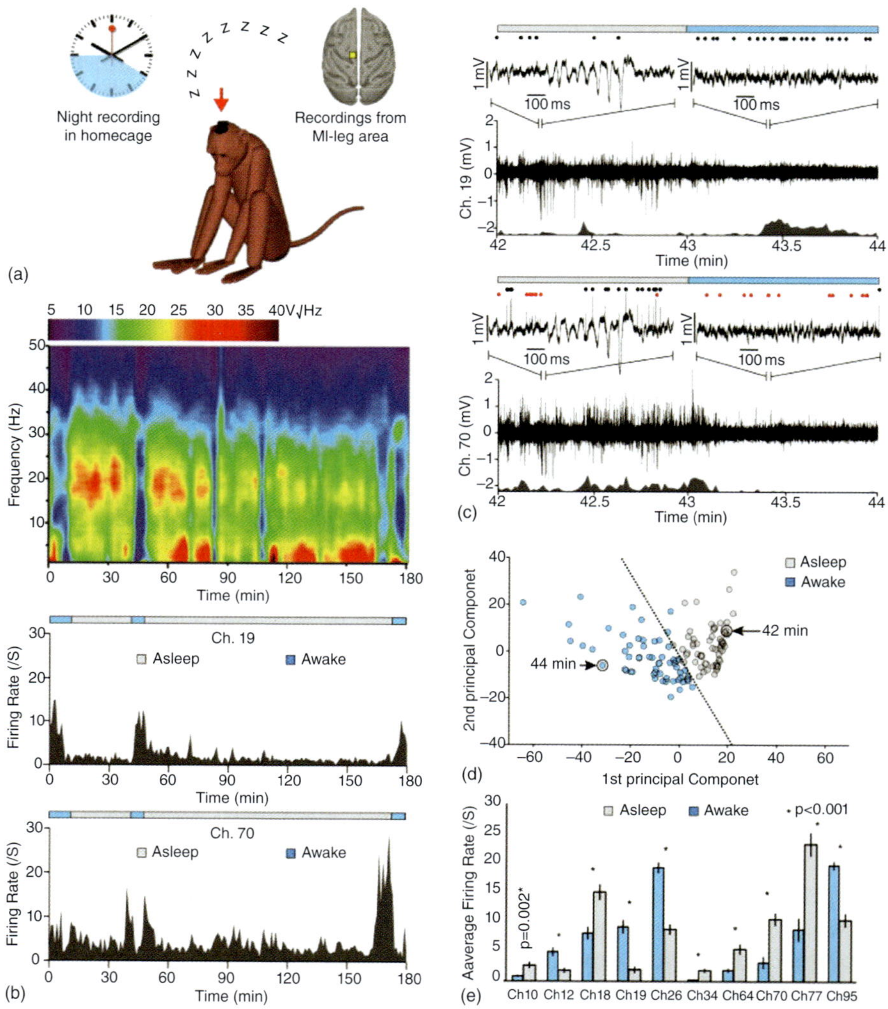

Figure 5.17 (a) Overnight recording from an untethered NHP in home cage with a four antenna SIMO receiver system. (b) Wireless data represented as 3 hours of recorded power spectrograms below 50 Hz (top) and neuron firing rates histograms of a single neuron from two channels as field potential and firing rate patterns across sleep (grey) and awake (blue) cycles monitored by IR camera video recordings. (c) Raw neural data and firing rate histograms (1 s bin size) from the two channels over a 2-min sleep-to-wake transition. (d) Classification of neural states based on the action potential signals during the particular 2-min sleep-to-wake transition using PCA. (e) Comparison of average firing rates between asleep and awake states for 10 neurons with dominant contribution to the PCA data. (Yin *et al.* 2014.)

the housing to support ethylene oxide sterilization, a standard for clinical sterilization. From a telemetry point of view for mobile subjects, this system has enabled design and configuration of RF links for wireless neural interfaces as a spatially distributed network of synchronized receivers that is scalable to hundreds of channels and vast environments. Mobile telemetry systems can suffer from signal degradation and loss due to multipath fading and electromagnetic interference, an issue of particular importance when planning wireless neural interfaces and prostheses for humans. We adapted the single-input-multiple-output (SIMO) telecommunication scheme analogous to approaches in cellular telecommunication to design a multiple receiver concept first using four antennas that simultaneously received the broadband neural data. The SIMO approach is scalable to a larger number of receiving antennas in order to achieve full spatial coverage of neural signal acquisition within vast and geometrically complex environments. For future human use, a network of antennas could be envisioned within a patient's home.

Related to scientific elements in Figure 5.17, continuous wireless broadband recording capability is essential in the study of natural sleep-wake transition with the subject in an unconstrained environment. Technological limitations have restricted previous studies on sleep-wake transitions to recordings of a few neurons (Zanos *et al.* 2011), or of neuronal populations under highly constrained conditions. Analysis of neuronal ensemble modulation over eight consecutive hours allowed identification of sleep-wake transitions against earlier literature using electroencephalography (EEG) and ECoG or spiking activity using microelectrodes. From the wireless data, shifts were found in spectral power of local field potentials (LFPs) between well-defined frequency bands, which corresponded to periods of sleep and awake states (Figure 5.17b), suggesting that the monkey was in a drowsy state. Occasional short-duration body movements during the initial transitory periods were reflected in spectral analysis of the Beta band (13–30 Hz) LFPs, generally associated with muscle contractions and suppression of movement. Conversely, during transition from shallow to deep sleep, the dominant frequency band shifted from beta to theta frequencies (Figures 5.17b and c). The power distribution of LFP signals abruptly de-correlated when the animal awoke (Figure 5.17c). Neural population dynamics during the sleep-wake transitions were also acquired across 96 channels during a representative 2-minute sleep-wake transition period (Figures 5.17c to e). PCA decoded from modulation of the population-based firing rates enabled first-order classification to UP-regulated and DOWN-regulated cortical states during sleep (Figure 5.17e).

5.8 SUMMARY AND PROSPECTS FOR HIGH DATA RATE BRAIN INTERFACES FOR NEURAL PROSTHESES

In this chapter we have described approaches to wireless neural interfaces through a prism of particular choices for technical solutions to high data rate recording microsystems. Although somewhat specific, the device and system approaches introduced in Section 5.4 should give the reader an assessment of the generic challenges of recording from ever larger numbers of neurons and their assemblies for interpretation as specific functional microcircuits. The platform approaches we have described were designed for broad use in research, clinical diagnostics, and applications such as neural prostheses, and are independent of whether the neural signals emanate from the brain, other parts of the central nervous system or extension to the peripheral nervous system.

It is important to note that the technology of wireless, high-performance neural recording interfaces for mobile subjects is still at an early stage. It is likely that the coming years will witness significant advances at many technical levels, ranging from development of new types of micro- and opto-electronic probe arrays to further decrease power consumption by sophisticated microelectronic components, as well as chipscale integration of tailored on-chip signal processing and storage functions.

At this writing, groundwork is being laid for translation to human neuroprosthetic use of the types of neural recording microsystems described above and the first clinical trials are likely in the near term given the many ongoing research and development efforts similar to those described in this chapter. The road to human use of any active medical device is, of course, complex and very costly, here perhaps even more so given the novelty and lack of precedents for implant technology for high data rate electronic interfaces to the brain for chronic use. Patient safety is, of course, paramount to which reliability, biocompatibility, hermeticity, electrical safety (Yin *et al.* 2013a), telemetry, etc. of the implant must conform to, while satisfying regulatory standards.

The platform wireless approaches reviewed here also offer the opportunity to enhance the functionality of the wireless interfaces by adding to the electronic payload while minimizing any cost to the device size, power consumption, etc. Among such functions is the implementation of patterned electrical (Dadarlat *et al.* 2015; Tabot *et al.* 2013) and/or optical micro-stimulation for enabling truly bidirectional communication opportunities with the brain. Development of dedicated microelectronic circuits for such multi-site stimulation are well under way in several academic and commercial laboratories and it is likely that some aspects of patterned wireless electrical micro-stimulation will be implemented in NHPs in the near term. Optical (wireless) micro-stimulation, exploiting developments in optogenetics in NHPs such as providing a proxy for tactile stimuli (May *et al.* 2014), is likewise on the cusp of transitioning from bench top to research for study of spatially and temporally tailored patterned cortical stimulation NHPs. Combination of multisite neural recording (such as from primary or premotor cortical microcircuits to read out the brain's motor plans) with multisite stimulation (such as in somatosensory cortex to provide sensory feedback) defines one ultimate goal of wireless neural interface technology to achieve truly closed-loop, biomimetic neural prostheses (such as sensory-motor assistive systems for persons with severe spinal cord injuries).

For realization of biomimetic, co-adaptive close-loop brain interfaces, implant device neurotechnology requires close engineering synergy with computational neuroscience in decoding increasingly rich sets of brain signals. Decoding strategies for tracking and interpreting neural population dynamics in real time have been advancing rapidly in their level of sophistication. A contemporary example is seen in the recent human clinical trials of subjects operating robotic limbs and communication devices based on neural population recordings from the motor and premotor cortical microcircuits (Chapter 9). Closed-loop sense/stimulate based algorithms have recently made considerable progress in deep brain stimulation (Afshar *et al.* 2013); however, the challenges encountered here, say, in a 100-channel *bidirectional* interface, require greatly increased sophistication and computational power for real-time implementation as prostheses. Here large amounts of real-time data, recorded from the nervous system, must be processed and decoded (Collinger *et al.* 2013; Truccolo *et al.* 2013) to develop electronic interfaces for persons with severe neurological disabilities to operate assistive devices ranging from telecommunication to advanced robotics. Moreover, given the inherent variability and statistical entropy-driven fluctuations in neural circuits and signal processing pathways, algorithms are likely to require inclusion of sophisticated machine-learning tools.

Finally, to complete a practical, user-friendly high data rate neural prosthesis system with flexibility and mobility, yet another set of technological innovations will be necessary. Presently, to implement the decoding algorithms and interface the "translated" brain signals with assistive electronic devices, sets of computers and other electronic hardware are used. As depicted in the concept image of Figure 5.18, it is desirable to compact entire sets of such firmware to a *wearable* module combining high computing power and telecommunication capability. Such a device system should be programmable, with a wide variety of sophisticated functions to enable research testing and end-user deployment of signal processing and decoding algorithms (including accommodation of anticipated parallel advances in fundamental brain sciences). Computational capabilities and throughput should support streaming of hundreds of channels of bidirectional broadband signals (say, 12 bit data per channel sampled at 30 ksps), which are recorded from and applied as stimulating therapies into the brain and/or other areas of the body.

Citing one early entry, Rapoport *et al.* (2012) have reported a low-power decoding architecture running on a FPGA for a fully *implantable* decoding system which, however, was optimized for low power at considerable expense of decoding flexibility. There are efforts under way at many institutions to create new types of architecture and much higher performance yet low power *wearable* external modules by adapting concepts of embedded digital systems, while exploiting the explosive growth in processing power of modern microprocessors. State–of-the-art microprocessors (at the 22 nm silicon CMOS process node) have at this writing reached performance measured in trillions of operations per second. In our laboratories, we have reached a prototype level of a low-power wireless embedded processor platform targeting the support of a full suite of current and investigational signal processing and decoding approaches being investigated for neural prosthetic applications (Komar *et al.* 2014). Briefly, this prototype, utilizing the Xilinx Zynq-7045 SoC, is configured and has been demonstrated for the specific case of wirelessly receiving broadband cortical neural data as an approximate 100 Mbps input data stream from an NHP, corresponding to 200 channels of broadband neural data composed of a sum of action potential spikes and field potentials. The input neural data is processed in parallel, using an on-board FPGA. The digitally filtered features alongside the raw neural

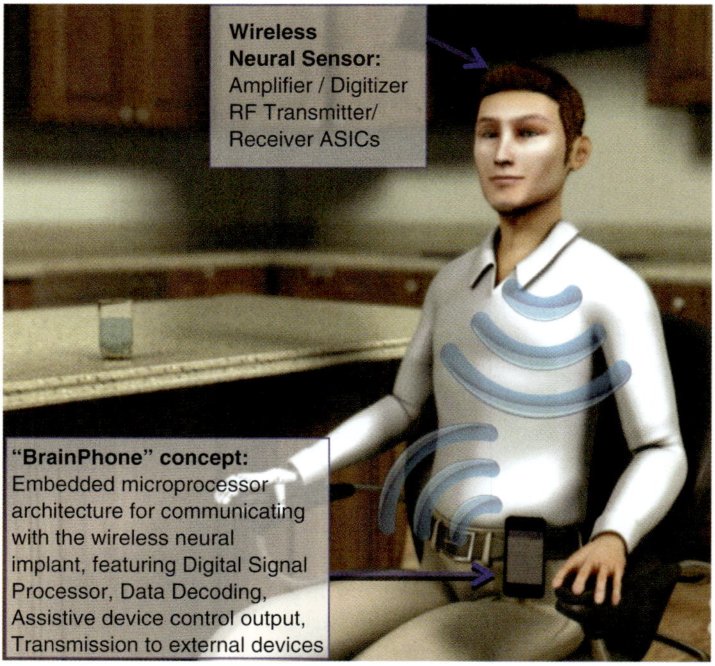

Figure 5.18 Futuristic concept of mobile, wireless neural prostheses, highlighting the functional roles of the subcutaneous wireless head implant together with the wearable external electronics for neural signal processing, decoding and wireless control of assistive neural prosthesis devices (the "BrainPhone" refers to an embedded microprocessor portable computing system, aimed at for real-time neural decoding by on-board algorithms). The wireless implant device is expected to also perform multichannel functional electrical stimulation.

data are forwarded to a server via Wi-Fi or Ethernet for data viewing and long-term storage. The low-power usage of the prototype system ($\approx$2 W) allows for battery operation in wearable and mobile settings. The deterministic scheduling of the FPGA logic in conjunction with a medical-grade real-time operating system provides for inclusion of real-time neural decoding or other complex and computationally intensive algorithms across hundreds of channels.

While forecasting the future is uncertain at best, it is perhaps even more so in the case of bidirectional high-performance wireless implants and associated neurotechnologies in terms of the "when" for their implementation in human use. Much of the technical and associated innovation to date has taken place in academic laboratories, so that in addition to the critical regulatory approvals to translate technologies such as described in this chapter, the role of medical device industry will be pivotal. Industry, in turn, must gauge the size of relevant markets so that careful engineering planning of platform approaches to the wireless implants to apply them to multiple models of neurological injuries and diseases is especially important. Epilepsy (Truccolo *et al.* 2014) and stroke rehabilitation are perhaps examples of reasonable-sized markets, to augment the use of this type of technology in cases of spinal cord injury.

ACKNOWLEDGEMENTS

The authors are very grateful to many members, past and present, in their laboratory. These include Y.-K. Song, William R. Patterson III, Kristina Davitt, Naubahar Agha, Farah Laiwalla, Sunmee Park, Jacob Komar and Chris Heelan among others. At Brown University, we have been fortunate to be part of a wonderful larger brain science and neurotechnology effort with colleagues such as John Donoghue, Leigh Hochberg, David Rosler, John Simeral and Carlos Vargas-Irwin, whom we thank for their continuing input and expertise. Special thanks are extended to Krishna Shenoy and his laboratory at Stanford for providing inspirational leadership in the field.

All animal procedures referred to in this chapter were conducted conforming to the National Research Council's Guide for the Use and Care of Laboratory Animals (1996), and according to protocols approved by Institutional Animal Care and Use Committee (IACUC) at each institution. Research in the authors' laboratory was supported by the US National Institutes of Health, Defense Advanced Projects Agency, and the National Science Foundation.

REFERENCES

Aceros, J., Yin, M., Borton, D., Patterson, W.R. and Nurmikko, A.V. (2011) A 32-channel fully implantable wireless neurosensor for simultaneous recording from two cortical regions, *IEEE EMBC Conference Proceedings*, pp 2300–2306.

Achtman, N., Afshar, A., Santhanam, G., Byron, M.Y., Ryu, S.I. and Shenoy, K.V. (2007) Free-paced high-performance brain-computer interfaces. *J. Neural. Eng.*, **4**: 336–347. DOI:10.1088/1741-2560/4/3/018

Afshar, P., Khambhati, A., Stanslaski, S., Carlson, D., Jensen, R. *et al.* (2013) A translational platform for prototyping closed-loop neuromodulation systems. *Frontiers in Neural Circuits*, **6**: 1

Afshar, P., Moran, D., Rouse, A., Wei, X. and Denison, T. (2011a) Validation of chronic implantable neural sensing technology using electrocorticographic (ECoG) based brain machine interfaces, *5th Int. IEEE/EMBS Conf. on Neural Eng,*, pp 704–707. DOI:10.1109/NER.2011.5910645

Afshar, A., Santhanam, G., Yu, B.M., Ryu, S.I., Sahani, M. and Shenoy, K.V. (2011b) Single-trial neural correlates of arm movement preparation. *Neuron*, **71**(3): 555–564.

Agha, N.S., Komar, J., Yin, M., Borton, D.A. and Nurmikko, A. (2013) A fully wireless platform for correlating behavior and neural data from an implanted, neural recording device: Demonstration in a freely moving swine model, *Proc. 6th Int. IEEE/EMBS Conf. on Neural Eng.*, pp 989–992.

Barretto, R.P.J., Ko, T.H., Jung, J.C., Wang, T.J., Capps, G. *et al.* (2011) Time-lapse imaging of disease progression in deep brain areas using fluorescence microendoscopy. *Nat. Med.*, **17**: 223–228.

Blabe, C., Gilja, V., Chestek, C.A., Shenoy, K.V., Anderson, K. and Henderson, J.M. (2015) Assessment of brain-machine interfaces from the perspective of people with paralysis. *J. Neural Eng.*, **12**(4): 043002. doi: 10.1088/1741-2560/12/4/043002. Epub 2015 Jul. 14.

Borton, D.A., Yin, M., Aceros, J. and Nurmikko, A. (2013) An implantable wireless neural interface for recording cortical circuit dynamics in moving primates. *J. Neural Eng.*, **10**: 026010-026010.

Chae, M.S., Yang, Z., Yuce, M.R., Hoang, L. and Liu, W. (2009) A 128-channel 6 mW wireless neural recording IC with spike feature extraction and UWB transmitter. *IEEE Trans. Neural Syst. Rehabil. Eng.*, **17**: 312–321.

Chestek, C., Gilja, V., Nuyujukian, P., Kier, R.J., Solzbacher, F. *et al.* (2009) HermesC: low-power wireless neural recording system for freely moving primates. *IEEE Trans. Neural Syst. Rehabil. Eng.*, **17**: 330–338. DOI:10.1109/TNSRE.2009.2023293

Collinger, J.L., Wodlinger, B., Downey, J.E., Wang, W., Tyler-Kabara, E.C. *et al.* (2013) High-performance neuroprosthetic control by an individual with tetraplegia. *Lancet*, **381**: 557–564.

Dadarlat, M.C., O'Doherty, J.E. and Sabes, P.N. (2015) A learning-based approach to artificial sensory feedback leads to optimal integration. *Nature Neurosci.*, **18**: 138–144.

Donaldson, P.E.K. (1996) The essential role played by adhesion in the technology of neurological prostheses. *Int. J. Adhesion and Adhesives*, **16**(2): 105–107.

Edell, D. (2002) Insulating Biomaterials N01-NS-9-2323, First Quarter Progress Report October–December, National Institute of Health. On-line; http://www.ninds.nih.gov/qpr/electrode/N01-NS-9-2347QPR01.pdf.

Gao, H., Walker, R.M., Nuyujukian, P., Makinwa, K.A.A., Shenoy, K.V. *et al.* (2012) HermesE: A 96-channel full data rate direct neural interface in 0.13 um CMOS. *IEEE J. Solid State Circuits*, **47**(4): 1043–1054.

Gilja, V., Nuyujukian, P., Ryu, S., Greger, B., Solzbacher, F. and Shenoy, K.V. (2009) Wireless neural recording with single low-power integrated circuit. *IEEE Trans. Neural Syst. Rehab. Eng.*, 17(**4**: 322–329.

Greenwald, E., Mollazadeh, M., Hu, C., Tang, W., Culurciello, E. and Thakor, N.V. (2011) A VLSI neural monitoring system with ultra-wideband telemetry for awake behaving subjects. *Biomedical Circuits and Systems, IEEE Trans.*, **5**: 112–119.

Guangqiang, J. and Zu, D.D. (2010) Technology advances and challenges in hermetic packaging for implantable medical devices, in *Implantable Neural Prostheses 2, Biological and Medical Physics, Biomedical Engineering*, Zhou, D.D. and Greenbaum, E. (eds), Springer-Verlag, New York.

Haessler, C., Boretius, T. and Stieglitz, T. (2011) Polymers for neural implants. *J. Polymer Sci.: Part B: Polymer Physics*, **49**: 18–33,

Harrison, R.R. and Charles, C. (2003) A low-power low-noise CMOS for amplifier neural recording applications. *IEEE J. Solid-State Circuits*, **38**: 958–965. DOI:10.1109/JSSC.2003.811979

Harrison, R.R. (2007) Designing efficient inductive power links for implantable devices. *Proc. IEEE Int. Symp. on Circuits and Systems (ISCAS '07)*, New Orleans LA, pp 2080–2083.

Harrison, R.R., Kier, R.J., Chestek, C.A., Gilja, V., Nuyujukian, P. *et al.* (2009) Wireless neural recording with single low-power integrated circuit. *IEEE Trans. Neural Syst. Rehab. Eng.*, **17**(4): 322–329.

Ho, J.S., Yeh, A.J., Neofytou, E., Kim, S., Tanabe, Y. *et al.* (2014) Wireless power transfer to deep-tissue micro-implants, *PNAS*, **111**, 7974–7979.

Hochberg, L.R., Serruya, M.D., Friehs, G.M., Mukand, J.A., Saleh, M. *et al.* (2006) Neuronal ensemble control of prosthetic devices by a human with tetraplegia. *Nature*, **442**: 164–171.

Hochberg, L.R., Bacher, D., Jarosiewicz, B., Masse, N.Y., Simeral, J.D. *et al.* (2012) Reach and grasp by people with tetraplegia using a neurally controlled robotic arm. *Nature*, **485**: 372–375.

Jackson, A., Mavoori, J. and Fetz, E.E. (2007) Correlations between the same motor cortex cells and arm muscles during a trained task, free behavior, and natural sleep in the macaque monkey. *J. Neurophysiol.*, **97**: 360–374.

Joung, Y.H. (2013) Development of implantable medical devices: from an engineering perspective. *Int. Neurol. J.*, **17**: 98–106.

Kim, S., Bhandari, R., Klein, M., Negi, S., Rieth, L. *et al.* (2009) Integrated wireless neural interface based on the Utah electrode array. *Biomedical Microdevices*, **11**(2): 453–466.

Kohler, F., Kiele, P., Ordonez, J.S., Stieglitz, T. and Schuettler, M. (2014) A polymer-metal two step sealing concept for hermetic neural implant packages. *Conf. Proc. IEEE Eng. Med. Biol. Soc.*, **2014**: 3981–3984.

Komar, J., Heelan, C., Simeral, J.D. and Nurmikko, A.V. (2014) A programmable 192-channel wireless neural signal processing and decoding platform for wearable mobile applications. *IEEE Eng. Biol. Med., Brain Grand Challenge Conference*, Washington DC, November.

May, T., Ozden, I., Brush, B., Borton, D., Agha, N. *et al.* (2014) Strong optically induced sensation in hand/digit areas of somatosensory cortex in non-human primates. *PLoS One*, **9**(12): e114529.

Mestais, C., Charvet, G., Sauter-Starace, F., Foerster, M., Ratel, D. and Benabid, A.L. (2015) WIMAGINE: Wireless 64-channel ECoG recording implant for long term clinical applications. *IEEE Trans. Neural Syst. Rehab. Eng.*, **23**: 10–21.

Miranda, H., Gilja, V., Chestek, C.A., Shenoy, K.V. and Meng, T.H. (2010) HermesD, a high-rate long-range wireless transmission system for simultaneous multichannel neural recording applications. *IEEE Trans. Biomed. Circ. Syst.*, **4**: 181–191.

Morrell, M. (2011) Responsive cortical stimulation for the treatment of medically intractable partial epilepsy. *Neurology*, **77**: 1296.

Nikita, K.S. (2014) Inductive coupling, in *Handbook of Biomedical Telemetry*, Ghovanloo, M. and Kiani (eds), Chapter 7, John Wiley & Sons, Inc., Hoboken NJ.

Nurmikko, A.V., Donoghue, J.P., Hochberg, L., Patterson, W.R., Song, Y.K. *et al.* (2010) Listening to brain microcircuits for interfacing with external world – progress in wireless implantable microelectronic neuroengineering devices. *Proc. IEEE*, **98**: 375.

Park, S., Borton, D.A., Kang, M., Nurmikko, A.V. and Song, Y.K. (2013) An implantable neural sensing microsystem with fiber-optic data transmission and power delivery. *Sensors*, **13**(5): 6014–6031.

Rapoport, B.I., Sarpeshkar, R, Turicchia, L., Wattanapanitch, W. and Davidson, T.J. (2012) Efficient universal computing architectures for decoding neural activity. *PLoS ONE*, **7**(9): e42492. doi:10.1371/journal.pone.0042492.

Rizk, M., Bossetti, C.A., Jochum, T.A., Callender, S.H., Nicolelis, M.A. *et al.* (2009) A fully implantable 96-channel neural data acquisition system. *J. Neural Eng.*, **6**: 026002.

Robinson, D.A. (1968) The electrical properties of metal microelectrodes. *Proc. IEEE*, **56**: 1065.

Rouse, A.G., Stanslaski, S.R., Cong, P., Jensen, R.M., Afshar, P. *et al.* (2011) A chronic generalized bi-directional brain-machine interface. *J. Neural. Eng.*, **8**: 036018.

Saleh, M., Takahashi, K. and Hatsopoulos, N.G. (2012) Encoding of coordinated reach and grasp trajectories in primary motor cortex. *J. Neurosci.*, **32**: 1220–1232.

Shahrokhi, F., Abdelhalim, K., Serletis, D., Carlen, P.L. and Genov, R. (2010) The 128-channel fully differential digital integrated neural recording and stimulation interface. *IEEE Trans. Biomed. Circ. Syst.*, **4**(3): 149–161.

Simeral, J., Kim, P., Black, M., Donoghue, J. and Hochberg, L. (2011) Neural control of cursor trajectory and click by a human with tetraplegia 1000 days after implant of an intracortical microelectrode array. *J. Neural. Eng.*, **8**: 025027. DOI:10.1088/1741-2560/8/2/025027

Sodagar, A.M., Wise, K.D. and Najafi, K. (2009) A wireless implantable microsystem for multichannel neural recording. *IEEE Trans. Microw. Theory Tech.*. **57**: 2565–2573.

Song, Y.K., Patterson, W.R., Bull, C.W., Beals, J., Hwang, N.J. *et al.* (2005) Development of a Chipscale integrated microelectrode/microelectronic device for brain implantable neuroengineering applications. *IEEE Trans. Neural. Rehab. Eng.*, **13**: 220. DOI:10.1109/TMTT.2009.2029957

Song, Y.K. *et al.* (2007) A brain implantable microsystem with hybrid RF/IR telemetry for advanced neuroengineering applications. *Proc. 29th Ann. Int. Conf. IEEE EMBS*, Lyon, France, August 23–26: 445–448.

Song, Y.K., Borton, D.A., Park, S., Patterson, W.R., Bull, C.W. *et al.* (2009) Active microelectronic neurosensor arrays for implantable brain communication interfaces. *IEEE Trans. Neural. Rehab. Eng.*, **17**: 339–345.

Szuts, T.A. *et al.* (2011) A wireless multi-channel neural amplifier for freely moving animals. *Nature Neurosci.*, **14**: 263–269.

Thomas, S., Harrison, R.R., Leonardo, A. and Reynolds, M.S. (2012) A battery-free multi-channel digital neural/EMG telemetry system for flying insects. *IEEE Trans. Biomed. Circ. Syst.*, **6**(5): 424–436.

Tabot, G.A., Dammann, J.F., Berg, J.A., Tenore, F.V., Boback, J.L. *et al.* (2013) Restoring the sense of touch with a prosthetic hand through a brain interface. *Proc. Nat. Acad. Sci. USA*, doi:10.1073/pnas.1221113110.

Truccolo, W., Friehs, G.M., Donoghue, J.P. and Hochberg, L.R. (2008) Primary motor cortex tuning to intended movement kinematics in humans with tetraplegia. *J. Neurosci.*, **28**: 1163–1178.

Truccolo, W., Ahmed, O.J., Harrison, M.T., Eskandar, E.N., Cosgrove, G.R. *et al.* (2014) Neuronal ensemble synchrony during human focal 738 seizures. *J. Neurosci.*, **34**(30): 9927–9944.

DOI:10.1038/nn.2730 Van Noorden, R. (2014) A better battery. *Nature*, **507**: 26.

Van Schuylenbergh, K. and Puers, R. (2009) *Inductive Powering Basic Theory and Application to Biomedical Systems*, Springer, Dordrecht, Netherlands.

Viventi, J., Kim, D.H., Vigeland, L., Frechette, E.S., Blanco, J.A. *et al.* (2011) Flexible, foldable, actively multiplexed, high-density electrode array for mapping brain activity *in vivo*. *Nature Neurosci.*, **14**: 1599.

Wattanapanitch, W., Fee, M, and Sarpeshkar, R. (2007) An energy-efficient micropower neural recording amplifier. *IEEE Trans. Biomed. Circ. Syst.*, **1**: 136.

Wenger, N., Moraud, E.M., Gandar, J., Capogrosso, M., Milekovic, T. *et al.* (2015) Electronic dura mater for long-term multimodal neural interfaces. *Science*, **347**: 159–163.

Wodlinger, B., Downey, J.E., Tyler-Kabara, E.C., Schwartz, A.B., Boninger, M.L. and Collinger, J.L. (2015) Ten-dimensional anthropomorphic arm control in a human brain–machine interface: difficulties, solutions, and limitations. *J. Neural Eng.*, **12**: 016011

Yin, M. and Ghovanloo, M. (2011) A low-noise clockless simultaneous 32-channel wireless neural recording system with adjustable resolution. *Analog Integ. Circ. Sign. Proc.*, **66**(3): 417–431.

Yin, M., Borton, D.A., Aceros, J., Patterson, W.R. and Nurmikko, A.V. (2013a) A 100-channel hermetically sealed implantable device for chronic wireless neurosensing applications. *IEEE Trans. Biomed. Circ. Syst.*, **7**: 115–128.

Yin, M., Li, H., Bull, C., Borton, D.A., Aceros, J., Larson, L. and Nurmikko, A.V. (2013b) An externally head-mounted wireless neural recording device for laboratory animal research and possible human clinical use. *IEEE Eng. Med. Biol. Soc. Conf.*, pp 3109–3114.

Yin, M., Borton, D.A., Komar, J., Agha, N., Lu, Y. *et al.* (2014) Wireless neurosensor for full-spectrum electrophysiology recordings during free behavior. *Neuron,* **84**(6): 1170–1182.

Yu, B.M., Cunningham, J.P., Santhanam, G., Ryu, S.I., Shenoy, K.V. and Sahani, M. (2009) Gaussian-process factor analysis for low-dimensional single-trial analysis of neural population activity. *J. Neurophysiol.*, **102**: 614–635.

Zanos, S., Richardson, A.G., Shupe, L., Miles, F.P. and Fetz, E.E. (2011) The Neurochip-2: an autonomous head-fixed computer for recording and stimulating in freely behaving monkeys. *IEEE Trans. Neur. Syst. Rehab. Eng.*, **19**: 427–435.

Ziv, Y., Burns, L.D., Cocker, E.D., Hamel, E.O., Ghosh, K.K. *et al.* (2013) Long-term dynamics of CA1 hippocampal place codes. *Nat. Neurosci.*, **16**: 264.

6

PRECLINICAL TESTING OF NEURAL PROSTHESES

DOUGLAS MCCREERY

Neural Engineering Program, Huntington Medical Research Institutes, Pasadena, CA, USA

6.1 INTRODUCTION

The development and certification of neural prostheses and devices intended for neuromodulation requires *in vitro* and *in vivo* tests in order to maximize the likelihood that the device will deliver the expected medical benefits with minimal risks. Neuroprostheses and devices intended for neuromodulation are regulated as Class III devices, a category which includes cardiac pacemakers and defibrillators and which support important or life-critical functions. Class III devices are subject to the most stringent controls by regulators agencies and the final stage of the pre-clinical evaluations must include testing of the actual device to be used in the clinical trials. Pre-clinical testing also may be required when a device that has previously been approved for clinical use is being considered for a novel application, or for use with a novel stimulation protocol or stimulus parameters.

Development of neural implants proceeds in stages towards the final clinical device and application. The earlier studies are intended to inform the physical configuration of the final device, its component materials, and the range of stimulus parameters that will provide the desired clinical benefits with minimal risk. This process continues towards development and the certification of the final device by the regulatory agencies, and entails an integrated sequence of *in vitro* and *in vivo* studies whose staging is dictated by the advantages and limitations of each. It also is dictated by the interplay

Neurobionics: The Biomedical Engineering of Neural Prostheses, First Edition.
Edited by Robert K. Shepherd.

of regulations affecting the use of animals in research and those governing the certification of medical devices. Regulations governing the use of animals in research require use of non-living systems, cell cultures and phylogenetically less advanced living systems whenever possible. However, regulations governing the development and certification of clinical devices (and especially of neural implants) generally require that prior to clinical trials, devices that are not substantially equivalent to existing approved devices must be evaluated in animal models that approximate the intended clinical use. When animal studies of the actual clinical device are impractical, appropriately configured surrogates may be used. Once animal studies are initiated, *in vivo* and *in vitro* studies frequently continue in parallel and are mutually informative. In the USA, medical devices are certified and clinical trials are approved by the Division of Devices and Radiological Health of the Food and Drug Administration (FDA). During the approval and certification process, the FDA Examiner uses as a guide the standards set forth by the International Standards Organization (ISO), as set forth in its guidance documents (Northup 1999).

Many of the engineering principles employed in the development of active neural implants are discussed in Chapter 5. As Class III devices, they are subject to regulation throughout their development, and after the completion of clinical trials but prior to sale they must receive Pre-market approval (PMA) by the FDA. The policy employed by the US FDA is to recommend, rather than to specifically require that device developers and manufactures adhere to the procedures described in its guidance documents. (http://www fda.gov/MedicalDevices/DeviceRegulationandGuidance/). It is then the prerogative of the FDA examiner to determine if the documentation provided by the developer or manufacturer specify procedures that insure the device's safety and efficacy, or that the reasons for excluding certain procedures are well justified. With some variations, these policies and procedures of the US FDA are similar to those of the regulatory agencies of other developed countries (Chapter 10).

6.2 BIOCOMPATIBILITY TESTING OF NEURAL IMPLANTS

Biological evaluation of medical devices ("Biocompatibility" in its broadest sense) is performed to determine the potential toxicity resulting from contact of the component materials with the body. The device materials should not, either directly or through the release of their material constituents: (i) produce adverse local or systemic effects; (ii) be carcinogenic; or (iii) produce adverse reproductive and developmental effects. Therefore, evaluation of any device intended for human use requires data from systematic testing to ensure that the benefits provided by the final product will exceed any potential risks. The US FDA uses the *in vitro* and *in vivo* tests described in ISO10993-1 as the basis of its certification of biocompatibility of medical implants (http://www.fda.gov/downloads/MedicalDevices/DeviceRegulationandGuidance/GuidanceDocuments/UCM348890.pdf)

While biocompatibility is a direct consequence of a device's component materials, the FDA makes its clearance or approval decision for a medical device in its final finished form. The Agency does not clear or approve individual materials that are used in the fabrication of medical devices, but will consider the information included in the master files developed by the material's manufacture. However, the biocompatibility of a final device depends not only on the materials but also on the processing of the materials, manufacturing methods (including the sterilization process), and the manufacturing residuals that may be present on the final device. When selecting the appropriate tests for biological evaluation of a medical device, the designer and manufacturer must consider the chemical characteristics of device materials and the nature, degree, frequency and duration of exposure to the body. The tests include: *in vitro* cytotoxicity; acute, sub-chronic and chronic toxicity; irritation; sensitization; hemocompatibility; genotoxicity; carcinogenicity; and effects on reproduction, including developmental effects. The specific tests for a particular device are specified in the ISO10993-1 matrix, and are determined by the site of implantation and the duration of implantation. Neural prostheses and devices for neuromodulation are classified as permanently implanted devices, which require the most biocompatibility comprehensive testing, including extended testing *in vivo*. Also, the broader mandate to maximizing both safety and efficacy dictates that the clinical device or an appropriate surrogate should be evaluated in the active state, and that it should function for the intended duration of the implant.

As outlined in the document, *Use of International Standard ISO10993, Biological Evaluation of Medical Devices Part 1: Evaluation and Testing – Draft Guidance for Industry and Food and Drug Administration Staff*, the US FDA strongly recommends that the biocompatibility testing be performed by an entity that is independent of the device developer, and a number of companies are certified to perform the ISO10993-1 testing. However, the tests described in ISO10993-1 are by design quite general with respect to the effects on a range of organ systems and may not be sufficient to demonstrate the biocompatibility of a device for a specific application. Additional tests for specific target organ toxicity, such as neurotoxicity and immunotoxicity, may be necessary. Most of the companies that perform ISO10993-1 testing will perform implantations into particular organ systems, including the brain. These special tests will complement the histologic and physiologic testing that is used to verify the safety and efficacy of the electrical stimulation, as discussed in Section 6.5 of this chapter. The early-stage studies related to safety and efficacy often are conducted in non-commercial research laboratories whose staff have the special expertise required. The US FDA strongly recommends that examples of the actual clinical implant be used for the biocompatibility testing. If the final product cannot be used, it may be possible to fabricate test samples (e.g. coupons) that are representative of the final product. If there are differences between the final product and the test sample, additional testing may be necessary to justify use of the test samples instead of the final product. This testing may include data to demonstrate that the test sample materials elute chemical leachants of the same type and relative quantity compared to the final product. In addition, exhaustive extraction and surface characterization techniques may be requested by the FDA examiner to support use of the representative test samples.

A critical adjunct to biocompatibility is the ability of the device to be rendered free of infectious microorganisms ("Sterilizeability"). The FDA's guidelines are described in the guidance document http://www.fda.gov/MedicalDevices/DeviceRegulationandGuidance/ucm109884. The evaluations should be conducted with the final device. Implantable devices must demonstrate that they can be sterilized to SAL 10^{-6}. A pre-determined dose of a test organism is applied to the device. Not more than 10^{-6} of the organisms must remain viable after the sterilization treatment. It is recognized that "traditional" methods of sterilization (e.g. dry or moist heat at elevated temperatures) often are not applicable to neural implants, and alternatives including gas sterilization must be used. If Ethylene Oxide sterilization is employed, the residual levels of ETO after the outgassing cycle should be measured and determined to be in accordance with ISO 10993-7:1995 "Biological Evaluation of Medical Devices – Part 7: Ethylene Oxide sterilization."

Although the scope of ISO10993-1 specifically excludes biological hazards arising from mechanical failure or degradation of the device, the FDA believes this potential risk is important to consider when designing biocompatibility studies. For example, if coating particles are released from a coated device, those particles could lead to a biological response because of their material properties, such as geometric and/or physicochemical properties. In addition, coating delamination could expose the biological system to leaching of different chemicals or to an increased level of chemicals from the substrate material. For devices intended to be implanted in the body for many years, these issues can be addressed by performing at least some the testing with devices that have undergone accelerated aging, as described below.

6.3 TESTING FOR MECHANICAL AND ELECTRICAL INTEGRITY

In keeping with mandate that a Class III device will deliver the expected medical benefits with maximum reliability and minimum risks, neural implants must demonstrate that they will function according to their specification when implanted at the end of their specified shelf life, after mishandlings that may occur during packaging, unpackaging and handling during the implant surgery. Procedures and acceptance criteria for many of these requirements are specified in ISO14708 *(International Organization for Standardization. Implants for surgery –Active implantable medical devices – Part 1. General requirements for safety, marking and for information to the provided to the manufacturer ISO14708-1:2000.Geneva, (Switzerland).* There are additional specifications for specific categories of devices. For example, the specifications for implantable stimulators are specified in ISO14708-3 and for cochlear implants in ISO14708-7.

Certification of neural prostheses and neuromodulation devices includes testing to verify their tolerance to the *in vivo* environment. The testing is performed on samples of the implant's components throughout development. After the device has been certified for clinical trials and then for marketing, testing is performed to ensure that individual devices comply with specifications (e.g. functionality of electrical components and embedded software, helium-leak tests for hermetic sealing at cable feed-throughs into electronic packages). Many of these are discussed in Chapter 1.

6.4 *IN VITRO* ACCELERATED TESTING AND ACCELERATED AGING OF NEURAL IMPLANTS

During development of neural implants, tests are employed to accelerate the processes that are responsible for device degradation and possible failure in the body. These include lifetime testing of cables that may be subject to repeated flexion and torqueing after implantation, to ensure that the cable's conductors do not fracture or that the electrical insulation does not fail.

Accelerated aging studies that are intended to yield quantitative estimates of the device's lifetime in the body at 37°C should be conducted in an environment that simulates the environment *in vivo* as closely as possible. These tests serve to evaluate the potential for functional failure of the implanted devices and to provide samples of partially deteriorated devices for biocompatibility testing, as described in the previous section. There are many potential pitfalls in accelerated aging testing (Meeker and Escobar 1998). However, since neural prostheses and devices for neuromodulation must remain functional for many years or decades, accelerated testing must occupy a central role in their development. The electrolyte composition and other constituents of the medium may strongly affect degradation of the stimulating electrodes and other components. For example, the rate of dissolution of platinum stimulation electrodes is reduced by more than 2 orders of magnitude (vs. the rate in inorganic saline) in a medium containing 0.2 mg/ml of serum albumin (Robblee *et al.* 1980). While testing *in vitro* often affords better control of the testing environment, it is never possible to duplicate exactly the chemical or physical environment *in vivo.*

For neuroprosthesis and neuromodulation devices, the physical composition of the device often requires that the testing be conducted with the device powered and operated as during its intended use as a medical device. The lifetime of the silicon dioxide and silicon nitride films used to insulate the electrical traces of silicon substrate neural probes is affected strongly by the electric field strength, and the effect of the electric field strength on the film's time to failure can be nearly two orders of magnitude (Ordonez *et al.* 2012; Vanhoestenberghe and Donaldson 2013). This also appears to be true of some organic insulating films used in neural implants, including parylene-C (li *et al.* 2011). Insulating films always exhibit some residual conductivity. The rate of degradation of inorganic passivation films, including silicon oxides and nitride, is proportional to $(1/J^2)$, where J is the current density through the film (Vanhoestenberghe and Donaldson 2013). J is proportional to the applied voltage, so the need to perform the testing under electrically active conditions is obvious. Also, metal dissolution and corrosion in a chloride-containing medium may be greatly accelerated under anodic bias (Medizza and Milner 1970).

Accelerated aging at elevated temperatures has been used (and sometimes misused) to estimate the lifetime of a device in the body. A key tenant of accelerated aging testing is that the outcome will be most informative as to what will occur *in vivo* if the chemical and physical processes that occur *in vivo* are accelerated, but failure mechanisms that do not occur *in vivo* at body temperature are not recruited or suppressed (Hu *et al.* 1993; Meeker and Escobar 1998). When these conditions are met, the rate of degradation of the device during accelerated testing often can be extrapolated with some confidence to the rate of degradation *in vivo* at normal body temperature. The rate of most chemical reactions increases with temperature, and the rate of a chemical process governed by first-order kinetics is given by the Arrhenius equation:

$$K = Ae^{Ea/RT} \tag{6.1}$$

where K is the reaction rate of the chemical reaction at absolute temperature T (in degrees Kelvin), A is a constant, E_a is the energy of activation of particular chemical process and R is the gas constant. Since accelerated aging studies are performed *in vitro* where the intended biological function of the devices cannot be monitored, the metric of outcome p(t,T) often is a quantity that is known (or assumed) to be linked closely to the device's functionality *in vivo*. In general, p(t,T) will be a function of time (t) and temperature (T). For a neuroprosthesis or neuromodulation device, p(t,T) might be the leakage of stimulus current through extraneous pathways as characterized by impedance measurements (Ii *et al.* 2011: Han *et al.* 2012; Prasad *et al.* 2012), the amount of cross-talk between channels of a multichannel device which would indicate leakage pathways between electrode channels (Gunalan *et al.* 2009), or the percentage of channels to loose electrical conductivity, indicative of corrosion of electrical conductors.

If it can be established that the rate of aging of a device, as inferred from p(t,T), is governed by the Arrhenius equation ("Arrhenius behavior'), and if the chemical and physical conditions of the testing closely approximate those *in vivo*, it is likely that the same chemical processes are responsible for the aging *in vitro* and *in vivo* over the specified range of temperatures. If the temperature range of Arrhenius behavior includes 37°C, the rate of aging of the device, or of similar devices in the body, can be estimated with some confidence.

The Arrhenius equation (1) can be written as:

$$\mathrm{Ln}\,(K) = (-E_a/R) \times (1/T) + \mathrm{Log}(A) \tag{6.2}$$

where Ln(K) is the natural logarithm of K. Two conditions must be satisfied in order for the behavior of p(t,T) to satisfy Equation 2 and thus to be classified as Arrhenius. First, if K is the rate of p(t,T) at temperature T, then K is the slope of the plot of p(t,T) vs. t, and the slope must be shown to be constant at each temperature T. Second, the plot of Log(K) vs. 1/T (in degrees Kelvin) must be a straight line, or a close approximation thereof. Verification requires data from at least 3 temperatures. The slope of the Arrhenius plot for a chemical process governed by first-order kinetics is $(-E_a/R)$, and if the plot of Ln(K) vs. 1/T can be fitted by a straight line, the activation energy E_a is similar or identical across the range of temperatures tested. The concordance of activation energy provides some assurance, but not conclusive proof, that the same chemical processes are operating across the full range of temperatures. Some practical considerations related to this requirement are discussed below.

The study by Shen and Burgess (2012) of the rate of release of Dexamethasone from an implant illustrates the rigorous procedure for establishing Arrhenius behavior. Figure 6.1A shows plots of the cumulative release over time, p(t,T), of the drug

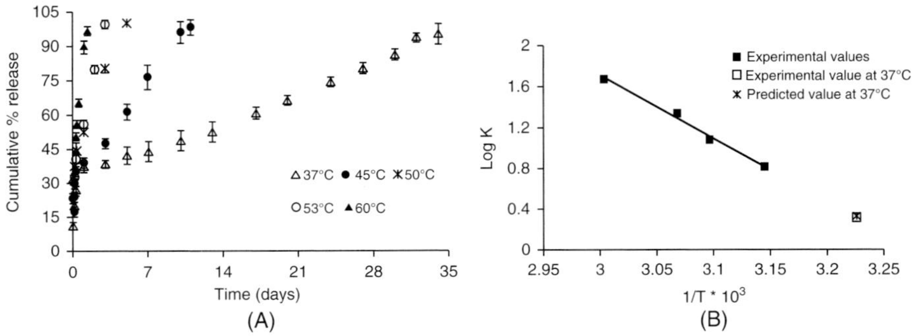

Figure 6.1 Example of Arrhenius behavior of an implant. (A) Cumulative release of Dexamethasone from subcutaneous implants at five temperatures. The rate constant K is the slope of the plots after the initial fast release phase. (B) Arrhenius plot of Log (K) vs. 1/T, where T is the absolute temperature. The rate constant measured at 37°C (310°Kelvin) corresponds to the value obtained by extrapolation of the plot obtained at elevated temperatures (reprinted from Shen and Burgess 2012, with permission from Elsevier).

from a sample of implants. p(t,T) is characterized by an early fast release followed by a slow release phase with nearly constant slope over the duration of the test. The early and late release phases of p(t,T) have very different slopes (K, rates), indicating that they are governed by different processes. The two phases must be handled separately in order to determine if the behavior is Arrhenius during either phase. Implantable devices exhibiting a comparable rapid early change in p(t,T) probably would be classified as early failures governed by mechanisms different from those responsible for the subsequent slower rate of degradation and would be deleted from the sample in order to evaluate for possible Arrhenius behavior of the survivors. Figure 6.1B shows that after the initial phase of fast drug release, Ln(K), the slopes of the plots in Figure 6.1A at each T, is approximately linear when plotted against 1/T (T in degrees Kelvin) and therefore, the rate of drug release after the initial fast release is approximately constant over time and approximately Arrhenius between 310 and 333°K (37°C and 60°C).

Also, the plot generated at 45, 47, 53 and 60°C passes though the value determined empirically at 37°C. Figure 6.1B also illustrates the practical issues related to the requirement for constant K over time, for each value of T. In order to accurately extrapolate the data from elevated temperatures to 37°C, the variance in Log(K) over time (and within the sample of devices) must be small compared to the range of Log(K) over the range of elevated temperatures. Devices fabricated from laminated Polyimide exhibit Arrhenius behavior up to at least 57°C (Ordonez *et al.* 2012), which represents a 4× acceleration of aging relative to 37°C.

Since the rate of an Arrhenius process approximately doubles with each 10°C increase in temperature, there are obvious advantages to conducting accelerated aging tests at high temperatures. Verification of Arrhenius behavior does convey confidence in the estimates of the device's lifetime *in vivo*. However, there remains the paradox that rigorously establishing Arrhenius behavior requires testing of a representative sample of devices at multiple temperatures spanning the full range of testing, including 37°C (or slightly above 37°C). In reliability testing, it is common to define a value $p_{failure}(T)$, of the test metric p(t,T), for which the device is no longer considered to be functional or reliable. The mean time required for a sample of devices reaching this state is the mean time to failure (MTTF) at temperature T. If the behavior of p(t,T) is Arrhenius, the slope of p(t,T) is K(T) and is constant, so $K(T) = (p_{failure}(T) - p(0,T)/MTTF$ where p(0,T) is p(t,T) at the start of the test (Wang *et al.* 2012). In this way the values of Ln(K) can be obtained from MTTF at three or more elevated temperatures, and those values used to extrapolate to the value of log(K) and thus of K at body temperature, as in Figure 6.1B, and thereby estimate MTTF at body temperature. In order to expedite the soak tests, the outcome may be based on a metric shown to signal impending device failure, so that it is not necessary to wait for complete functional failure to occur. If the electrical conductors of an implant are insulated from the tissue by organic or inorganic dielectric layers, the various failure mechanisms of these films can be classified into three groups: failures due to mechanical stress and mechanical defects, failures due to chemical, physicochemical and electrochemical reactions, and failures due to the interaction of mechanical stress and chemical stress (Schmitt *et al.* 1999). All of these failure modes are characterized by various amounts of fracture, dissolution and delamination of the insulations as the failure progresses (Figure 6.2).

For multichannel devices, the onset of inter-channel crosstalk in a multichannel device is indicative of delamination of the electrical insulation and the accompanying saline intrusion under the insulation (Aceros *et al.* 2012: Gunalan *et al.* 2009). The development of pinholes, stress fractures and corrosion of the insulating layers

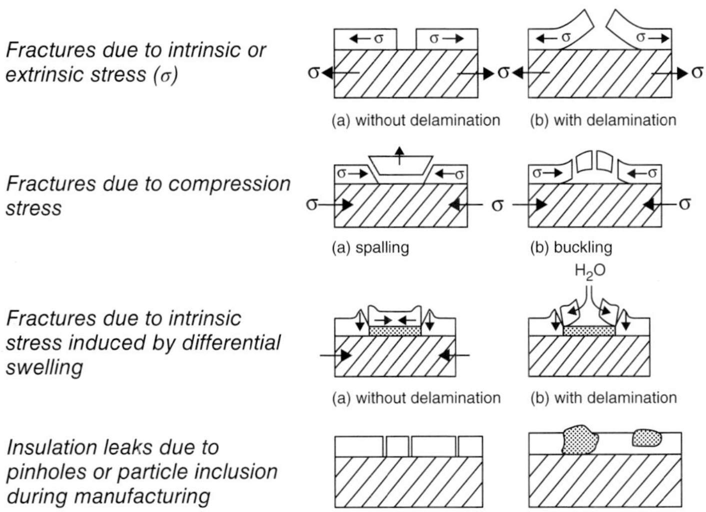

Figure 6.2 Examples of failure modes of an active implant caused by internal and external stresses, and by degradation of the electrical insulation in an aquaeous invironment (from Schmidt *et al.* 1999, with permission from Elsevier).

is accompanied by a reduction in the electrical impedance of the individual channels (Sim and Lawson 1979; Vanhoestenberghe and Donaldson 2013). Also, repeated measurements of the impedance spectrum of each channel reveals changes in channel impedance across a range of frequencies, and can lend insights into the nature of the insulation failure (Prasad *et al.* 2012; Prasad and Sanchez 2012; Streit *et al.* 2012).

After Arrhenius behavior has been verified, testing of representative samples of the device at the elevated temperature can be used to monitor quality control during subsequent fabrication runs, and to validate modifications to the device and its fabrication procedures, with the proviso that the alterations are of a type that is unlikely to introduce new failure modes. However, even after Arrhenius behavior has been established, the lifetime predictions must be regarded with some caution, since it is never possible to exactly duplicate the environment *in vivo*. For example, there is a growing literature detailing the interrelationships of brain inflammation and the generation of reactive oxygen and nitrogen species (Valko *et al.* 2005) and some studies (Streit *et al.* 2012) suggest a link between recording failure of parylene-insulated microwire electrodes and reactive oxygen species in the surrounding tissue.

Accelerated aging testing under validated Arrhenius behavior does provide the most reliable estimate of the device's lifetime *in vivo*. However, the procedure is time-consuming and requires a relatively large number of devices just to establish that the behavior is Arrhenius. A much less rigorous approach is to assume that the mechanism(s) responsible for failure are Arrhenius across the range of elevated temperatures tested and also at 37°C, but without actually obtaining data at 37°C. The Arrhenius plot (Figure 6.1B) then is generated from the data at three or more elevated temperatures and the plot of log(K) is extrapolated to determine K at 37°C. This approach has been used to estimate the lifetime of devices or test samples whose lifetime *in vivo* is projected to be many years and for which a rigorous verification of Arrhenius is deemed to be impractical. Using this approach, the MTTF of a test package for a retinal prosthesis insulated with Parylene-C was estimated to be greater than 20 years, based on soak testing in saline at 77°C and 90°C (li *et al.* 2011). However, this approach will yield invalid lifetime estimates if the chemical processes responsible for failure differ across the range of temperatures (Hu *et al.* 1993; Meeker and Escobar 1998). Thus, if the activation energy of the failure mechanism(s) operating at elevated temperatures are different from those at or near 37°C, the slope of the Arrhenius plot generated from the high temperature data will be erroneous, causing the extrapolated value of log(K) and of K at 37°C to be erroneous, and thus the predicted time to failure will be erroneously long or short (Meeker and Escobar 1998).

Accelerated aging at high temperature without formally establishing Arrhenius behavior may be more useful as a means of establishing a ranking of device configurations, materials, and fabrication procedures, but with the understanding that this type of testing may not yield reliable estimates of the devices' lifetime *in vivo*. Harsh accelerated testing assumes that devices that better withstand these test conditions will survive longer in the body, but this assumption is subject to the same concerns regarding the commonalty of failure modes in the harsh environment and at body temperature. In this regard, the outcomes convey greater confidence if the device exhibits similar deterioration and failure in the harsh testing environment and *in vivo*. Takmanov *et al.* (2012) soaked devices at high temperature (85°C) in a mixture of normal saline and 2 to 10 mM hydrogen peroxide, which was included in the test medium to simulate degradation of the device by reactive oxygen species *in vivo* (Streit *et al.* 2012: Valko *et al.* 2005). The investigators monitored changes in the device's electrical impedance spectrum (1Hz to 1 MHz) and documented physical damage to the devices and their electrical insulation by scanning electron microscopy. Devices soaked in the harsh environment for 2 weeks were compared to identical devices implanted in animals' brains for 1 year. The changes in the devices' impedance spectra and the appearance of the physical deterioration as documented by the scanning electron microscopy were similar for both test conditions, suggesting that the type and severity of the damage induced by the *in vivo* and *in vitro* tests was generally similar. It is possible that this type of harsh accelerated testing can be a valuable aid for screening materials and refining fabrication procedures in the early stages of device development.

6.5 *IN VIVO* TESTING OF NEURAL PROSTHESES

In vivo evaluation of a neural prosthesis or neuromodulation device in a suitable animal model usually is integrated with the regimens of *in vitro* testing. *In vivo* studies in laboratory animals are conducted to evaluate the safety and efficacy of a novel device or nascent technology, or to evaluate the functionality and safety of an existing device that is being considered for a novel clinical application. *In vivo* testing for function and safety of a neural prosthesis is performed at multiple stages of its development and typically includes functional measurements (e.g. recording of neuronal activity that is evoked or modulated by the electrical stimulation) and/or recording of the animals' behavioral responses to the stimulation, combined with histologic studies of the implant sites. While it sometimes is possible to record neural and behavioral responses that are quantitative markers of the evoked neural activity, these responses often are incomplete or ambiguous markers of the intended clinical function. Histologic studies of the tissue responses at the implant site, together with the recorded neuronal and behavior responses, are likely to provide the most complete information regarding the safety and functional stability of the implant, and of the regimen of electrical stimulation.

For a neural prosthesis or a device intended for neuromodulation, it is important to evaluate the tissue effects that are due to the device alone, and also the effects caused by the neural stimulation, and the definitive testing should be conducted with the device implanted at the site of the intended clinical use. Prior to granting an investigation device exemption to begin clinical trials, the FDA certifies the entire device including the stimulator, its imbedded software and the stimulation protocol that will be used in the clinical trial and in the final stages of pre-clinical animal testing, the actual clinical stimulator must be used. However, for the earlier stages of animal testing, it is essential to use a device for which the parameters of the electrical stimulation that are closely linked to safety and efficacy (including residual direct current (DC) in the electrical stimulation) can be monitored. Also, animals testing of all components of the final clinical device may not be practical (e.g. a cochlear implant, due to differences in the size of the cochlea in humans and available animal models) and an appropriately configured surrogate device may be substituted. The studies must be designed with the recognition that neural injury and/or dysfunction can occur during device implantation and also during its subsequent long-term residence in the body, and that tissue injury may be caused by the electrical stimulation.

Since any tissue damage resulting from the electrical stimulation will be superimposed on the injury due to device implantation and its long-term residence in the body, the study will be better controlled and will have maximum statistic power if the testing conditions with and without the neural stimulation are as similar as possible, so that the various damage mechanism can be isolated without introducing confounding interactions between the active and inactive conditions. Thus, the active and passive devices should be as similar as possible, both in their configuration and materials, and in their history of fabrication and handling. To control for co-variances arising from difference in surgical technique, anatomical sites and inter-animal variations, the electrically active and inactive electrode sites might be included in the same integrated electrode array, or they might be implanted into mirroring sites in the same animal (or at least into animals of the same species), and be implanted by the same surgeon. Modern statistical packages readily accommodate these designs as hierarchical crossed or nested data structures.

The design of the studies of the safety and efficacy of the electrical stimulation will vary somewhat, depending on whether the goal is to evaluate the final clinical device for a specific clinical application, or to gather data that will inform the ongoing design and development of a device or class of devices. For the latter, the studies likely will encompass a greater range of stimulus parameters and perhaps a range of configurations of the implant. Also, if evaluation of the final clinical device in an animal model is not practical, it is important to have data that can inform which features of the clinical device must be duplicated (or scaled appropriately) in the surrogate device that is used in the animal testing. An investigation concerning the interactions of charge density (QD) and charge per phase (Q) in the induction of neural damage by electrical stimulation of the cerebral cortex (McCreery *et al.* 1990) was designed to inform the design and development of a class of neuroprostheses and neuromodulation devices that employ simulating electrodes implanted on the surface of the cerebral cortex.

Platinum electrodes ranging in size from 0.002 to 0.5 cm^2 were implanted over the parietal cortex of adult cats. Geometric charge density and charge per phase of the electrical stimulation were varied independently by varying the size of the electrodes. The electrodes were pulsed continuously for 7 hours using charge-balanced, controlled current pulse pairs, 400 μs per phase in duration, at a repetition rate of 50 Hz. The histologic evaluation of the stimulated cortex showed that charge density (as measured at the surface of the stimulating electrode), and charge per phase interact to determine the threshold of stimulation-induced neural injury. Figure 6.3A shows that this interaction between charge density and charge per phase occurred over a

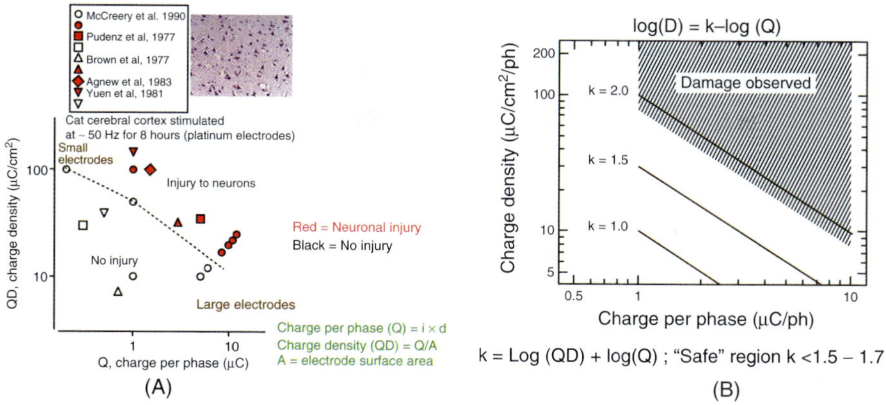

Figure 6.3 (A) Relation between charge density, charge per phase and neuronal injury after 8 hours of electrical stimulation applied through platinum electrodes implanted on cat cerebral cortex. Damaged neurons appeared dark, with shrunken somas, as shown in the inset (adapted from McCreery *et al.* 1990). (B) Numeric representation of the region of the graph shown in A, for which injury to neuron did or did not occur ("safe" region) (from Shannon 1992, with permission from IEEE).

wide range of QD (0.1–8 $\mu C/cm^2$) and Q (1–100 μC per phase). For the larger electrodes, damage to neurons occurred at charge densities much lower than the reversible charge density for platinum, which is approximately 50 to 150 $\mu C/cm^2$ depending on the stimulus pulse duration (Rose and Robblee 1990). This demonstrates that adhering to the reversible charge injection limits of a stimulating electrode (Chapter 3) is a necessary but not always sufficient requirement for safe neural stimulation. Shannon (1992) converted the data from this study into a numeric form:

$$k = \text{Log}(QD) + \text{Log}\,Q \tag{6.3}$$

where $k < 1.7$ defines combinations of Q and QD that did not induce neural damage in the cat cerebral cortex. Over the ensuring years, the "Shannon criterion" has proven to be a useful starting point for more comprehensive studies of the safety of individual neural prostheses and neuromodulation devices (Merrill *et al.* 2005). However, the "Shannon criterion" is based on the tissue responses to short-term stimulation, and studies employing long duration stimulation are required in order to insure the safety of individual devices, prior to human trials.

For devices intended for long-term implantation, there is an emerging consensus that the device and the host neural tissue should be viewed as an integrated system whose long-term function is dictated by the interactions and interplay of "biotic" and "abiotic" factors (Prasad *et al.* 2012; Streit *et al.* 2012), as depicted in Figure 6.4. Thus, the mechanical and electrical integrity of an implant (abiotic factors) is affected by its evolving biological environment ("biotic factors"), while the process of implanting the device and its subsequent long-term residence in the tissue induces biotic changes in the host tissue that affect its interactions with the implant.

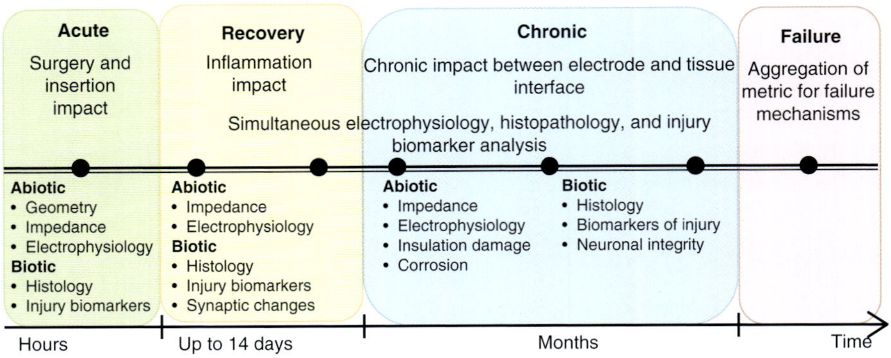

Figure 6.4 A suggested protocol for monitoring the evolution of "biotic" and "abiotic" factors that contributes to the failure of a neural implant. The scheme incorporates metrics of the functional status of the implant, metrics linked to its functional status, and also metrics of the biological changes and status in the adjacent tissue that may affect the device's function (from Prasad *et al.* 2012, with permission from IOP Publishing).

There have been many studies of the tissue responses to neural implants, including those that penetrate into the parenchyma of the brain or spinal cord (Chapter 2). As noted above, the details of histologic studies of implanted devices will be dictated by the study objectives. If the goal is to validate the safety of the final configuration of a neural prosthesis or neuromodulation device, for which only minor modifications are anticipated prior to clinical studies, the histologic studies might be restricted to a quantization of neuronal density around the implant, and semi-quantitative measures of mechanical injury to the adjacent tissue, including hemorrhage and tissue scaring (Agnew *et al.* 1983; Cohen 2007; Edell *et al.* 1992; McCreery *et al.* 1992, 2010). However, if the objective is to develop or refine a novel device or technology, the histologic evaluations will be more comprehensive. It is now apparent that intraparenchymal implants that induce a persisting inflammatory response may produce persisting changes in nearby neurons, including "pruning" of the neuron's dendritic trees (McConnell *et al.* 2009). While these neurons may remain viable, it is likely that their responses to the electrical stimulation as well as their other intrinsic function are altered significantly. Therefore it seems prudent for the pre-clinical histologic evaluations of all neural implants to include markers of persisting inflammation, as well as markers for loss of neurons and axons, hemorrhage and tissue scarring.

A universal concern in late-stage pre-clinical testing of devices is the appropriate duration of the animal studies. Animals studies of the final device should be designed so that all damage mechanism(s) that have been identified from prior experiences with similar devices can be isolated and identified. The histologic evaluation of the stimulated tissue should include at least two time points that will allow identification of early, non-progressive injury, and also injury mechanism(s) that are ongoing or progressive. This is especially important if the device and its implant site suggest that multiple damage mechanism may be operational, such as those associated with device implantation, its subsequent residence in the tissue, and the ongoing electrical stimulation. The US FDA's Medical devices advisor committee has accepted implant durations of 3 to 6 months for the final device. During the development of the NeuroPace, Inc. RNS stimulator system for treating epilepsy, the devices were implanted into 11 sheep for 33 to 200 days, with a mean implant time of 131 days. Seven sheep had implant durations of 167 or more days. The findings from these studies were accepted by the Medical devices advisory committee of the US FDA in granting PMA for the NeuroPace RNS system (www.fda.gov/downloads/.../UCM340258).

In deference to the time and expense required to conduct a study of the safety of a novel device or application, the question often arises as to whether findings from previous studies are applicability to the novel device or application. Factors that need to be considered are the similarities and differences of the implant site and the neural substrate, the differences in the novel device's physical configuration, including the electrode materials, and the differences in stimulus parameters. Some studies of the safety of electrical stimulation have been designed to elucidate the biological processes that are responsible for the tissue damage, and can help to inform these decisions. The likelihood that these mechanisms might be similar or different in the novel device or application, should be central to the decision regarding the applicability of data from earlier studies. For example, microelectrodes with very small surface areas must operate at high charge densities that may induce voltage gradients that are sufficient to induce electroporation of neural membranes (Weaver 1993,1995; Butterwick *et al.* 2007). This damage mechanism is unlikely with the larger electrodes and the lower charge densities used in the studies depicted in Figure 6.3.

To address concerns related to prolonged intracortical stimulation with microelectrodes, we conducted an animal study of prolonged intracortical microstimulation with penetrating microwire microelectrodes (McCreery *et al.* 2010). This work was part of a sequence that illustrates how animal studies and early clinical trials are integrated during the development of a neural prosthesis. The microelectrode arrays were similar to those being designed for a cortical-level visual prosthesis (Bradley *et al.* 2005; Troyk *et al.* 2003, 2005). In our study (McCreery *et al.* 2010), arrays of 16 activated iridium microelectrodes were implanted for 450–1282 days into the cerebral cortex of 7 adult domestic cats. Half of the microelectrodes in each array were pulsed for 240 h (8 h per day for 30 days) at 50 Hz. The selection of stimulus parameters was informed by data from an earlier clinical study using similar microelectrodes implanted into the primary visual cortex of a blinded person, in which the threshold of the stimulation-induced visual percepts was ≤ 4 nC/phase for about 80% of the microelectrodes (Schmidt *et al.* 1996). That human trial had in turn been preceded by animal studies showing that the microelectrodes could be implanted into the brain without excessive damage to nearby neurons (Salcman and Bak 1976). In our study of the safety of prolonged microstimulation, 240 hours of stimulation through the microelectrodes (8 h per day for 30 days) at 50 pps and 2 nC/phase, and at a geometric charge density of 100 $\mu C/cm^2$ produced no detectable change in neuronal density in the tissue surrounding the microelectrode tips. However, continuous stimulation at 50 pps and 4 nC/phase with a geometric charge density of 200 $\mu C/cm^2$ induced loss of cortical neurons over a radius of at least 150 μm from the electrode tips (Figure 6.5B). The same stimulus regimen, but with a

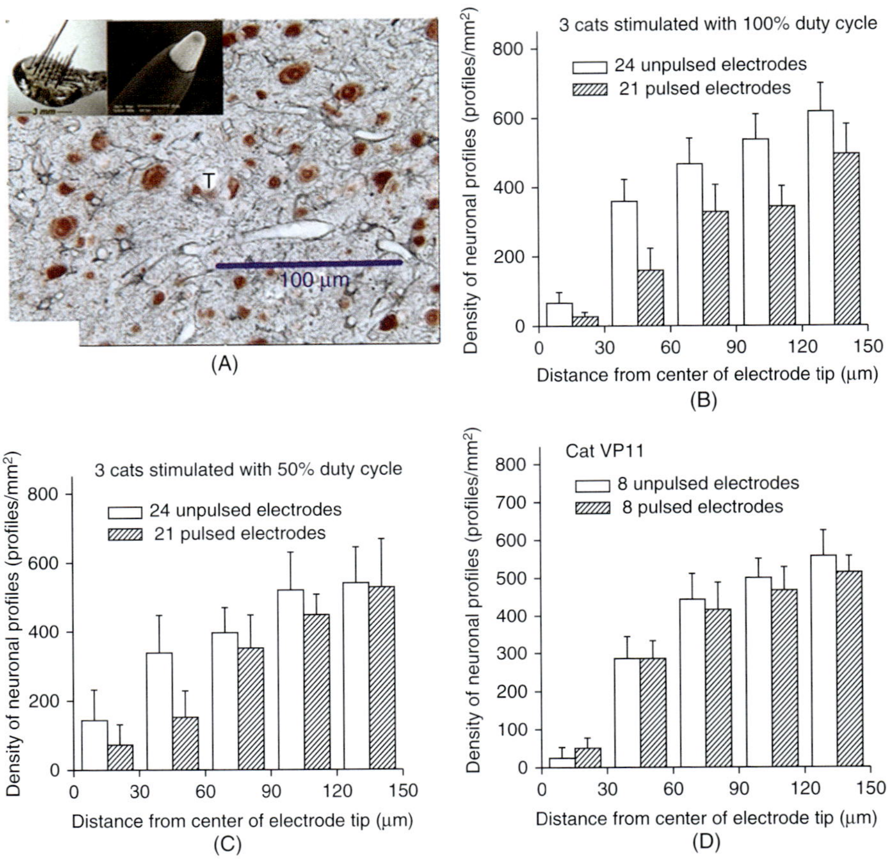

Figure 6.5 (A) Light micrograph of a tissue section through the site of a microelectrode implanted chronically in cat cerebral cortex. The inset shows a photo of the microelectrode array and a scanning electron micrograph of one of the microelectrode tip sites. Neurons identified by the neuron-specific label NeuN are shown as red-brown profiles against the background of astrocyte processes stained with the astrocyte marker GFAP. (B) The density of cerebral cortex neurons surrounding 24 unstimulated microelectrode sites, and 21 sites from the same arrays stimulated for 240 h at 4 nC/phase, 50 Hz and ~200 μC/cm^2 and with 100% duty cycle. (C) The density of neurons around 24 electrode sites stimulated with a 50% duty cycle, also at 50 pps, 4 nC/phase and 200 μC/cm^2. (D) Neuron density around 8 microelectrodes pulsed with 100% duty cycle, 50 pps, 2 nC/phase and ~100 μC/cm^2 (from McCreery 2010, with permission from IOP Publishing).

duty cycle of 50% (1 s of stimulation, and then 1 s without stimulation, repeated for 8 h) produced neuronal loss within a smaller radius, extended approximately 60 μm from the center of the electrode tips (Figure 6.5C). Comparable temporal modulation of the stimulus probably would be used in a clinical visual prosthesis, since continuous stimulation through any one microelectrode would convey no information.

Stimulation at 2 nC/phase with a 100% duty cycle produced no detectable loss of neurons (Figure 6.5D), but the data from the clinical study cited above indicates that stimuli of this amplitude might be insufficient for a cortical level visual prosthesis. There also was some loss of neurons adjacent to the unpulsed microelectrodes, presumably a sequela of mechanical injury during their implantation and long-term residence in the tissue (450 days or more). The strong radial gradient of neuronal loss around the unpulsed microelectrodes showed that this loss of neurons was not strongly related to a global effect from the ensemble of stimulated electrodes. Taken together, the animal study and the earlier clinical study suggest that a cortical level visual prosthesis based on intraparenchymal microstimulation would be possible, although the dynamic range of the visual percepts might be reduced somewhat by the loss of neurons very close to the microelectrodes. The findings also suggest that additional animal studies related to selecting the size of the electrodes sites on microstimulating electrodes may be warranted.

It is not surprising that the severity of the damage to cortical neurons would be related to the short-interval duty cycle of the stimulus. Another study of the safety of electrical stimulation of a peripheral nerve illustrates how the stimulus pulse rate may interact with other stimulus parameters, and also illustrate how the interrelations of stimulus parameters can lend insights into the biological processes that are responsible for the damage inflicted by the electrical stimulation (Agnew *et al.* 1990b; McCreery *et al.* 1995). The stimulus was delivered through bipolar cuff electrode arrays implanted chronically on the nerve (Figure 6.6A). Injury to the myelinated axons was detected as early axonal degeneration (EAD) in which the axons' myelin sheath prolapses into the axons space, and reduces or even obliterates the profile of the enclosed axon, as illustrated by the dark profiles in the light micrographs (Figure 6.6B, right). For each axonal profile, the area of each of the darkly-stained myelin and of the light-colored axon profile was computed using commercial image analysis software. In unstimulated nerves implanted with the same type of array, the area of 98.85% of the axon profiles was greater than $A_{normal} = 0.2 \times (MA - 5.8\ \mu m^2)$ where MA, the myelin area, is in units of μm^2. Profiles whose axon profiles were smaller than A_{normal} were counted as shrunken axonal profiles (SAP) and considered to be undergoing EAD (McCreery *et al.* 1997). In Figure 6.6D the stimulus amplitude is shown on the abscissa and is expressed as multiples of the amplitude required to fully recruit the α component of the nerve's compound action potential (CAP). The α component of the CAP represents the synchronized activation of the nerves' large myelinated axons (Patton and Woodbury 1960) and a stimulus amplitude of $1 \times \alpha$ is sufficient to excite all or most of these large axons (Figure 6.6C). Figure 6.6D shows the relation between the prevalence of EAD in the stimulated nerve, the stimulus pulse rate and stimulus amplitude for 17 sciatic nerves from 11 cats. The graph's ordinate spans the low range of the prevalence of EAD in order to reveal the threshold of neural damage. Continuous stimulation for 8 hours at 20 pulses per second (Hz) at high amplitude (8 x α) did not induce damage to the myelinated axons, but stimulation at a higher rate and at a much lower amplitude was damaging. Also, the threshold stimulus amplitude for inducing neural damage in the form of EAD was very close to

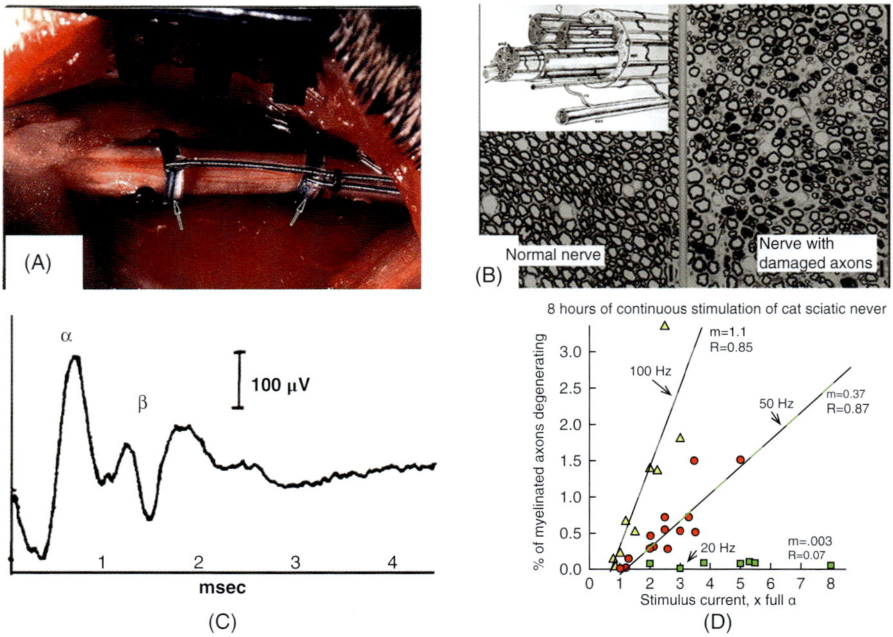

Figure 6.6 Injury to the myelinated axons of a peripheral nerve (cat sciatic nerve) induced by 8 h of continuous electrical stimulation. (A) The electrode array and nerve at the time of the implant surgery. (B) Light micrographs of cross-sections of a normal unstimulated sciatic nerve and of a stimulated nerve, showing the appearance of normal myelinated axons and damaged axons undergoing EAD. (C) Compound action potential recorded from a cat's sciatic nerve in response to electrical stimulation with the cuff electrode. The α component represents the activity in the largest and most rapidly conducting axons. (D) Plots of the relation between stimulus pulse amplitude and the % of axons undergoing EAD, with stimulus pulse rate as a parameter. The stimulus current is expressed as multiples of α, the stimulus amplitude required to excite all of the nerves' large myelinated axons (B,D from McCreery *et al.* 1995, with permission from IFMBE. C from Agnew *et al.* 1990b, with permission from Butterworth-Heinemann for BES).

$1 \times \alpha$. These interactions of stimulus parameters led us to hypothesize that the neural damage was a consequence of the neuronal activity induced by the stimulation, rather than a mechanism that is related specifically to the electrical stimulation, such as membrane electroporation. It is possible that when the homeostatic capacity of the nerve or the individual axons is exceeded, axonal injury may occur (Huang *et al.* 1998).

This hypothesis is supported by an earlier study in which we showed that blocking the action potentials in most of the axons with local anesthetics (10% procaine or 2% lidocaine) almost completely prevented axonal degeneration during high-rate, high amplitude electrical stimulation (Agnew *et al.* 1990a). Also, partial suppression of the induced neural activity by a smaller dose of local anesthetic that blocked the action potentials in only some of the axons afforded nearly complete protection of all axons. If the axonal injury was indeed a consequence of the induced activity exceeding the homeostatic capacity of the sciatic nerve, it is possible that the range of non-injurious stimulus parameters determined for a particular nerve would not apply to a different nerve that has evolved to accommodate greater (or less) neuronal activity. However, the general principle of the interactions of stimulus amplitude and stimulus rate does seem to apply. Other studies of electrical stimulation of the auditory nerve have shown prolonged reduction of nerve excitability at high stimulation rates and at high intensities (Huang *et al.* 1998; Huang and Shepherd 1999, 2000; Tykocinski *et al.* 1995, 1997). Findings such as this emphasize the importance of evaluating the final clinical device in a site that accurately represents the target for the clinical device and of using a stimulation protocol as similar as possible to what will be used in the clinical application.

When evaluating the safety and efficacy of neural stimulation, it is important to consider how the parameters of the electrical stimulation may alter the efficacy of the stimulation through mechanisms other than neuronal death. One such phenomenon is prolonged but reversible depression of neuronal excitability (elevation of a neuron's threshold for excitation), which we have described as stimulation-induced depression of neural activity (SIDNE). The name and acronym was chosen to describe the phenomenon without conveying inferences regarding the responsible biological processes. SIDNE is characterized by reversible depression of the excitability of individual neurons, as characterized by an elevation in their threshold for excitation by electrical stimulation. The effect may persist for many days after termination of the stimulus (McCreery *et al.* 1992, 1997, 2000, 2002). By its definition, neurons do eventually recover from SIDNE, but the prolonged depressed of excitability (3 to 17 days or more after termination of the stimulation) suggests a phenomenon that is very different from a transient accommodation to the stimulus. The prolonged depression of excitability does imply that the electrical stimulation imposed marked stress on the neurons and/or the local homeostatic mechanisms. Also, when SIDNE raises the threshold for exciting the neurons, attempting to compensate for the elevated threshold by increasing the stimulus amplitude is likely to compound the problem by producing more depression and increasing the risk of producing irreversible tissue injury.

In one study, we examined the changes in neuronal excitability induced in neurons of the cat cerebral cortical during prolonged microstimulation (McCreery *et al.* 2002). The findings illustrate how the performance of a neural prosthesis employing intracortical microstimulation might be affected by the overall stimulation protocol as well as by the stimulus parameters for the individual microelectrodes. The microstimulation was delivered through arrays of intracortical microwire microelectrodes implanted chronically into the postcruciate gyri of cats. Seven hours of intracortical stimulation with individual microelectrodes, or stimulation with 4 of the 16 microelectrodes at 50 Hz and at 4 nC/phase, induced little or no change in the neurons' electrical thresholds. However, SIDNE that persisted for 2 to 18 days occurred when all 16 microelectrodes were pulsed simultaneously. The prevalence of SIDNE increased further and persisted longer when all 16 microelectrodes were pulsed in the interleaved mode, which will increase the rate of stimulation of neurons that are activated by multiple adjacent electrodes. Stimulation at a higher amplitude (15 nC/ph) induced much more depression of excitability. These features of the SIDNE imply that multiple processes mediate the stimulation-induced depression of neuronal excitability, and that one of the mechanisms is prolonged high rate excitation of neurons. The study also demonstrated that the SIDNE can be minimized by employing a stimulus regimen in which the spatial resolution of the array is maximized (e.g. by pulsing the array's electrodes sequentially rather than simultaneously, and restricting the stimulus amplitude to the range for which there is minimal spatial overlap of the effective stimulus). Since SIDNE is by definition reversible, it is instructive to speculate whether the phenomenon might serve to predict that the stimulus, if prolonged, is likely to induce irreversible neuronal injury or death. While induction of reversible SIDNE does imply that the stimulus regimen is a significant stressor, its relation to neuronal loss after long-duration stimulation remains unclear and it is obvious that a neuroprosthesis or neuromodulation device should be restricted to a range of stimulus parameters that induce neither reversible SIDNE nor neural injury. At this time the precise relation of these phenomena remains unclear, but it does illustrate the importance of having a good understanding of the physiology of the target neural population.

An obvious limitation of histologic evaluations of the sites of neural stimulation is that the findings only reveal the state of the tissue at a single point in time. However, much can be learned by observing the state of the stimulated tissue at multiple time points throughout a regimen of prolonged stimulation, which will convey information regarding the relations between stimulus parameters, device configuration, and the time of onset of neural injury. Also, there are distinct advantages to observing the state of the stimulated tissue soon after the onset of the alterations induced by the stimulation, since analysis of tissue fixed days or weeks post-insult could be distorted by gliosis, phagocytosis, or dead neurons and tissue regeneration (Bolon *et al.* 2006; Garman 2006). Optical coherence tomography (OCT) is an optical signal acquisition and processing method that can provide near real-time data related to the tissue's response to electrical stimulation. OCT captures micrometer-resolution, three-dimensional images from within optical scattering media (e.g. biological tissue). An *in vitro* study of the safety of epi-retinal stimulation by Cohen *et al.* (2011) illustrates the utility of this technology, as well as some of its limitations. To obtain preliminary data for the safety of an epi-retina visual prosthesis, the authors employed a frequency domain OCT procedure (Bizheva *et al.* 2006) to monitor the tissue response throughout a regimen of localized electrical stimulation of the retinal ganglion cells.

During OCT imaging, 50 Hz trains of biphasic current pulses 1 ms/phase (23–749 $\mu C/cm^2$) were applied to the retinal surface for 5 min. After imaging, the stimulated regions were stained with the dye propidium iodide (PI) to reveal cytotoxic damage (Figure 6.7C). The damage seen in the OCT images matched the pattern observed in histological sections and the pattern of stain with the PI dye (Figures 6.7B,C). With pulse trains of 442 $\mu C/cm^2$, rapid increases in the reflectivity of the IPL could be observed under the stimulus electrode. The duration of the experiments was relatively short (30 min in most cases), and the quantitative relations between stimulus parameters and retina damage probably do not inform the standards for prolonged safe and

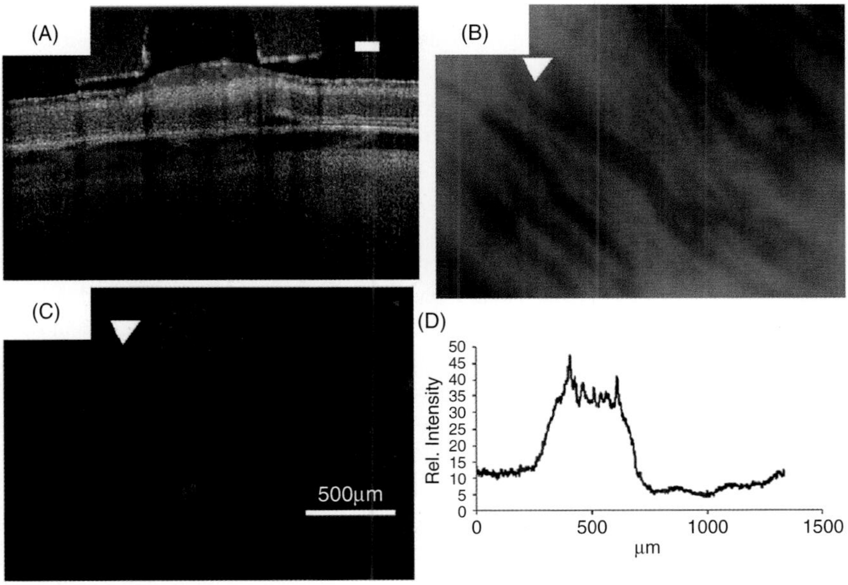

Figure 6.7 Comparison of OCT image of retinal swelling with other indicators of cytotoxic damage. (A) OCT scan of retina region stimulated at 442 μC cm^2 shows swelling of the ILM and IPL under the electrode. (B) Low power light micrograph of the retinal surface after the stimulation electrode was removed, showing only a barely visible circular retinal opacity (arrow). (C) The same region as in (B), now stained with the vital dye PI. The fluorescence image reveals a disk-shaped region of damage at the location of the stimulating electrode, with bright specks at the retinal surface indicating that the local neurons were damaged. (D) Intensity profile of the florescence across the region shown in C (modified from Cohen *et al.* 2011, with permission from IOP Publishing).

effective stimulation with an epi-retinal visual prostheses, but could guide the design of future studies in living animals. The extent to which OCT may be applicable to the early development of other neural prostheses and neural modulation devices is unclear. The view of the site of the stimulation must not be obstructed by the stimulating electrodes, or by optically dense connective tissue (limiting its utility in studies involving chronically implanted devices) or deep sites with intervening neural tissue. Perhaps the most restrictive limitation would be the difficulty of imaging stimulation sites deep in the brain.

6.6 CONCLUSION

Pre-clinical testing of implantable medical devices, including neurobionics, is an ever-evolving field that encompasses and draws upon the basic neurosciences, neuropharmacology, toxicology, the material sciences and instrumentation. As the standards for conducting clinical trials have evolved in recent decades and have become incorporated into the formal process of certifying medical devices, the pre-clinical testing of medical devices also has evolved as an indispensible part of bringing these devices to clinical use. Certainly, we must expect that the data obtained during the staged clinical trials that follow the pre-clinical testing will continue to inform the device's place in clinical practice, and how it can be used most effectively in that service. It is the role of pre-clinical testing to ensure that the first human trials will entail a minimal risk of injury and discomfort to the participants, and that they can be objectively informed of the medical benefits that they can expect and what risks that they may incur. The preclinical testing should provide evidence that the device will have a reasonably chance of providing the intended medical benefit, if not to the earliest participants, then at least by the end of the clinical trials. This obligation is especially acute for Class III devices such as neurobionics. To this end, the pre-clinical testing encompasses a wide gamut of studies, ranging from those that inform the development of classes of devices, to the testing of the final clinical device for a specific clinical application.

REFERENCES

Aceros, J., Yin, M., Borton, D., Patterson, W., Bull, C. and Nurmikko, A. (2012) Polymeric packaging for fully implantable wireless neural microsensors. *Conf. Proc. IEEE Eng. Med. Biol. Soc.*, pp 743–746.

Agnew, W., Yuen, T. and McCreery, D. (1983) Morphologic changes after prolonged electrical stimulation of the cat's cortex at defined charge densities. *Exp. Neurol.*, **79**(2): 397–411.

Agnew, W., McCreery, D., Yuen, T. and Bullara, L. (1990a) Local anaesthetic block protects against electrically-induced damage in peripheral nerve. *J. Biomed. Eng.*, **12**(4): 301–308.

Agnew, W., McCreery, D., Yuen, T. and Bullara, L. (1990b) Effects of prolonged electrical stimulation of peripheral nerve, in *Neural Prostheses: Fundamental Studies*. Agnew, W. and McCreery, D. (eds), Englewood Cliffs NJ, Prentice Hall.

Bizheva, K., Pflug, R., Hermann, B., Povazay, B., Sattmann, H. *et al.* (2006) Optophysiology: depth-resolved probing of retinal physiology with functional ultrahigh-resolution optical coherence tomography. *Proc. Natl. Acad. Sci. USA*, **103**(13): 5066–5071.

Bolon, B., Garman, R., Jensen, K., Krinke, G., Stuart, B. and S. T. P. S. Ad Hoc Working Group of the C. Regulatory Policy (2006) A "best practices" approach to neuropathologic assessment in developmental neurotoxicity testing – for today. *Toxicol. Pathol.*, **34**(3): 296–313.

Bradley, D.C., Troyk, P.R., Berg, J.A., Bak, M.J., Cogan, S. (2005) Visuotopic mapping through a multichannel stimulating implant in primate V1. *J. Neurophysiol.*, **93**(3): 1659–1670.

Butterwick, A., Vankov, A., Huie, P., Freyvert, Y. and Palanker, D. (2007) Tissue damage by pulsed electrical stimulation. *IEEE Trans. Biomed. Eng.*, **54**(12): 2261–2267.

Cohen, E.D. (2007) Prosthetic interfaces with the visual system: biological issues. *J. Neural. Eng.*, **4**(2): R14–31.

Cohen, E., Agrawal, A., Connors, M.,. Hansen, B., Charkhkar, H. and Pfefer, J. (2011) Optical coherence tomography imaging of retinal damage in real time under a stimulus electrode. *J. Neural. Engin.*, **8**(5): 1–13.

Edell, D.J., Toi, V.V, McNeil V.M. and Clark, L.D. (1992) Factors influencing the biocompatibility of insertable silicon microshafts in cerebral cortex. *IEEE Trans. Biomed. Eng.*, **39**(6): 635–643.

Garman, R.H. (2006) The return of the dark neuron. A histological artifact complicating contemporary neurotoxicologic evaluation. *Neurotoxicology*, **27**(6): 1126.

Gunalan, K., Warren, D.J., Perry, J.D., Normann, R.A. and Clark, G.A. (2009) An automated system for measuring tip impedance and among-electrode shunting in high-electrode count microelectrode arrays. *J. Neurosci. Meth.*, **178**(2): 263–269.

Han, M., Manoonkitiwongsa, P.S., Wang, C.X., and McCreery, D.B. (2012) *In vivo* validation of custom-designed silicon-based microelectrode arrays for long-term neural recording and stimulation. *IEEE Trans. Biomed. Eng.*, **59**(2): 346–354.

Hu, J.M., Barker, D., Dasguta, A. and Arora, A. (1993) Role of failure-mechanism identification in accelerated testing. *J. Inst. Environ. Sci.*, **36**(4): 39–45.

Huang, C.Q. and Shepherd, R.K. (1999) Reduction in excitability of the auditory nerve following electrical stimulation at high stimulus rates. IV: Effects of stimulus intensity. *Hear Res.*, **132**(1–): 60–68.

Huang, C.Q. and Shepherd, R.K. (2000) Reduction in excitability of the auditory nerve following electrical stimulation at high stimulus rates: V: Effects of electrode surface area. *Hear Res.*, **146**(1–2): 57–71.

Huang, C.Q., Shepherd, R.K., Seligman, P.M. and Clark, G.M. (1998) Reduction in excitability of the auditory nerve following acute electrical stimulation at high stimulus rates: III: Capacitive versus non-capacitive coupling of the stimulating electrodes. *Hear Res.*, **116**(1-2): 55–64.

li, W., Rodger, D.C., Meng, E., Weiland, J.D., Humayun, M.S. and Tai, Y.C. (2101) Wafer-level parylene packaging with integrated RF electroncis for wireless retinal prostheses. *J. Micromech. Syst.***19**(4): 735–741.

McConnell, G.C., Rees, H.D., Levey, A.I., Gutekunst, C.A., Gross, R.E. and Bellamkonda, R.V. (2009) Implanted neural electrodes cause chronic, local inflammation that is correlated with local neurodegeneration. *J. Neural. Eng.*, **6**(5): 056003.

McCreery, D.B., Agnew, W.F., Yuen, T.G. and Bullara, L. (1990) Charge density and charge per phase as cofactors in neural injury induced by electrical stimulation. *IEEE Trans. Biomed. Eng.*, **37**(10): 996–1001.

McCreery, D.B., Yuen, T.G., Agnew, W.F. and Bullara, L.A. (1992) Stimulation with chronically implanted microelectrodes in the cochlear nucleus of the cat: Histologic and physiologic effects. *Hear Res.*, **62**(1): 42–56.

McCreery, D.B., Agnew, W.F., Yuen, T.G. and Bullara, L.A. (1995) Relationship between stimulus amplitude, stimulus frequency and neural damage during electrical stimulation of sciatic nerve of cat. *Med. Biol. Eng. Comput.*, **33**(3 Spec No): 426–429.

McCreery, D.B., Yuen, T.G., Agnew, W.F. and Bullara, L.A. (1997) A characterization of the effects on neuronal excitability due to prolonged microstimulation with chronically implanted microelectrodes. *IEEE Trans. Biomed. Eng.*, **44**(10): 931–939.

McCreery, D.B., Yuen, T.G. and Bullara, LA. (2000) Chronic microstimulation in the feline ventral cochlear nucleus: physiologic and histologic effects. *Hear Res.*, **149**(1–2): 223–238.

McCreery, D.B., Agnew, W.F. and Bullara, L.A. (2002) The effects of prolonged intracortical microstimulation on the excitability of pyramidal tract neurons in the cat. *Ann. Biomed. Eng.,*. **30**(1): 107–119.

McCreery, D., Pikov, V. and, Troyk, P.R. (2010) Neuronal loss due to prolonged controlled-current stimulation with chronically implanted microelectrodes in the cat cerebral cortex. *J. Neural. Eng.*, **7**(3): 036005.

Medizza, A, and Milner, P.C. (1970) Electrochemistry and protection of surfaces, in *Materials Technology*. Prentice Hall, Englewood Cliffs NJ, pp 197–255.

Meeker, W.Q. and Escobar, L.A. (1998) Pitfalls of accelerated testing. *IEEE Trans. Reliability*, **47**(2): 114–118.

Merrill, D.R., Bikson, M. and Jefferys, J.G. (2005) Electrical stimulation of excitable tissue: design of efficacious and safe protocols. *J. Neurosci. Meth.*, **141**(2): 171–198.

Northup, S.J. (1999) Safety evaluation of medical devices: US Food and Drug Administration and International Standards Organization guidelines. *Int. J. Toxicol.*, **18**(4): 275–283.

Ordonez, J.S., Boehler, C., Schuettler, M. and Stieglitz, T. (2012) Improved polyimide thin-film electrodes for neural implants. *Conf. Proc. IEEE Eng. Med. Biol. Soc.*, pp 5134–5137.

Patton, H.D. and Woodbury, J.W. (1960). Special properties of nerve trunks and tract, in *Medical Physiology and Biophysics*. Ruch, T. and Fulton, J.F. (eds), W.B. Saunders, Philidelphia and London, pp 66–95.

Prasad, A. and Sanchez, J.C. (2012).Quantifying long-term microelectrode array functionality using chronic *in vivo* impedance testing. *J. Neural. Eng.*, **9**(2): 026028.

Prasad, A., Xue, M.Q.S., Sankar, V., Nishida, T., Shaw, G. *et al.* (2012) Comprehensive characterization and failure modes of tungsten microwire arrays in chronic neural implants. *J. Neural. Eng.*, .**9**(5): 1–21.

Robblee, L.S., McHardy, J., Marston, J.M. and Brummer, S.B. (1980) Electrical stimulation with Pt electrodes. V: The effect of protein on Pt dissolution. *Biomaterials*, **1**(3): 135–139.

Rose, T.L. and Robblee, L.S. (1990) Elect–rical stimulation with Pt electrodes. VIII: Electrochemically safe charge injection limits with 0.2 ms pulses. *IEEE Trans. Biomed. Eng.*, **37**(11): 1118–1120.

Salcman, M. and Bak, M.J. (1976) A new chronic recording intracortical microelectrode. *Med. Biol. Eng.*, **14**(1): 42–50.

Schmidt, E.M., Bak, M.J., Hambrecht, F.T., Kufta, C.V., O'Rourke, D.K. and Vallabhanath, P. (1996) Feasibility of a visual prosthesis for the blind based on intracortical microstimulation of the visual cortex. *Brain*, **119** (2): 507–522.

Schmitt, G., Schultze, J.W., Fassbender, F., Buss, G., Luth, H. and Schoning, M.J. (1999) Passivation and corrosion of microelectrode arrays. *Electrochimica Acta*, **44**: 3865–3883.

Shannon, R.V. (1992) A model of safe levels for electrical stimulation. *IEEE Trans. Biomed. Eng.*, **39**(4): 424–426.

Shen, J. and Burgess, D.J. (2012) Accelerated *in vitro* release testing of implantable PLGA microsphere/PVA hydrogel composite coatings. *Int. J. Pharm.*, **422**(1–2): 341–348.

Sim, S. and Lawson, R. (1979). Insulating Biomaterials. *IEEE 17th Ann. Reliability Physics Symp.*, pp 103–112.

Streit, W., Xue, Q.S., Prasad, A., Sankar, V., Knott, E. *et al.* (2012) Electrode failure: tissue, electrical, and material responses. *IEEE Pulse*, **3**(1): 30–33.

Takmanov, P., Phillips, I.S., Isayeva, C.G., Welle, E., Civillico, E. and Krauthamer, V. (2012) *Accelerated Aging of Neural Implants: Fast Bench Test for Reliability of Brain-Machine Interface*. Pittsburg Conference, Orlando FL.

Troyk, P., Bak, M.J., Berg, J., Bradley, D., Cogan, S., Erickson, R. *et al.* (2003) A model for intracortical visual prosthesis research. *Artif. Organs*, **27**(11): 1005–015.

Troyk, P.R., Bradley, D., Bak, M.J., Cogan, S., Erickson, R. *et al.* (2005) Intracortical visual prosthesis research – approach and progress. *Conf. Proc. IEEE Eng. Med. Biol. Soc.*, **7**: 7376–7379.

Tykocinski, M., Shepherd, R.K. and Clark, G.M. (1995) Reduction in excitability of the auditory nerve following electrical stimulation at high stimulus rates. *Hear Res.*, **88**(1–2): 124–142.

Tykocinski, M., Shepherd, R.K. and Clark, G.M. (1997) Reduction in excitability of the auditory nerve following electrical stimulation at high stimulus rates. II: Comparison of fixed amplitude with amplitude modulated stimuli. *Hear Res.*, **112**(1–2): 147–157.

Valko, M., Morris, H. and Cronin, M.T. (2005) Metals, toxicity and oxidative stress. *Curr. Med. Chem.*, **12**(10): 1161–1208.

Vanhoestenberghe, A. and Donaldson, N. (2013) Corrosion of silicon integrated circuits and lifetime predictions in implantable electronic devices. *J. Neural. Eng.*, **10**(3): 031002.

Wang, J.-S., Tsai, C.-C., Liou, J.-S., Cheng, W.-C. *et al.* (2012) Mean-time-to-failure evaluations of encapsulation materials for LED package in accelerated thermal tests. *Microelectronics Reliability*, **52**(5): 813–817.

Weaver, J.C. (1993) Electroporation: A general phenomenon for manipulating cells and tissues. *J. Cell Biochem.*, **51**(4): 426-435.

Weave, J.C. (1995) Electroporation theory. Concepts and mechanisms. *Meth. Mol. Biol.*, **55**: 3–28.

PART III

CLINICAL APPLICATIONS

7

AUDITORY AND VISUAL NEURAL PROSTHESES

ROBERT K. SHEPHERD, PETER M. SELIGMAN AND MOHIT N. SHIVDASANI

Bionics Institute and Medical Bionics Department, University of Melbourne, East Melbourne, Victoria, Australia

7.1 INTRODUCTION

This chapter will briefly review sensory prostheses designed for use in patients with severe auditory or visual disorders. While there are some similarities in the architecture of each technology, there are also important differences. Significantly, these devices are at very different stages of development. The cochlear implant, the mainstay of auditory prostheses, is a commercially mature technology that has been established for over 30 years. In contrast, while vision prostheses using cortical stimulation underwent initial clinical trials as early as the 1960s (Chapter 1), only recently have these devices – in the form of retinal-based prostheses – obtained regulatory approval and been commercialised.

Unlike many other neurobionic devices, sensory prostheses must continuously monitor their ever-changing environment and transmit that information in real time to the user. The sensory input provided by these devices, while impressive, remains relatively rudimentary. For example, a cochlear implant with 22 electrodes, is attempting to replicate the fidelity of a normal cochlea containing nearly 15 000 sensory hair cells. Nevertheless, to date, these devices have been successful clinically, which is a testament to both the technology and to the plasticity of the human brain.

Neurobionics: The Biomedical Engineering of Neural Prostheses, First Edition.
Edited by Robert K. Shepherd.

7.2 AUDITORY PROSTHESES

7.2.1 The auditory system

The inner ear or cochlea transduces the mechanical vibrations of sound into nerve impulses that are transmitted along the central auditory pathway where they are perceived as sound. The sensory cells of the cochlea are located within the Organ of Corti, which is located on the basilar membrane (Figure 7.1). The mechanical characteristics of the basilar membrane allow the cochlea to resolve different frequencies at different spatial locations along its length; high frequencies produce maximum basilar membrane displacement in the base, while lower frequencies produce maxima at more apical sites. Thus the cochlea has a frequency-to-place or *cochleotopic* map along the length of the basilar membrane.

The human cochlea contains approximately 3500 inner hair cells and 11 000 outer hair cells located on the basilar membrane (Ashmore 2008). Each hair cell contains up to 150 stereocilia protruding from its apical surface. Located within these stereocilia

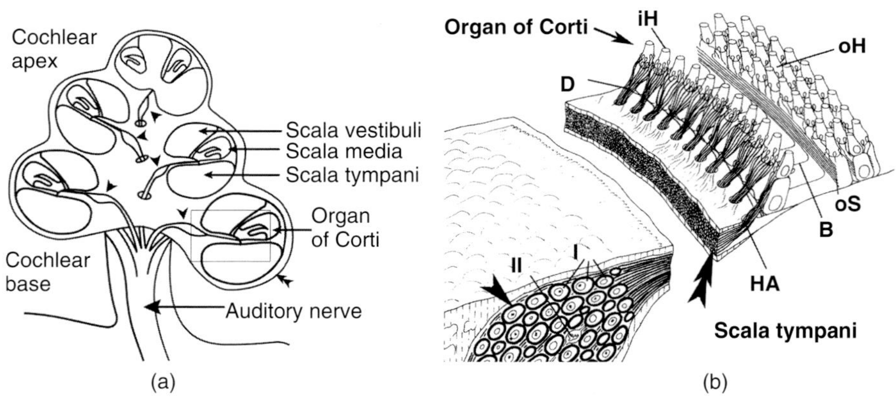

Figure 7.1 Transverse section of the mammalian cochlea illustrating the three fluid-filled cochlear ducts (scala vestibule, scala media and scala tympani) spiralling around the modiolus containing the auditory nerve. The soma of the auditory nerve fibres, the spiral ganglion neurons (SGN), are located in a small bony canal (Rosenthal's canal) that also spirals from the cochlear base to its apex (arrowheads). There are ~30 000 SGNs in a normal human cochlea. A cochlear implant electrode array is inserted into the scala tympani at the cochlear base either via the round window, or a membrane separating the scala tympani and the middle ear (not illustrated), or via a small hole drilled into the bone overlying (double arrowhead). The boxed area of the basal turn is illustrated in more detail in (b). (a) Schematic illustrating the afferent innervation pattern in the mammalian cochlea. SGN soma reside in Rosenthal's canal (arrowhead) and their peripheral processes (double arrowhead) project to the hair cells within the Organ of Corti. (b) Type I SGNs (I) make direct synaptic contact with a single inner hair cell (iH). In contrast, Type II SGNs (II) project across the floor of the basilar membrane (B) and make synaptic contact with numerous first, second and third row outer hair cells (oH). HA – Habenular perforata, OSL – osseous spiral lamina (adapted from Spoendlin 1984, reprinted with permission Elsevier Science).

are ion channels that are extremely sensitive to mechanical displacement generated by acoustic stimuli. On displacement of the stereocilia, the hair cell depolarises, resulting in the generation of an action potential within the auditory nerve fibers in synaptic contact with it.

There are approximately 30 000 auditory neurons termed spiral ganglion neurons (SGNs) that innervate the hair cells within the Organ of Corti (Figure 7.1b). The cochlea contains two SGN populations; a large Type I SGN (~90% of the total population) and the smaller Type II SGN population (Kimura *et al.* 1979; Spoendlin and Schrott 1989). Type I SGNs innervate their closest inner hair cell, while Type II SGNs innervate several outer hair cells.

Projecting from the cochlea, the auditory nerve makes connection with the central auditory pathway, a complex structure containing over a dozen auditory processing centres. Like the cochlea, each centre within the auditory pathway is cochleotopically organised.

7.2.2 Hearing loss

Over 5% of the world's population have a disabling hearing loss, making it a major health and economic burden on society (WHO 2012). Sensorineural hearing loss is the most common form of deafness and is due to a loss of hair cells. Widespread hair cell loss results in a severe-to-profound hearing loss that cannot be effectively aided by amplification via a hearing aid; cochlear implants provide the only therapeutic intervention for these patients.

The SGN undergo gradual degenerative changes in response to hair cell loss, eventually resulting in significant reductions in the number of surviving neurons (Hardie and Shepherd 1999; Leake and Hradek 1988; Nadol 1990) (Figure 7.2). Despite these pathological changes, surviving SGNs remain capable of initiating and propagating action potentials elicited via an electrical stimulus (Shepherd and Javel 1997).

7.2.3 Cochlear implants

Commercial cochlear implants were developed in the early 1980s as a therapeutic intervention for patients with a severe-profound bilateral sensorineural hearing loss (Clark 2003). Designed as an "aid to lip-reading", these devices have proven to be very successful, with many users being able to communicate in a near normal fashion in an environment with low background noise. By 2014, over 330 000 cochlear implants had been implanted worldwide. As clinical performance has improved, patient selection criteria have expanded to include children as young as 6 months of age, and patients with some level of residual hearing.

7.2.3.1 Implant technology Cochlear implant systems are comprised of an external sound processor and an implanted receiver-stimulator (Figure 7.3). These components are coupled by a transcutaneous inductive link that conveys power and data to the implant. Worn behind the ear, the sound processor contains directional microphones that are typically incorporated in a beam forming configuration to minimise

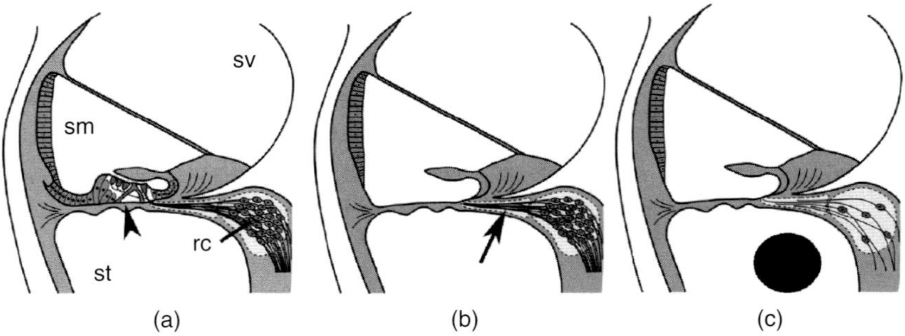

Figure 7.2 Schematic diagram of a single cochlear turn illustrating the degenerative changes that occur following loss of the Organ of Corti. (a) Normal cochlea, illustrating the three fluid filled chambers (scala tympani, st; scala media, sm; scala vestibule, sv), the Organ of Corti containing the sensory hair cells (arrowhead), and Rosenthal's canal (rc) containing the spiral ganglion neuron (SGN) soma and their peripheral processes projecting to the Organ of Corti to synapse with the sensory hair cells. (b) The most common forms of SNHL target the hair cells of the Organ of Corti, eventually resulting in the degeneration of the Organ of Corti. Hair cell loss initially induces loss of SGN peripheral processes (arrow) that would normally innervate the Organ of Corti. (c) More gradual degenerative changes following hair cell loss includes extensive loss of peripheral processes and gradual loss of SGNs. This is typically the status of the cochleae receiving a cochlear implant. The typical position of the electrode array within the scala tympani is illustrated in this panel (adapted from Shepherd *et al.* 2013b, with permission).

unwanted noise from specific directions. The microphone signal is analysed to select important spectral features while reducing the effects of noise. Sound-processing algorithms are based on a spectral analysis performed by a Fourier transform or a bank of parallel band-pass filters that typically cover a frequency range of 0.2–8.0 kHz (McDermott *et al.* 1992; Wilson *et al.* 1991). Each frequency band is assigned to a single electrode in the cochlea. High frequency bands are assigned to electrodes near the base of the cochlea, where high-frequency components of sounds in normal hearing result in mostly neural activity, whereas lower frequency bands are assigned to more-apical electrodes. Variations in amplitude in each band result in corresponding variations in the stimulus level delivered to each electrode along the cochlea. This information is specifically coded for each patient to match the sensitivity and dynamic range of his/her electrodes.

The power to supply implantable electronics can take a number of different forms including: (i) an implanted battery or primary cell commonly used in heart pacemakers; (ii) implanted rechargeable battery, charged inductively from an externally worn power source, used in fully implanted hearing aids and cochlear implants; and (iii) electromagnetic induction power though a pair of coils, one outside the body and the other implanted commonly used in cochlear implants (Shepherd *et al.* 2014). The receiver-stimulator, the implantable component of the cochlear implant system, receives power via the external sound processor through this radio frequency (RF) inductive link, avoiding the need for an implanted battery and the periodic surgical

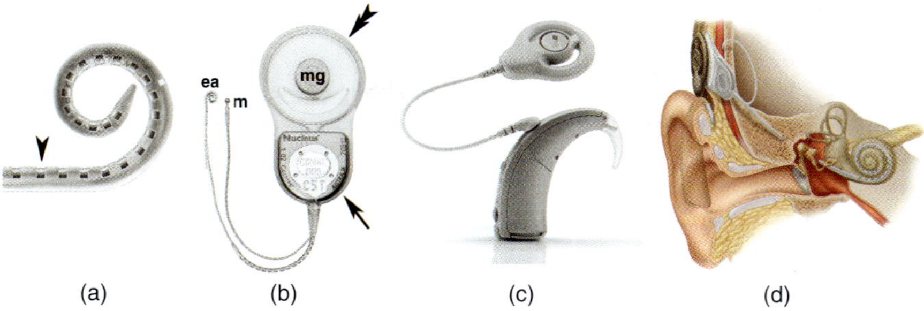

Figure 7.3 A modern cochlear implant system illustrating the (a) scala tympani electrode array containing platinum electrode contacts (arrow); (b) implant receiver-stimulator, the electrode array (ea); a monopolar return electrode (m), a hermetically sealed titanium case containing the stimulator electronics (arrow) and a magnet (mg) to align the external radio frequency (RF) coil (double arrowheads). The implant package is ~15 mm across and 7 mm thick; (c) the external sound processor with the external RF coil and magnet; and (d) a schematic illustrating the surgical placement of the system together with the external speech processor and RF coil. Images courtesy of Cochlear Ltd.

procedure required to replace it. Importantly, this also results in a smaller device suitable for implantation in young children.

The receiver-stimulator electronics are protected from the biological environment by placing them in a hermetically sealed container. The industry standard sealing technique is to encase the electronics within a titanium capsule that is closed using a laser welder (Amanat *et al.* 2010). Some cochlear implants use ceramic packages to seal the electronics; however, this approach is less common as ceramic is brittle and can be damaged. The implant package also has an electrical feed-through – a hermetic interface that provides an electrical path between the hermetically sealed implant electronics, the leadwire/electrode array and the receiver coil (Patrick *et al.* 1990) (Chapter 1). The receiver coil is located outside the titanium metal package so that the electro-magnetic field is not attenuated (Figure 7.3b).

The receiver-stimulator contains a low-power Application Specific Integrated Circuit (ASIC) designed to decode information received by the coil and convert it into current controlled pulses suitable for safe stimulation. The implanted electronics are generally designed to be as simple as possible in the interest of reliability, leaving the complicated electronics outside the body where it can be repaired or upgraded as often as necessary. Apart from the ASIC, there are a small number of capacitors and diodes which cannot be integrated onto the chip.

The receiver-stimulator is connected via a flexible leadwire to the electrode array. The leadwire consists of insulated platinum-iridium micro-wires; each micro-wire is connected to an individual electrode on the array. The leadwire passes from the receiver-stimulator through a drilled recess in the mastoid bone to the electrode array inserted into the cochlea (Figure 7 3d).

The electrode array consists of a flexible silicone carrier containing between 12 and 22 platinum electrodes at a pitch of between 0.75 and 2.5 mm (Figure 7.3a).

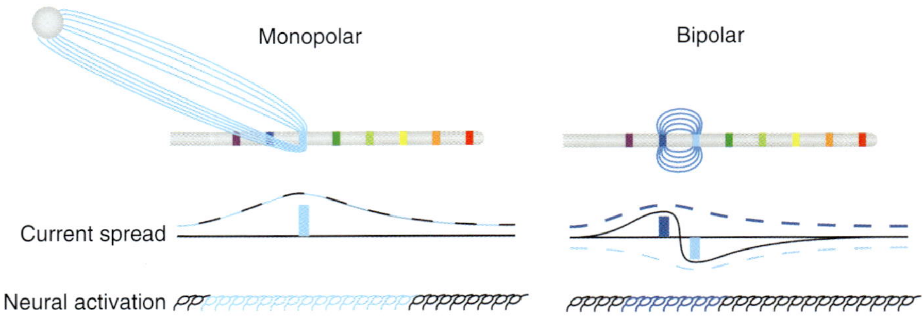

Figure 7.4 Schematic illustrating the two most common electrode geometries used in cochlear implants. In a monopolar configuration, a scala tympani electrode proximal to the SGN population is stimulated against a remote large surface area electrode. In contrast, bipolar electrodes incorporate a single adjacent return electrode. Although monopolar stimulation is efficient, it also results in greater current spread compared with bipolar stimulation. Nevertheless, the majority of patients can identify electrodes via an orderly change in pitch using monopolar stimulation (images courtesy of Fallon *et al.* 2009, with permission).

The electrode array is inserted up to 25 mm into the scala tympani of the cochlea via the round window or an ~1 mm diametre hole drilled through the bony wall of the cochlea. The preferred mode of stimulation in cochlear implants is monopolar; the active intra-cochlear electrodes use a remote extra-cochlear electrode – typically a platinum ball – as a current return path (Figure 7.4). Bipolar stimulation can be implemented in most commercial devices; however, monopolar stimulation is far more efficient, requiring an order of magnitude less charge/phase than bipolar stimulation to evoke an auditory percept, without a decrement in clinical performance.

Cochlear implants typically stimulate the auditory nerve using sequential stimulation, delivering a current pulse to a single channel along the electrode array at a time, rather than the simultaneous stimulation of multiple channels. This is done to eliminate any untoward effects of current interaction (Shannon 1983). The cycle time or stimulation rate for sequential stimulation is of the order of 1 ms, during which a subset of electrodes are stimulated, depending on the sound coding strategy (McDermott *et al.* 1992; Wilson *et al.* 1991).

Although cochlear implants use a small number of frequency channels compared to the normal cochlea, this is sufficient for good speech recognition in quiet environments. However, they are limited in the representation of music (McDermott 2004) and voice pitch in tonal languages such as Mandarin, although due to redundancy in the speech signal, cochlear implants have generally still been very successful in countries where a tonal language is spoken (Lee *et al.* 2002).

7.2.3.2 Device fitting and clinical performance Cochlear implants do not restore normal hearing; learning and adapting to the device is required for a good clinical outcome. Before surgery, the existing auditory, medical and psychological status of each prospective patient must be carefully assessed. Counselling is key to the success of the procedure; it is important that patients, their families and carers are fully informed so that they have realistic expectations (Chapter 11).

After a recovery period of up to three weeks following surgery, the patient's device is "switched on" by an implant clinician who establishes threshold and comfortable stimulus levels for each electrode on the array. This device fitting procedure is important to ensure optimal clinical performance (Holden *et al.* 2013). Attention is paid to ensure that any unpleasant, painful or non-auditory sensations are eliminated from the patient's individual speech processing map. The fitting procedure may take several hours; when completed, the map is written to the patient's sound processor and will translate acoustic sound into patterns of electrical stimulation delivered to specific electrodes along the array in a manner customised for that patient. Each patient may receive a number of maps to trial and continues to return to the clinic for periodic check-ups and remapping (if required) at less frequent intervals.

While almost all cochlear implant recipients obtain substantial benefit from their device using measures of clinical safety and efficacy that include: sensitivity to sound; speech perception; speech production; adverse effects of treatment; health-related quality of life and educational outcomes (Bond *et al.* 2009), the range in clinical performance across cochlear implant users is very wide (Blamey *et al.* 2013; Holden *et al.* 2013; Lazard *et al.* 2012). The evaluation of factors affecting performance in adult implant subjects demonstrate that among many factors, the duration of auditory experience – either pre-implantation through normal or aided hearing, or post-implantation via the use of the cochlear implant – positively correlates with clinical performance. In contrast, duration of deafness negatively correlates with performance (Blamey *et al.* 2013; Lazard *et al.* 2012). Individuals that have been profoundly deaf for periods in excess of 30 years tend to show far less benefit with cochlear implant use (Bond *et al.* 2009). Moreover, while late-implanted prelingually deaf subjects show clinical improvement with device use over an extended period of time (Schorr *et al.* 2005), their speech perception remains much poorer than subjects implanted at an early age. Reductions in clinical performance associated with long durations of deafness and age at implantation are presumably associated with a reduction in the plastic capacity of the central auditory pathway (Blamey *et al.* 2013; Kral and Sharma 2012; Lazard *et al.* 2012) (Section 7.4).

Clinical performance associated with paediatric cochlear implantation is less variable. Results include improvements in academic performance, with greater benefits obtained with earlier cochlear implantation and shorter durations of deafness (Bond *et al.* 2009). Moreover, implanting children in the 6–12 month age group significantly increases the likelihood of those children obtaining language growth rates *equal* to their normal hearing peers (Dettman *et al.* 2007).

7.2.4 Central auditory prostheses

There are some clinical conditions where deaf patients are not candidates for a cochlear implant because they have no surviving auditory nerves, their cochleae contain extensive fibrous tissue and/or new bone as a result of the disease process (e.g. meningitis) or the cochlea is malformed, preventing the implantation of an electrode array. Under these conditions a central auditory prosthesis becomes a viable alternative. The human central auditory pathway consists of a series of auditory nuclei that are cochleotopically organised, making many of these potential sites for the development of a central auditory prosthesis capable of evoking pitch percepts based on electrode position (Figure 7.5).

7.2.4.1 Auditory brainstem implants The auditory brainstem implant was designed for patients that were not considered a candidate for a cochlear implant (Colletti *et al.* 2009; McCreery 2008). Initially developed in the late 1970s for patients with tumours on both auditory nerves, these devices consist of up to 21×0.7 mm platinum disc electrodes mounted on a silicone carrier that is placed on the surface of the cochlear nucleus during surgery to remove the tumour. Clinical experience has shown this array to be stable for periods of more than 10 years (Otto *et al.* 2002; Shannon *et al.* 1993).

Approved by the FDA in 2000, it is estimated that over 1100 adults with bilateral tumours on the auditory nerve have received this device (R.V. Shannon, personal communication). Although non-auditory sensations can be evoked, the offending electrodes can be removed from the sound processor map (Otto *et al.* 2002). Using a cochlear implant receiver-stimulator fitted with the modified electrode array, and an identical speech processor to that developed for cochlear implants, stimulation of auditory brainstem electrodes can evoke electrode-specific pitch sensations (Otto *et al.* 2002). Surface stimulation of the human cochlear nucleus with these electrodes provides a level of electrode-specific pitch percepts important for sound awareness and environmental sound discrimination; however, only a few of these patients have achieved word and sentence recognition without the aid of lip reading (Colletti *et al.* 2009; Otto *et al.* 2002; Shannon *et al.* 1993). In contrast, non-tumour patients receiving these devices, including aetiologies such as temporal bone fracture, cochlear ossification and congenital malformations of the cochlea or auditory nerve, consistently perform significantly better than tumour patients (Colletti *et al.* 2009); many are able to understand speech without the aid of lip reading. This success has resulted in auditory brainstem implants now being successfully used in children (Colletti *et al.* 2009).

The limited clinical performance of the early auditory brainstem implant users was thought to be affected by the relatively poor spatial selectivity associated with surface electrodes. In response, researchers developed auditory brainstem implants containing penetrating electrodes of different lengths designed to take advantage of the cochleotopic organisation of the cochlear nucleus (McCreery 2008). While results from a clinical trial demonstrated reduced thresholds and more selective stimulation, speech perception scores were not significantly better than that of patients using the surface arrays (Shannon 2012), and the approach has been abandoned.

7.2.4.2 Auditory midbrain and cortical implants A clinical trial was recently performed in deaf patients using an electrode array inserted along the cochleotopic axis of the inferior colliculus (Lim *et al.* 2008) (Figure 7.5). Although each patient reported pitch and loudness cues with stimulation on some electrodes, more than

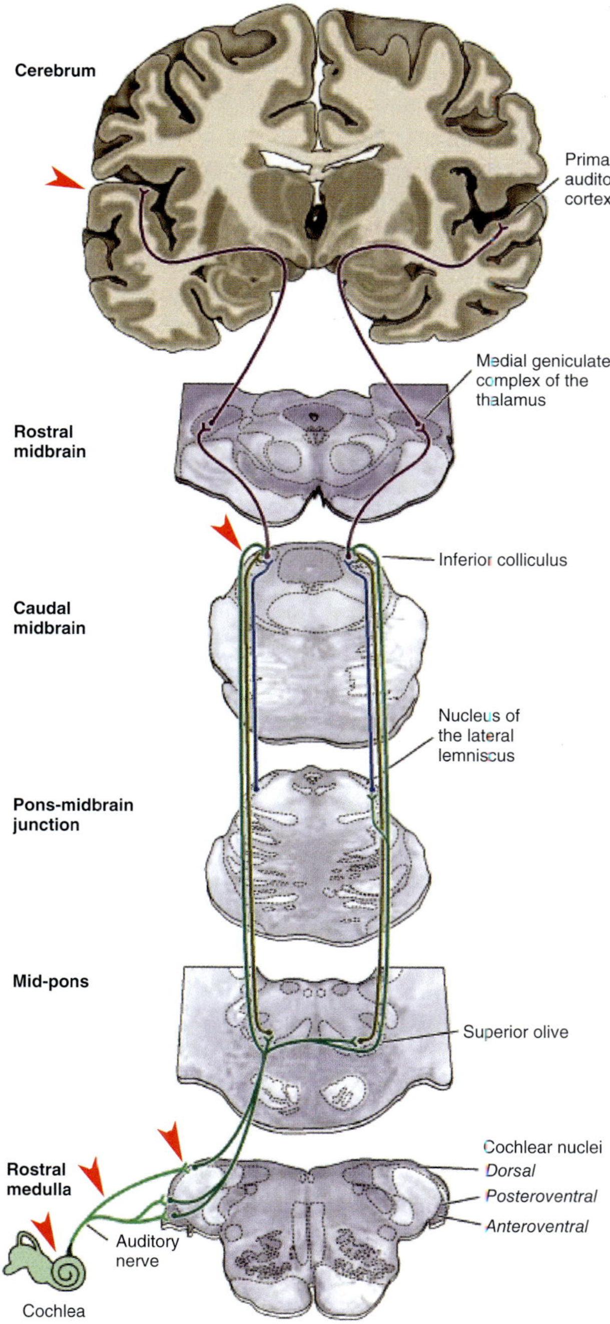

Figure 7.5 Schematic diagram of the human auditory pathway. Five stimulation sites have been trialed clinically over the course of auditory prosthesis development (arrowheads) including: (i) the cochlea; (ii) the auditory nerve; (iii) the cochlear nucleus; (iv) the inferior colliculus; and (v) the primary auditory cortex. To date, only cochlear and cochlear nucleus based devices have received regulatory approval for clinical use (adapted from Purves *et al.* 2001, with permission from Sinauer Associates, Inc.).

half the electrodes evoked non-auditory sensations. In addition, these subjects had difficulty in processing temporal cues important for speech understanding (McKay *et al.* 2013). While some users showed awareness of environmental sounds, open-set speech perception without lip reading was not achieved (Lim *et al.* 2008).

Electrical stimulation of the primary auditory cortex is known to evoke auditory sensations including buzzing, whistling and knocking, while stimulation of secondary auditory areas evoke more complex sensations including voices and music (Penfield and Perot 1963). Although there has been some interest in auditory prostheses based on cortical stimulation (Dobelle *et al.* 1973), to date no commercial device has been developed. An issue common to all cortical based sensory prostheses is that these devices bypass potentially important neural processing from more peripheral nuclei within the pathway.

7.2.5 Combined electric and acoustic stimulation

The excellent speech perception scores achieved by some cochlear implant recipients have led to a relaxation of the criteria for candidacy for implantation. Patients with significant levels of low-frequency hearing, who previously would only have been treated with hearing aids, are now being considered for a cochlear implant (Lorens *et al.* 2008). These patients receive auditory information from both acoustic and electric stimulation and are said to have "electric-acoustic hearing". The increase in information provided by the combined stimulation results in an improved aesthetic quality of sound for many patients as well as improved speech perception, particularly in the presence of background noise (Gantz and Turner 2003; Simpson *et al.* 2009; Turner *et al.* 2008).

There are several factors that affect the performance of patients with combined electric and acoustic stimulation, the most obvious of which is the amount of residual hearing the patient has. There have been a number of studies focused on the preservation of hearing following implant surgery demonstrating that it is possible to preserve auditory function in an implanted ear over long periods of time (Adunka *et al.* 2004; Simpson *et al.* 2009; Turner *et al.* 2008). While promising, further research is required to develop techniques that will ensure greater levels of hair cell preservation when combining electric and acoustic hearing.

7.2.6 Bilateral cochlear implants

By 2010, approximately 5% of cochlear implant recipients were fitted bilaterally (Peters *et al.* 2010). Bilateral cochlear implants provide users with the ability to partially restore spatial hearing abilities, resulting in improved sound localisation ability and some improvement in speech understanding in noise (Kan and Litovsky 2014). Although bilateral cochlear implants do not preserve the interaural timing differences required for optimal binaural hearing (Shannon 2012), there is good evidence, at least in the developing auditory pathway, that the auditory brainstem exhibits improved binaural processing once fitted with bilateral implants (Gordon *et al.* 2012). Research to develop improved binaural processing strategies is ongoing and likely to include the implants working in concert rather than independently.

7.2.7 Future directions

7.2.7.1 ***Advanced stimulation strategies*** Cochlear implants aim to stimulate discrete regions of the cochlea; however, neural excitation via a cochlear electrode is spatially broad due to the conductive nature of the fluids of the inner ear (van den Honert and Stypulkowski 1987). As a consequence, the neural populations activated by each electrode are not independent but overlap considerably. Implant recipients therefore perceive relatively low resolution spectral information. This broad spatial activation significantly reduces sound quality and speech perception, particularly in discriminating speech in noise and the perception of music and tonal languages that are rich in spectral and temporal information (Skinner *et al.* 1994; Sucher and McDermott 2007). Research has focused on developing more targeted or localised activation of SGNs using novel electrode geometries and current focusing techniques (Bierer and Middlebrooks 2002; George *et al* 2014, Saoji *et al.* 2013; van den Honert and Kelsall 2007). A number of these techniques use simultaneous stimulation from multiple electrodes; although promising (Marozeau *et al.* 2015; Smith *et al.* 2013), the main limitation is associated with increased thresholds compared to monopolar stimulation resulting in increased power requirements.

7.2.7.2 ***Totally implantable cochlear implants*** As described above, cochlear implants consist of both an external sound processor and an implanted receiver-stimulator. From a user's point of view, this arrangement is not ideal; the sound processor is visible, and can be dislodged or damaged. In an attempt to minimise these disadvantages, a prototype totally implantable cochlear implant has been developed and trialed in three patients (Briggs *et al.* 2008). While these patients are still using their devices more than nine years later, a commercially available totally implantable device remains some years off.

A major technical challenge with the development of a totally implanted device is the requirement that a microphone implanted within the body be sensitive to sound originating from outside the body while rejecting internally generated sounds (e.g. breathing, heartbeats and muscle activity associated with movement). Until this problem is resolved, totally implanted cochlear implants will not match the performance of devices using an externally worn microphone.

7.3 VISUAL PROSTHESES

7.3.1 The visual system

The human retina is a delicate and intricate network of photo-sensitive cells and neurons lining the back of the eye (Figure 7.6). It transduces incident visible light, focused by the optics of the eye, into neural impulses, which form the brain's perception of vision. The retina consists of an outer layer of photoreceptors, several specialist neural layers and supporting architecture. Humans have two primary photoreceptors: rods and cones. The rods are optimised for low-light monochrome vision,

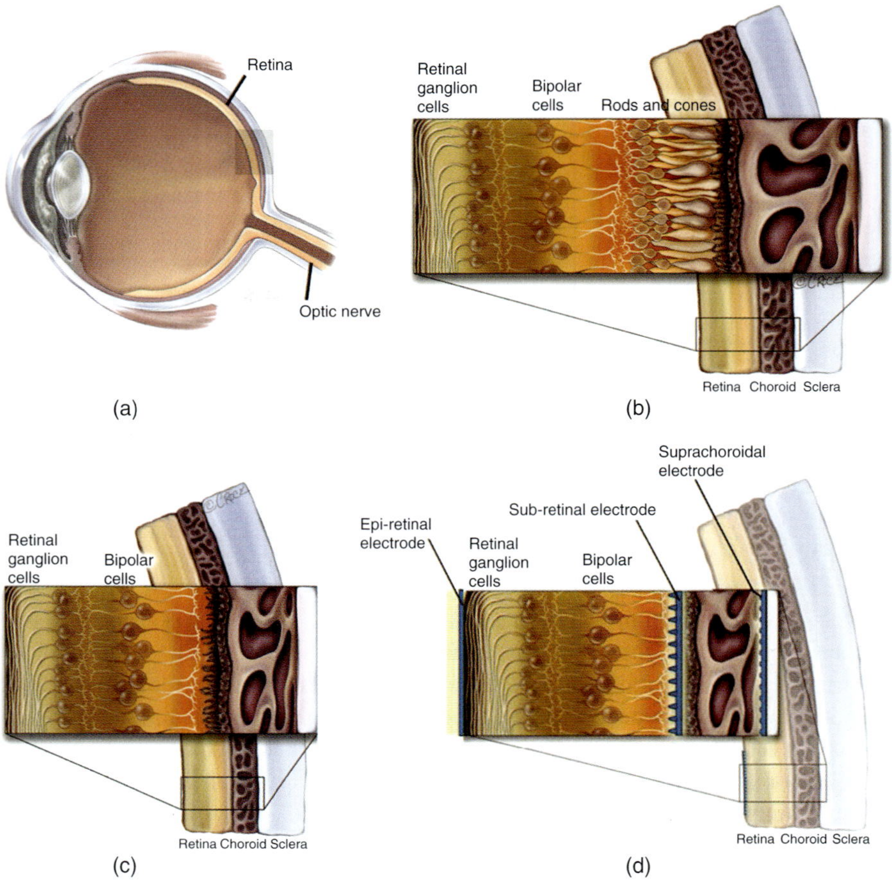

Figure 7.6 The anatomy of the eye in normal vision and following loss of photoreceptors. (a) Schematic of a normal eye. The boxed area of the retina is enlarged in panels b–d. (b) Schematic of a normal retina, choroid and sclera. The retina consists of several processing layers extending from the rods and cones of the outer retina through bipolar cells of the middle retina to the retinal ganglion cells (RGCs) that make up the inner retina. Axons of the RGCs project via the optic disc to form the optic nerve. (c) Schematic of a retina with widespread photoreceptor degeneration. This pathology also results in the remodelling of the neural architecture within the inner retina and shrinkage of the choroid. (d) Potential sites for the location of an electrode array for a retinal prosthesis including epiretinal, sub-retinal and suprachoroidal locations. Images courtesy of Bionic Vision Australia (image by C. Roce).

while cones are specialised for colour vision in brighter environments. The photoreceptors are highly metabolically active and are supplied by a rich network of blood vessels in the adjacent layer of the eye, known as the choroid (Figure 7.6b). The light-sensing photoreceptor cells initiate a cascade of neural activity that propagates via a convergent retinal network to retinal ganglion cells (RGCs) whose axons form the optic nerve (Figure 7.6a).

7.3.2 Vision loss

Retinitis pigmentosa (RP) and age-related macular degeneration (AMD) are two degenerative diseases of the retina that result in blindness, secondary to photoreceptor loss (Figure 7.6c). Because both pathologies affect the photoreceptors while sparing the majority of RGCs, these forms of blindness are targets for vision restoration via a retinal prosthesis.

RP represents a group of hereditary disorders that lead to blindness in midlife, as a result of a gradual degeneration of photoreceptors. The lesion is initially restricted to the peripheral retina, but over time the central macular region is also affected, resulting in severe blindness. RP has an incidence of 1 in 4000 with an estimated 1 million patients worldwide (Hartong *et al.* 2006). There is currently no other clinical treatment for RP.

AMD is a leading cause of vision loss in older adults; in Western countries it accounts for almost 50% of all severe visual impairment and blindness (Finger *et al.* 2011). The disease process gradually destroys the high resolution foveal region of the retina via a proliferation of blood vessels while typically leaving peripheral vision intact. At present it is possible to treat some forms of AMD using anti-vascular endothelial growth factors (anti-VEGF) (Lai and Landa 2015) to reduce the formation of new blood vessels; however, it is not yet possible to reverse the progression of the pathology.

Significant populations of RGCs survive following photoreceptor degeneration, although the loss of the photoreceptors produces retinal remodelling that includes the migration and rewiring of retinal circuitry, gliosis, ectopic neurite outgrowth and RGC degeneration (Jones *et al.* 2012). These alterations influence the sensitivity of RGCs to electrical stimulation as well as the neural processing through the retinal network (Jones *et al.* 2012). Importantly, electrical stimulation of the long-term blind retina evoke stable, retinotopically organised visual percepts (Ayton *et al.* 2014; Weiland *et al.* 2011; Zrenner *et al.* 2011).

7.3.3 Retinal prostheses

There are two general approaches to the design of retinal prostheses:

1. optical sensors such as an array of photodiodes that are implanted close to the retina, allowing the normal optical properties of the eye to focus light onto the photodiodes, the output of which is converted into electrical pulses to depolarise RGCs (Mathieson *et al.* 2012; Zrenner *et al.* 2011); and
2. an electrode array located at the level of the retina connected to a leadwire and implantable stimulator that receives power and data from an external video camera and vision processor via a transcutaneous telemetry link (Humayun *et al.* 2012; Shepherd *et al.* 2013a) (Figure 7.7).

There are several advantages in using a photodiode array, primarily associated with the absence of wires (Chow *et al.* 2002), which simplifies the implant design and

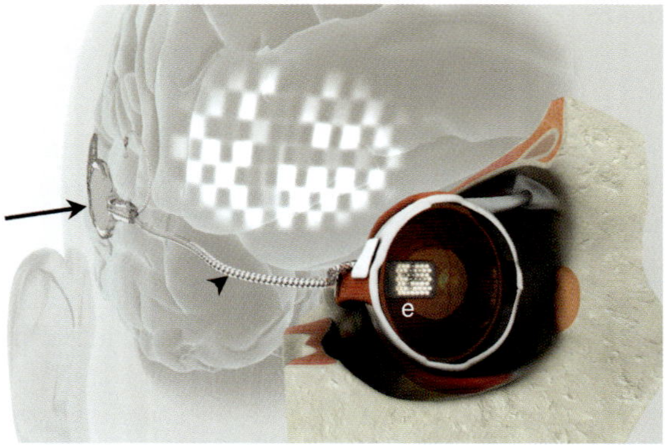

(a)

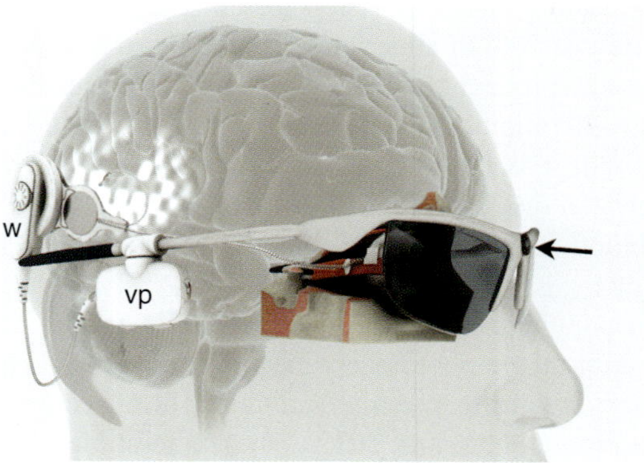

(b)

Figure 7.7 (a) Schematic diagram of a generic retinal prosthesis illustrating a receiver-stimulator unit implanted in the mastoid bone behind the ear (arrow), a lead-wire assembly (arrowhead) connecting the output of the stimulator to an array of electrodes (e) implanted in the retina. The electrode array can be tacked to the retina (epiretinal); inserted between the choroid and the retina (subretinal); or inserted between the sclera and the choroid (suprachoroidal). (b) Overall schematic of a retinal prosthesis that includes a video camera incorporated onto glasses (arrow), an external vision processor (vp) that provides both data and power across the skin via a wireless link (w) to the implanted receiver-stimulator, leadwire and electrode array illustrated in (a). The camera continuously feeds video signals to the vision processor that contains the patient's phosphene map, visual processing algorithms and stimulation strategies. Each frame of the input video generates a sequence of commands at the vision processor that defines the electrodes and stimulus parameters required to generate a prosthetic image of the scene. (Images by Jack Parry; courtesy of the Bionics Institute).

surgery. In addition, because a photodiode array utilises existing ocular optics and eye position to localise the visual field, eye tracking technology is not required. The major limiting factor with the use of photodiodes is their inability, at this stage of their development, to provide sufficient energy to achieve retinal stimulation using visible light (Fernandes *et al.* 2012). To overcome this limitation one photodiode-based device includes an additional external power source and a leadwire assembly (Stingl *et al.* 2013; Zrenner *et al.* 2011), while a second design proposes the use of pulsed infrared light to provide both sufficient power and visual information directly to photodiodes implanted in the retina (Mathieson *et al.* 2012).

The most common form of a visual prosthesis incorporates an electrode array located on or close to the retina using an architecture similar to a cochlear implant (Figure 7.7). There are a number of anatomical sites where these electrodes can be placed including the inner surface of the retina (epiretinal), between the retina and choroid (subretinal) or between the choroid and the sclera (suprachoroidal) (Figure 7.6d).

Placing an electrode array on the inner surface of the retina ensures the electrodes stimulate the output of the retina – the axons of the RGCs. Such proximity to RGCs results in low thresholds for neural activation, minimising the physical size required of individual electrodes and theoretically maximising the resolution and acuity of prosthetic vision. Surgically, the electrode array is attached to the inner retinal surface, typically by using one or more retinal tacks. There are several research groups and companies developing epiretinal devices (Hadjinicolaou *et al.* 2012; Humayun *et al.* 2012; Keseru *et al.* 2012; Klauke *et al.* 2011). Second Sight Medical Products received regulatory approval in both Europe (CE Mark in 2011) and the USA (FDA in 2013) to treat late-stage RP with their 60 electrode Argus II device. Two technical challenges associated with the epiretinal approach are related to the significant constraints on device size with this surgical approach, and the fixation and long-term mechanical stability of the electrode array on the delicate retinal surface.

Significant neural processing occurs within the outer and middle layers of the retina; epiretinal prostheses do not take advantage of this processing. Subretinal electrode arrays are designed to be positioned at the level of the outer retina where, in a healthy eye, photoreceptors would be located (Mathieson *et al.* 2012; Rizzo 2011; Stingl *et al.* 2013; Zrenner *et al.* 2011) (Figure 7.6d). While this is a logical choice for an implant whose function is to replace lost photoreceptors, it presents its own challenges. The surgical approach is difficult and the electrode array and associated electronics must be extremely thin (<400 μm) to minimise retinal damage or detachment. In addition, there is the potential for a subretinal electrode array to impede blood supply from the choroid to the surviving retina. A number of groups are developing this technology, with the Alpha-IMS device manufactured by Retina Implant AG recently gaining regulatory approval in Europe (CE Mark in 2013). There are technical challenges with manufacturing long-term hermetic encapsulation with the thin profile necessary for these devices. Some have chosen photodiodes as the sensing elements (Mathieson *et al.* 2012; Zrenner *et al.* 2011), while another group has developed an array of 256 electrodes driven by an implantable stimulator containing 256 current drivers (Rizzo *et al.* 2011).

While electrodes located in a tissue pocket between the choroid and sclera (the "suprachoroidal" position) are distal to their target neurons in the inner retina, this approach has been adopted by a number of groups. This placement offers a safe and simple surgical approach and a mechanically stable location (Ayton *et al.* 2014; Nayagam *et al.* 2014; Villalobos *et al.* 2013). Clinical complications are minimised with the suprachoroidal approach, as multiple layers of the eye do not have to be breached in order to position the electrode array (Ayton *et al.* 2014; Fujikado *et al.* 2011). A limitation of this approach is an increase in stimulus thresholds as a result of the greater distance between the electrode array and the retina compared with epi- and subretinal devices; however, experimental and clinical studies have demonstrated that the retina can be effectively stimulated at safe levels using this electrode position (Ayton *et al.* 2014; Fujikado *et al* .2011; Nayagam *et al.* 2014, Shivdasani *et al.* 2014). Recently, two groups have conducted successful clinical trials using this approach; both demonstrated that long-term severely blind patients can perceive discriminable visual percepts in response to electrical stimulation within safe stimulation limits (Ayton *et al.* 2014; Fujikado *et al.* 2011; Shivdasani *et al.* 2014).

7.3.4 Central visual prostheses

Retinal prostheses are not suitable for forms of blindness where the injury focus is central to the RGCs such as glaucoma, diabetic retinopathy, traumatic eye injury and central disorders. In these cases, electrode arrays that directly stimulate more central structures within the visual pathway have the potential to provide prosthetic vision (Figure 7.8). Although clinical trials of cortical based devices were undertaken early in the development of visual prostheses (Brindley and Lewin 1968; Dobelle *et al.* 1974) (Chapter 1), there is currently no commercial or regulatory approved central visual prosthesis.

Clinical trials have been performed using cuff electrodes on the optic nerve (Veraart *et al* 2004), via surface (Brindley and Lewin 1968; Dobelle *et al.* 1974) or penetrating microelectrodes in the visual cortex (Schmidt *et al* 1996). This research has recently been advanced by taking advantage of new technologies (Lewis *et al.* 2015; Normann *et al.* 2009; Troyk *et al.* 2003).

The primary visual cortex is thought to be an ideal location for a visual prosthesis for several reasons; it is well organised retinotopically, it has adequate space for multiple implanted components and is known to have a large area devoted to central vision. However, it is unclear whether a prosthesis that bypasses all visual processing occurring within the retina and the central visual pathway will contain sufficient information for higher-order brain centres to accurately recognise percepts generated from electrical stimulation. Clinical trials, proposed by a number of research groups, will determine the efficacy of this approach.

7.3.5 Perceptual effects of visual prostheses

Phosphenes are visual sensations produced by stimuli other than light, including electrical stimulation of the retina or brain (Marg 1991). The phosphenes evoked by electrical stimuli have been described as flashes of light, often amorphous but sometimes with a well-defined shape. Complex phosphenes, including shapes darker than the patient's naturally perceived background, have also been described (Keseru *et al.* 2012). The brightness, shape, size, duration and location of a phosphene can

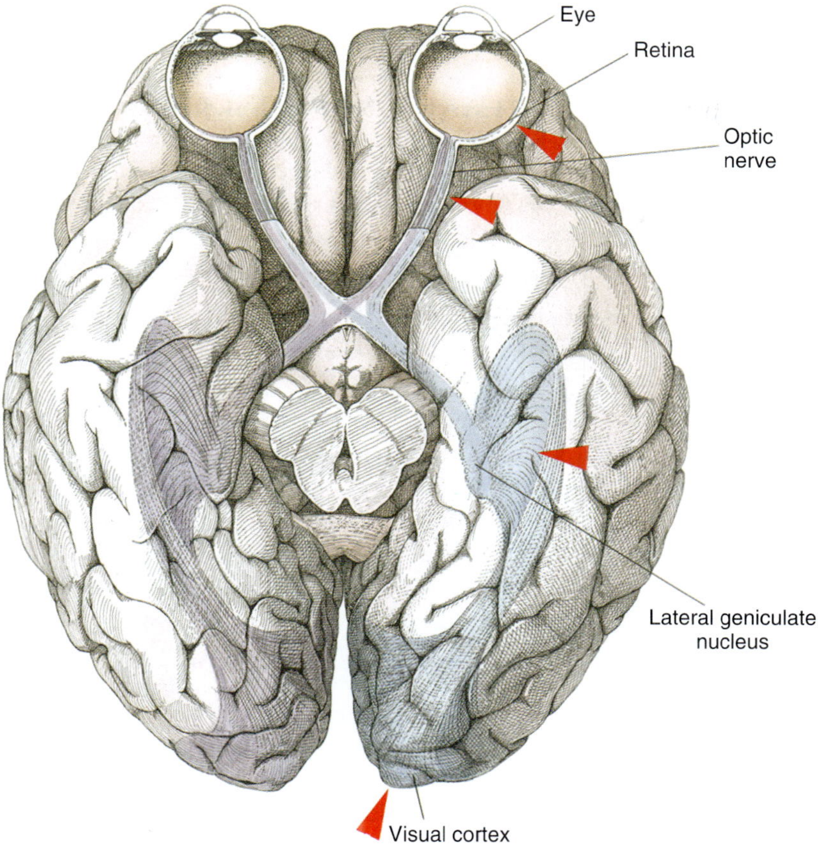

Figure 7.8 Overview of the visual pathway from the retina to the primary visual cortex in the human. Visual prostheses can potentially target several sites along this pathway (arrowhead) including the retina, the optic nerve, lateral geniculate nucleus and the visual cortex. The retina, optic nerve and visual cortex have been trialed clinically over the course of visual prosthesis development. The majority of these structures have a well organised topographic map of the retina (adapted from Hubel 1988, with permission).

be manipulated by varying the location and configuration of electrodes being stimulated, and the electrical waveforms used (Nanduri *et al* .2012; Perez Fornos *et al.* 2012; Wilke *et al.* 2011).

Despite the complexities of phosphenes, they form the basic building blocks of prosthetic vision. By stimulating appropriate electrodes on an array, it is possible to create retinotopically organised patterns and simple shapes such as lines of different orientations and geometric objects such as triangles and squares, regardless of the position of the electrode array (Ayton *et al.* 2014; Fujikado *et al.* 2011; Humayun *et al.* 2012; Wilke *et al.* 2011).

A video camera and an image processing algorithm are required to provide a visual representation of the subjects' surroundings (Figure 7.7). Subjects implanted with the Argus II epiretinal system and using an external camera to source their visual field can perform object localisation, motion discrimination and discrimination of oriented gratings with a best visual acuity to date of 20/1260 (Humayun *et al.* 2012). The Alpha-IMS subretinal device has demonstrated a best visual acuity of 20/546 (Stingl *et al.* 2013), allowing patients implanted with this device to be able to detect and localise light and motion, identify, localise and discriminate high contrast objects and read large font letters (Stingl *et al.* 2013). Finally, the prototype suprachoroidal retinal implant demonstrated a best visual acuity of 20/4451, allowing patients to detect and localise light and motion, detect objects and discriminate high contrast objects (Ayton *et al.* 2014).

7.3.6 Future directions

There remain considerable technical challenges that must be addressed before visual prostheses receive widespread clinical acceptance. These include effective electrode fixation and long-term mechanical and biological stability; a safe surgical technique that can be performed routinely by retinal or neurosurgeons; the maintenance of a healthy/undamaged electrode/neural interface; device miniaturization and hermetic sealing; MRI compatibility; low power consumption; safety associated with the simultaneous electrical stimulation of multiple electrodes; eye tracking technology to account for eye movements which significantly interfere with the locations of perceived phosphenes; and the use of sophisticated image processing techniques such as feature extraction. However, the single-most challenging issue will be associated with the optimal fitting and patient training using devices with a large number of electrodes (>40). Despite these challenges, visual prostheses are likely to become widely used in the clinical management of severely blind patients. Ultimately these devices may provide a level of prosthetic vision that will allow users to read and potentially recognise faces.

7.4 SENSORY PROSTHESES AND BRAIN PLASTICITY

It is clear from the evaluation of clinical performance of cochlear implant subjects that brain plasticity plays an important role in the clinical success. This is most clearly seen in the application of these devices in very young children; however, the great majority of elderly cochlear implant recipients also show ongoing improvements in clinical performance over time that can also be attributed to plasticity.

Importantly there are a number of factors that are known to contribute to clinical performance. A major negative determinant of clinical outcome is duration of deafness (Blamey *et al.* 1996), while auditory experience, including the duration of experience with a cochlear implant and hearing aid, is associated with improved outcomes (Blamey *et al.* 2013; Lazard *et al.* 2012). Both of these processes have been attributed to brain plasticity.

There are also a range of more subtle changes that occur within the auditory pathway as a result of deafness and subsequent reactivation with a cochlear implant.

Evidence of these changes comes from both clinical and animal studies and include changes in: cochleotopic organization (Fallon *et al.* 2009); temporal processing (Vollmer and Beitel 2011); binaural processing (Gordon *et al.* 2013; van Hoesel 2011); and the coupling between the primary auditory cortex and other auditory centres (Kral and Sharma 2012).

While the status of the visual cortex after long-term blindness is not as well understood (Weiland *et al.* 2011), it is clear from the existing clinical data that useful visual cues can be evoked via a visual prosthesis decades after the onset of blindness. Moreover, visual experience obtained through ongoing use of a device is expected to result in improved performance as a result of plasticity within the central visual pathway.

It is important to note that careful device fitting and ongoing training are key factors associated with the clinical success of these devices. In order to maximise the benefits of plasticity, training using everyday tasks – not just object recognition – is required (Dagnelie 2012). Optimal device fitting can be a time-consuming procedure that will also require ongoing monitoring; a task that, as noted above, is likely to become more problematic with high density electrode arrays.

7.5 CONCLUSIONS

In the 1970s, cochlear implants were branded as "an aid to lip reading". Over the subsequent 40 years, they have significantly exceeded these expectations and expanded the patient base from profoundly deaf adults to include both severely deaf adults and children and patients with usable levels of residual hearing. We can anticipate similar outcomes for visual prostheses over the following decades, as both the technology and clinical experience in managing patients using these devices become more sophisticated.

ACKNOWLEDGEMENTS

We acknowledge funding from the Australian Research Council, the National Health and Medical Research Council of Australia and the US National Institutes of Health. The Bionics Institute acknowledges the support it receives from the Victorian Government through its Operational Infrastructure Support Program. We thank Berenice Hale for administrative assistance.

REFERENCES

Adunka, O., Kiefer, J., Unkelbach, M.H., Lehnert, T. and Gstoettner, W. (2004) Development and evaluation of an improved cochlear implant electrode design for electric acoustic stimulation. *Laryngoscope,* **114**: 1237–1241.

Amanat, N., James, N.L. and McKenzie, D.R. (2010) Welding methods for joining thermoplastic polymers for the hermetic enclosure of medical devices. *Med. Eng. Phys.,* **32**: 690–699.

Ashmore, J. (2008) Cochlear outer hair cell motility. *Physiol Rev.,* **88**: 173–210.

Ayton, L.N., P.J. , Guymer, R.H., Luu, C.D., Nayagam, D.A. *et al.* (2014) First-in-Human trial of a novel suprachoroidal retinal prosthesis. *PLoS One,* **9**: e115239.

Bierer, J.A. and Middlebrooks, J.C. (2002) Auditory cortical images of cochlear-implant stimuli: dependence on electrode configuration. *J. Neurophysiol.,* **87**: 478–492.

Blamey, P., Arndt, P., Bergeron, F., Bredberg, G., Brimacombe, J. *et al.* (1996) Factors affecting auditory performance of postlinguistically deaf adults using cochlear implants. *Audiol. Neurootol.,* **1**: 293–306.

Blamey, P., Artieres, F., Baskent, D., Bergeron, F., Beynon, A. *et al.* (2013) Factors affecting auditory performance of postlinguistically deaf adults using cochlear implants: an update with 2251 patients. *Audiol, Neurotol.,* **18**: 36–47.

Bond, M., Mealing, S., Anderson, R., Elston, J., Weiner, G. *et al.* (2009) The effectiveness and cost-effectiveness of cochlear implants for severe to profound deafness in children and adults: a systematic review and economic model. *Health Technol., Assess.* **13**: 1–330.

Briggs, R.J., Eder, H.C., Seligman, P.M., Cowan, R.S., Plant, K.L. *et al.* (2008) Initial clinical experience with a totally implantable cochlear implant research device. *Otol. Neurotol.,* **29**: 114–119.

Brindley, G.S. and Lewin, W.S. (1968) The sensations produced by electrical stimulation of the visual cortex. *J. Physiol.,* **196**: 479–493.

Chow, A.Y., Pardue, M.T., Perlman, J.I., Ball, S.L., Chow, V.Y. *et al.* (2002) Subretinal implantation of semiconductor-based photodiodes: durability of novel implant designs. *J. Rehabil. Res. Dev.,* **39**: 313–321.

Clark, G.M. (2003) *Cochlear Implants: Fundamentals and Applications.* New York: Springer-Verlag, 830 pp.

Colletti, V., Shannon, R.V., Carner, M., Veronese, S. and Colletti L. (2009) Progress in restoration of hearing with the auditory brainstem implant. *Prog. Brain Res.* **175**: 333–345.

Dagnelie, G. (2012) Retinal implants: emergence of a multidisciplinary field. *Curr. Opin. Neurol.,* **25**: 67–75.

Dettman, S.J., Pinder, D., Briggs, R.J., Dowell, R.C. and Leigh, J.R. (2007) Communication development in children who receive the cochlear implant younger than 12 months: risks versus benefits. *Ear Hear.,* **28**: 11S–18S.

Dobelle, W.H., Mladejovsky, M.G. and Girvin, J.P. (1974) Artifical vision for the blind: electrical stimulation of visual cortex offers hope for a functional prosthesis. *Science,* **183**: 440–444.

Dobelle, W.H., Stensaas, S.S., Mladejovsky, M.G. and Smith, J.B. (1973) A prosthesis for the deaf based on cortical stimulation. *Ann. Otol. Rhinol. Laryngol.,* **82**: 445–463.

Fallon, J.B., Irvine, D.R. and Shepherd, R.K. (2009) Cochlear implant use following neonatal deafness influences the cochleotopic organization of the primary auditory cortex in cats. *J. Comp. Neurol.,* **512**: 101–114.

Fernandes, R.A., Diniz, B., Ribeiro, R. and Humayun, M. (2012) Artificial vision through neuronal stimulation. *Neurosci. Lett.,* **519**: 122–128.

Finger, R.P., Fimmers, R., Holz, F.G. and Scholl, H.P. (2011) Incidence of blindness and severe visual impairment in Germany: projections for 2030. *Invest. Ophthalmol. Vis. Sci.,* **52**: 4381–4389.

Fujikado, T., Kamei, M., Sakaguchi, H., Kanda, H., Morimoto, T. *et al.* (2011) Testing of semichronically implanted retinal prosthesis by suprachoroidal-transretinal stimulation in patients with retinitis pigmentosa. *Invest. Ophthalmol. Vis. Sci.,* **52**: 4726–4733.

Gantz, B.J. and Turner, C.W. (2003) Combining acoustic and electrical hearing. *Laryngoscope,* **113**: 1726–1730.

George, S.S., Wise, A.K., Shivdasani, M.N., Shepherd, R.K. and Fallon, J.B. (2014) Evaluation of focused multipolar stimulation for cochlear implants in acutely deafened cats. *J.Neural. Eng.,* **11**: 065003.

Gordon, K.A., Salloum, C., Toor, G.S., van Hoesel, R. and Papsin, B.C. (2012) Binaural Interactions develop in the auditory brainstem of children who are deaf: effects of place and level of bilateral electrical stimulation. *J. Neurosci.,* **32**: 4212–4223.

Gordon, K.A., Wong, D.D. and Papsin, B.C. (2013) Bilateral input protects the cortex from unilaterally-driven reorganization in children who are deaf. *Brain,* **136**: 1609–1625.

Hadjinicolaou, A.E., Leung, R.T., Garrett, D.J., Ganesan, K., Fox, K. *et al.* (2012) Electrical stimulation of retinal ganglion cells with diamond and the development of an all diamond retinal prosthesis. *Biomaterials,* **33**: 5812–5820.

Hardie, N.A. and Shepherd, R.K. (1999) Sensorineural hearing loss during development: morphological and physiological response of the cochlea and auditory brainstem. *Hear Res.,* **128**: 147–165.

Hartong, D.T., Berson, E.L. and Dryja, T.P. (2006) Retinitis pigmentosa. *Lancet,* **368**: 1795–1809.

Holden, L.K., Finley, C.C., Firszt, J.B., Holden, T.A. and Brenner, C. *et al.* (2013) Factors affecting open-set word recognition in adults with cochlear implants. *Ear Hear.,* **34**: 342–360.

Hubel, D.H. (1988) *Eye, Brain and Vision.* Scientific American Library.

Humayun, M.S., Dorn, J.D., da Cruz, L., Dagnelie, G., Sahel, J.A. *et al.* (2012) Interim results from the international trial of Second Sight's visual prosthesis. *Ophthalmology,* **119**: 779–788.

Jones, B.W., Kondo, M., Terasaki, H., Lin, Y., McCall, M. and Marc. R.E. (2012) Retinal remodeling. *Jpn. J. Ophthalmol.,* **56**: 289–306.

Kan, A. and Litovsky, R.Y. (2014) Binaural hearing with electrical stimulation. *Hear Res.,* **322**: 127–137.

Keseru, M., Feucht, M., Bornfeld, N., Laube, T., Walter, P. *et al.* (2012). Acute electrical stimulation of the human retina with an epiretinal electrode array. *Acta Ophthalmol.,* **90**: e1–8.

Kimura, R.S., Ota, C.Y. and Takahashi, T. (1979) Nerve fiber synapses on spiral ganglion cells in the human cochlea. *Ann. Otol. Rhinol. Laryngol.* – Suppl. **88**: 1–17.

Klauke, S., Goertz, M., Rein, S., Hoel, D., Thomas, U. *et al.* (2011) Stimulation with a wireless intraocular epiretinal implant elicits visual percepts in blind human. *Invest. Ophthalmol. Vis. Sci.,* **52**: 449–455.

Kral, A. and Sharma, A. (2012) Developmental neuroplasticity after cochlear implantation. *Trends Neurosci.,* **35**: 111–122.

Lai, K. and Landa, G. (2015) Current choice of treatments for neovascular AMD. *Expert Rev. Clin. Pharmacol.,* **8**: 135–140.

Lazard, D.S., Vincent, C., Venail, F., Van de Heyning, P., Truy, E. *et al.* (2012) Pre-, per- and postoperative factors affecting performance of postlinguistically deaf adults using cochlear implants: a new conceptual model over time. *PLoS One,* **7**: e48739.

Leake, P.A. and Hradek, G.T. (1988) Cochlear pathology of long-term neomycin induced deafness in cats. *Hear Res.,* **33**: 11–33.

Lee, K.Y., van Hasselt, C.A., Chiu, S.N. and Cheung, D.M. (2002) Cantonese tone perception ability of cochlear implant children in comparison with normal-hearing children. *Int. J. Pediatr. Otorhinolaryngol.,* **63**: 137–147.

Lewis, P.M., Ackland, H.M., Lowery, A.J. and Rosenfeld, J.V. (2015) Restoration of vision in blind individuals using bionic devices: a review with a focus on cortical visual prostheses. *Brain Res.,* **1595C**: 51–73.

Lim, H.H., Lenarz, T., Anderson, D.J. and Lenarz, M. (2008) The auditory midbrain implant: effects of electrode location. *Hear Res.,* **242**: 74–85.

Lorens, A., Polak, M., Piotrowska, A and Skarzynski, H. (2008) Outcomes of treatment of partial deafness with cochlear implantation: a DUET study. *Laryngoscope,* **118**: 288–294.

Marg, E. (1991) Magnetostimulation of vision: direct noninvasive stimulation of the retina and the visual brain. *Optom. Vis. Sci.,* **68**: 427–440.

Marozeau, J, McDermott, H.J., Swanson, B.A. and McKay, C.M. (2015) Perceptual interactions between electrodes using focused and monopolar cochlear stimulation. *J. Assoc. Res. Otolaryngol.,* **16**: 401–412.

Mathieson, K., Loudin, J., Goetz, G., Huie, P., Wang, L. *et al.* (2012) Photovoltaic retinal prosthesis with high pixel density. *Nat. Photonics,* **6**: 391–397.

McCreery, D.B. (2008) Cochlear nucleus auditory prostheses. *Hear Res.,* **242**: 64–73.

McDermott, H.J. (2004) Music perception with cochlear implants: a review. *Trends Amplif.,* **8**: 49–82.

McDermott, H.J., McKay, C.M. and Vandali, A.E. (1992) A new portable sound processor for the University of Melbourne/Nucleus Limited multi-electrode cochlear implant. *J. Acoust. Soc. Am.,* **91**: 3367–3371.

McKay, C.M., Lim, H.H. and Lenarz, T. (2013) Temporal processing in the auditory system: insights from cochlear and auditory midbrain implantees. *J. Assoc. Res. Otolaryngol.,* **14**: 103–124.

Nadol, J.B.Jr. (1990) Degeneration of cochlear neurons as seen in the spiral ganglion of man. *Hear Res.,* **49**: 141–154.

Nanduri, D., Fine, I., Horsager, A., Boynton, G.M., Humayun, M.S. *et al.* (2012) Frequency and amplitude modulation have different effects on the percepts elicited by retinal stimulation. *Invest. Ophthalmol. Vis. Sci.,* **53**: 205–214.

Nayagam, D.A., Williams, R.A., Allen, P.J., Shivdasani, M.N., Luu, C.D. *et al.* (2014) Chronic electrical stimulation with a suprachoroidal retinal prosthesis: a preclinical safety and efficacy study. *PLoS One,* **9**: e97182.

Normann, R.A., Greger, B., House, P., Romero, S.F., Pelayo, F. and Fernandez, E. (2009) Toward the development of a cortically based visual neuroprosthesis. *J. Neural. Eng.,* **6**: 035001.

Otto, S.R., Brackmann, D.E., Hitselberger, W.E., Shannon, R.V. and Kuchta, J. (2002) Multichannel auditory brainstem implant: update on performance in 61 patients. *J. Neurosurg.* **96**: 1063–1071.

Patrick, J.F., Seligman, P.M., Money, D.K. and Kuzma, J.A. (1990) Engineering, in *Cochlear Prostheses*, Clark, G.M., Tong, Y.C. and Patrick, J.F. (eds), Churchill Livingston, Edinburgh, pp 99–124.

Penfield, W and Perot, P. (1963) The brain's record of auditory and visual experience. *Brain,* **86**: 595–696.

Perez Fornos, A., Sommerhalder, J., da Cruz, L., Sahel, J.A., Mohand-Said, S. *et al.* (2012) Temporal properties of visual perception on electrical stimulation of the retina. *Invest. Ophthalmol. Vis. Sci.,* **53**: 2720–2731.

Peters, B.R., Wyss, J. and Manrique, M. (2010) Worldwide trends in bilateral cochlear implantation. *Laryngoscope,* **120**, Suppl 2: S17–S44.

Purves, P., Augustine, G.J., Fitzpatrick, D., Katz, L.C., LaMantia, A-S. *et al.* (2001) *Neuroscience.*, USA Sinauer, Sunderland MA.

Rizzo, J.F. 3rd. (2011) Update on retinal prosthetic research: the Boston Retinal Implant Project. *J. Neuroophthalmol.,* **31**: 160–168.

Rizzo, J.F. 3rd, Shire, D.B., Kelly, S.K., Troyk, P., Gingerich, M. *et al.* (2011) Development of the boston retinal prosthesis. *Conf. Proc. IEEE Eng. Med. Biol. Soc.*, pp 3135–3138.

Saoji, A.A., Landsberger, D.M., Padilla, M. and Litvak, L.M. (2013) Masking patterns for monopolar and phantom electrode stimulation in cochlear implants. *Hear Res.,* **298**: 109–116.

Schmidt, E.M., Bak, M.J., Hambrecht, F.T., Kufta, C.V., O'Rourke, D.K. and Vallabhanath, P. (1996) Feasibility of a visual prosthesis for the blind based on intracortical microstimulation of the visual cortex. *Brain,* **119**: 507–522.

Schorr, E.A., Fox, N.A., van Wassenhove, V. and Knudsen, E.I. (2005) Auditory-visual fusion in speech perception in children with cochlear implants. *Proc. Natl. Acad. Sci. USA,* **102**: 18748–18750.

Shannon, R.V. (1983) Multichannel electrical stimulation of the auditory nerve in man. II: Channel interaction. *Hear Res* **12**: 1–16.

Shannon, R.V. (2012) Advances in auditory prostheses. *Curr. Opin. Neurol.,* **25**: 61–66.

Shannon, R.V., Fayad, J., Moore, J., Lo, W.W., Otto, S. *et al.* (1993) Auditory brainstem implant: II: Postsurgical issues and performance. *Otolaryngol. Head. Neck. Surg.,* **108**: 634–642.

Shepherd, R.K. and Javel, E. (1997) Electrical stimulation of the auditory nerve. I: Correlation of physiological responses with cochlear status. *Hear Res.,* **108**: 112–144.

Shepherd, R.K., Shivdasani, M.N., Nayagam, D.A., Williams, C.E. and Blamey, P.J. (2013a) Visual prostheses for the blind. *Trends Biotechnol.*, **31**: 562–571

Shepherd, R.K., Wise, A.K. and Fallon, J. (2013b) Cochlear implants, in *Disorders of Peripheral and Central Auditory Processing*. Celesia, G.G. (ed.), Elsevier, New York, pp 315–331.

Shepherd, R.K., Fallon, J.B. and McDermott, H. (2014) Medical bionics, in *Comprehensive Biomedical Physics*. A. Brahme (ed.), Elsevier, Amsterdam, pp 327–341.

Shivdasani, M.N., Sinclair, N.C., Dimitrov, P.N., Varsamidis, M., Ayton, L.N. *et al.* (2014) Factors affecting perceptual thresholds in a suprachoroidal retinal prosthesis. *Invest. Ophthal. Vis. Sci.,* **55**: 6467–6481.

Simpson, A., McDermott, H.J, Dowell, R.C., Sucher, C. and Briggs, R.J. (2009) Comparison of two frequency-to-electrode maps for acoustic-electric stimulation. *Int. J. Audiol.,* **48**: 63–73.

Skinner, M.W., Clark, G.M., Whitford, L.A., Seligman, P.M., Staller, S.J. *et al.* (1994) Evaluation of a new spectral peak coding strategy for the Nucleus 22 Channel Cochlear Implant System. *Am. J..Otol.,* **15**(S–uppl 2): 15–27.

Smith, Z.M., Parkinson, W.S. and Long, C.J. (2013) Multipolar current focusing increases spectral resolution in cochlear implants. *IEEE EMBS.*, pp 2796–2799.

Spoendlin, H. (1984) Primary neurons and synapses, in *Ultrastructural Atlas of the Inner Ear*. Butterworths, London, pp 133–164.

Spoendlin, H. and Schrott, A. (1989) Analysis of the human auditory nerve. *Hear Res.,* **43**: 25–38.

Stingl, K., Bartz-Schmidt, K.U., Besch, D., Braun, A., Bruckmann, A. *et al.* (2013) Artificial vision with wirelessly powered subretinal electronic implant alpha-IMS. *Proc. Biol. Sci.*, **280**: 20130077.

Sucher, C.M. and McDermott, H.J. (2007) Pitch ranking of complex tones by normally hearing subjects and cochlear implant users. *Hear Res.*, **230**: 80–87.

Troyk, P., Bak, M., Berg, J., Bradley, D., Cogan, S. *et al.* (2003) A model for intracortical visual prosthesis research. *Artif. Organs,* **27**: 1005–1015.

Turner, C., Gantz, B.J. and Reiss, L. (2008) Integration of acoustic and electrical hearing. *J. Rehab. Res. Dev.,* **45**: 769–778.

van den Honert, C. and Kelsall, D.C. (2007) Focused intracochlear electric stimulation with phased array channels. *J. Acoust. Soc. Am.*, **121**: 3703–3716.

van den Honert, C. and Stypulkowski, P.H. (1987) Single fiber mapping of spatial excitation patterns in the electrically stimulated auditory nerve. *Hear Res.,* **29**: 195–206.

van Hoesel, R. (2011) Bilateral cochlear implants In *Auditory prostheses: New horizons*, Zeng, F.G., Popper, A.N. and Fay, R.R. (eds), Springer, New York, pp 13–57.

Veraart, C., Duret, F., Brelen, M. and Delbeke, J. (2004) Vision rehabilitation with the optic nerve visual prosthesis. *Conf. Proc. IEEE Eng. Med. Biol. Soc.,* **6**: 4163–4164.

Villalobos, J., Nayagam, D.A., Allen, P.J., McKelvie, P., Luu, C.D. *et al.* (2013) A wide-field suprachoroidal retinal prosthesis is stable and well tolerated following chronic implantation. *Invest. Ohthal. Vis. Sci.,* **54**: 3751–3762.

Vollmer, M. and Beitel, R.E. (2011) Behavioral training restores temporal processing in auditory cortex of long-deaf cats. *J. Neurophysiol.*, **106**: 2423–2436.

Weiland, J.D., Cho, A.K. and Humayun, M.S. (2011) Retinal prostheses: current clinical results and future needs. *Ophthalmology,* **118**: 2227–2237.

WHO (2012). Global estimates on prevalence of hearing loss. http://www.who.int/pbd/deafness/WHO_GE_HL.pdf

Wilke, R., Gabel, V.P., Sachs, H., Bartz Schmidt, K.U., Gekeler, F. *et al.* (2011) Spatial resolution and perception of patterns mediated by a subretinal 16-electrode array in patients blinded by hereditary retinal dystrophies. *Invest. Ophthalmol. Vis. Sci.,* **52**: 5995–6003.

Wilson, B.S, Finley, C.C., Lawson, D.T., Wolford, R.D., Eddington, D.K. and Rabinowitz, W.M. (1991) Better speech recognition with cochlear implants. *Nature,* **352**: 236–238.

Zrenner, E., Bartz-Schmidt, K.U., Benav, H., Besch, D., Bruckmann, A. *et al.* (2011) Subretinal electronic chips allow blind patients to read letters and combine them to words. *Proc. Biol. Sci.,* **278**: 1489–1497.

8

NEUROBIONICS: TREATMENTS FOR DISORDERS OF THE CENTRAL NERVOUS SYSTEM

HUGH McDERMOTT
Bionics Institute and Medical Bionics Department, University of Melbourne, East Melbourne, Victoria, Australia

8.1 INTRODUCTION

The application of electric stimulation to treat disorders of the central nervous system (CNS) has expanded remarkably over the past several decades. Such disorders include neurological and psychiatric conditions. At present, various devices are in use to alleviate disability and ill health associated with epilepsy, Parkinson's disease (PD) and other movement disorders, obsessive-compulsive disorder (OCD), depression, and a range of related conditions. In addition, application of CNS stimulation to relieve severe, chronic pain is currently increasing, and there are numerous further medical indications for which therapeutic electric stimulation is being trialed.

This chapter focuses particularly on selected conditions that have been shown to respond beneficially to electric stimulation delivered via electrodes implanted in or on the brain. In some studies, alternative types of device which deliver electric or magnetic stimulation to the CNS from sources that are entirely external to the body have been used to treat certain neurological and psychiatric disorders. Examples of such devices are transcranial direct-current stimulators and magnetic stimulators (Hoy and Fitzgerald 2010; Lefaucheur 2009; Nitsche *et al.* 2009). However, the scope of the overview presented below is limited to therapeutic systems that have at least some surgically implanted components. The conditions treated with electric stimulation using these systems are discussed under four main headings: psychiatric conditions,

Neurobionics: The Biomedical Engineering of Neural Prostheses, First Edition.
Edited by Robert K. Shepherd.

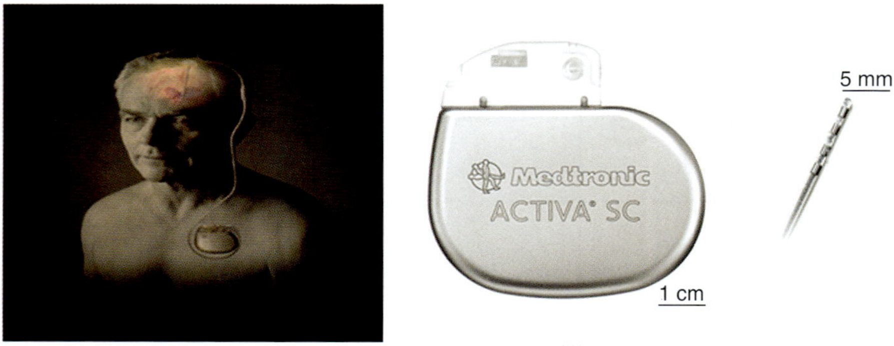

(A) (B) (C)

Figure 8.1 A typical configuration of a DBS system: (A) sketch showing the main implanted components; (B) implantable pulse generator; (C) electrode array with four stimulation contacts (reprinted with the permission of Medtronic, Inc. © 2014).

movement disorders, epilepsy and pain. The chapter concludes with some brief comments about how electric stimulation provides benefit and what developments can be expected in this field in the near future.

A typical configuration of a Deep Brain Stimulation (DBS) system is illustrated in Figure 8.1. It includes multiple individual electrodes arranged in groups on several insulating carriers (often referred to as "leads"). For instance, one widely-used DBS electrode array comprises 4 discrete platinum rings, each with a length of 1.5 mm mounted on a 1.27 mm-diameter carrier at a spacing of 1.5 mm. Each ring is connected by a separate insulated wire to an implantable pulse generator (IPG). The IPG is usually located at a site remote from the DBS electrodes, such as below the clavicle or in the abdomen. Subcutaneous extension cables are used to connect the electrodes to the IPG. Most existing types of IPG can drive at least two DBS electrode arrays (i.e. 8 electrodes), and also provide a separate current-return electrode for monopolar stimulation. In addition, the IPG contains either a primary or a rechargeable battery, enabling the system to function autonomously without any external components. The electrical parameters of the stimulation are selected using an external, wireless programmer that allows clinicians to choose which electrodes to activate and the magnitude of the stimulus pulses as well as the rate at which pulses are delivered to the electrodes. In most devices it is possible to select either a bipolar electrode configuration (i.e. two closely spaced electrodes in the brain conduct the stimulation) or a monopolar configuration (i.e. one electrode in the brain conducts the stimulation using a remote electrode to return the current). Optimizing the selection of the active electrodes and the settings of stimulation parameters is an ongoing objective of DBS research (Kuncel *et al.* 2006).

The goal of DBS treatment is to alleviate or control severe symptoms that are refractory to other forms of therapy while avoiding intolerable side effects. Obtaining a satisfactory outcome requires, among other factors, that the electrodes are placed appropriately. Although the details of the surgical procedures vary among clinicians, in general the first step is to select the anatomical target or targets in the brain. As discussed below, the choice of target depends on the condition to be treated, although in many cases it is unclear which neuroanatomical region (if any) will provide optimal results for each patient when electrically stimulated. Once the target has been selected, images of the brain are obtained, typically using both computed tomography and magnetic resonance imaging. The images are aligned with each other and with predetermined locations on the patient's head using markers that are attached to the skull while images are acquired. Specialized computer software processes the images to provide coordinates in three dimensions to guide the insertion of the electrode arrays during surgery. Accuracy in setting the intended electrode trajectory is often achieved by use of a stereotactic frame that is attached to the patient's head, although frameless systems are sometimes used instead, as they may be less uncomfortable for the patient. With either method, the surgery involves drilling a burrhole through which the electrode array is inserted. Using the stereotactic system, a cannula is inserted to a location close to the anatomical target. A microelectrode is then inserted through the cannula to record electrical activity in the brain.

Such activity has characteristics that indicate the position of the microelectrode tip, assisting the surgeon to verify the exact location of the target. Subsequently the microelectrode is removed and the DBS array inserted, with an internal stylet maintaining its rigidity during the insertion procedure. Once the tip of the array is believed to be at the desired location, an external stimulator may be connected temporarily to verify that electrical stimulation can relieve symptoms without inducing unacceptable side effects. In general, the patient will have received only local anesthesia and mild sedation during the procedure, and is therefore able to report directly on the subjective effects of stimulation to clinicians. Finally, the stylet is removed from the electrode lead, and a cap is placed in the burrhole to fix the electrode lead securely in position. The IPG may then be implanted, usually in a surgically created pocket inferior to the clavicle, and connected to the electrodes via extension leads that are tunneled subcutaneously along the neck. In some clinics, implantation of the IPG is delayed for up to two weeks after the DBS electrode insertion procedure. During that time, an external stimulator can be connected to the electrodes to ensure that DBS is effective in relieving the patient's symptoms. This helps to avoid the expense, inconvenience and possible complications associated with implantation of a complete DBS system in patients for whom the treatment is ineffective or causes intolerable side effects. Once the DBS system is permanently implanted, adjustments to the stimulation can be carried out at any time using an external programming unit. Unless complications arise, further surgery related to the DBS implant will be required only when the battery needs to be replaced.

8.2 PSYCHIATRIC CONDITIONS

8.2.1 Obsessive-compulsive disorder

OCD is a chronic anxiety disorder that affects approximately 2% of the general population (Denys *et al.* 2010). It often commences in adolescence or early adulthood.

OCD is characterized by recurrent obsessions (persistent thoughts or images that are experienced as intrusive and irrational) and compulsions (intentional behaviors performed in response to obsessions in a stereotyped and excessive way). A common obsession observed in people with OCD is fear of contamination, and the compulsions that may be associated with it include an excessive response to potential sources of contamination, such as vigorous and prolonged washing after shaking hands with another person. In severe cases, people with OCD may encounter extreme difficulty in carrying out everyday tasks in ordinary situations. For example, they may avoid any activities or circumstances that could possibly lead to contamination, and devote disproportionate time and effort to compulsive behaviors. Not only is the person with OCD effectively disabled in some instances, but also their family and other members of their social group can be greatly affected. Comorbidities such as depression are common, and in extreme cases suicide might be attempted. Existing treatments for OCD include medications and cognitive behavior therapy (Burdick *et al.* 2009; Greenberg *et al.* 2010; Patel and Simpson 2010; Read and Greenberg 2009). However, in at least 10% of cases, little or no improvement is observed, even after all available non-surgical treatments have been applied and evaluated.

Since the 1940s, surgical procedures such as cingulotomy (creating lesions in the tissue of the cingulate gyrus of the frontal lobe) have been carried out on selected patients with OCD when other therapies have failed (Benabid *et al.* 2009; Burdick *et al.* 2009). In part, because of the relatively good outcomes reported from several studies using ablative neurosurgical procedures, there has been an increasing trend in recent years towards application of DBS to treat a small proportion of patients with severe, intractable OCD. It is generally accepted that DBS may produce beneficial effects similar to lesions, but with the advantage of being more selective, controllable and reversible.

A number of different neuroanatomical regions have been targeted with DBS electrodes for treatment of OCD (Figure 8.2). These include the nucleus accumbens (NAc), ventral caudate/ventral striatum (VC/VS), the anterior limb in the internal capsule (ALIC) and the subthalamic nucleus (STN). In a relatively recent study (Denys *et al.* 2010), NAc stimulation was used in 16 patients and outcomes were assessed mainly using the standard Y-BOCS instrument (i.e. the Yale-Brown Obsessive Compulsive Scale). Patients who obtained at least a 35% decrease on the Y-BOCS measure when DBS was activated were classified as responders. On this basis, 9 of the 16 patients were responders. They had an average decrease in the Y-BOCS score of 72%. In the double-blind, sham-controlled phase of the study, there was a statistically significant difference in the mean Y-BOCS score between the conditions with stimulation enabled and disabled. No patients experienced a worsening of their OCD with active DBS, and side effects of the stimulation were reported as generally mild and well tolerated.

Comparable outcomes have been reported from studies in which other targets were stimulated. In one study involving 8 patients with DBS of the VC/VS region who were assessed over 3 years, 4 were classified as responders according to the above criterion, and a further 2 had decreases in Y-BOCS scores of 25–35% (Greenberg *et al.* 2006). A randomized, blinded study of 16 OCD patients with DBS of the STN found that

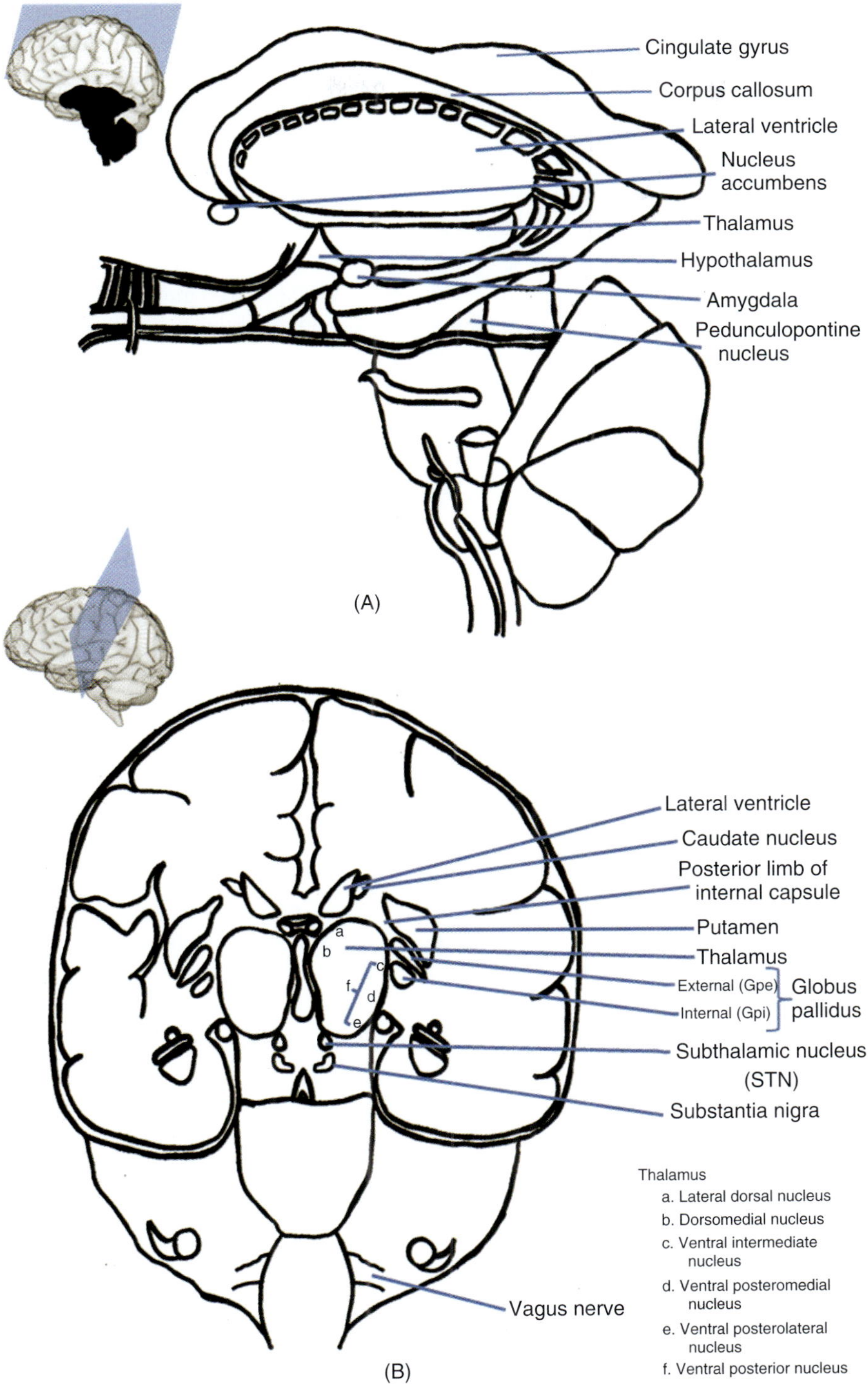

Figure 8.2 Sketch showing approximate positions of the principal anatomical targets used in DBS (A) part of sagittal section as indicated in the inset picture at upper left; (B) coronal section as indicated in the inset.

12 patients had at least a 25% reduction in their Y-BOCS score (Mallet *et al.* 2008). Reviews of additional studies are provided in Mian *et al.* (2010) and Lipsman *et al.* (2013b); in general, both the proportion of responders and the mean magnitudes of Y-BOCS score reductions are similar. Based on the limited evidence available so far, it is not clear whether a particular brain target for DBS should be preferred for treatment of OCD (de Koning *et al.* 2011). It is also unclear how to optimize DBS settings for each patient, and why some patients fail to gain benefit from the stimulation. Nevertheless, the overall findings suggest that DBS is an effective and relatively safe treatment for OCD in at least some cases where other therapies have proven unsatisfactory. The US Food and Drug Administration (FDA) approved the use of DBS as a treatment option for OCD via a Humanitarian Device Exemption in 2009.

8.2.2 Major depression

Major depressive episodes, particularly when recurrent, can lead to a chronic condition that results in a loss of occupational function, reduced ability to perform everyday activities, and impaired social relationships. In addition to suffering from depressed mood, people with major depression experience diminished interest or pleasure in many ordinary activities most of the time. Associated characteristics frequently include insomnia, weight changes, chronic fatigue, inability to concentrate, and feelings of worthlessness. In extreme cases, suicide may be attempted. Approximately 2–10% of adults have major depression, and about 20–35% of people diagnosed with recurrent major depression have a chronic condition that causes severe impairment. A range of treatment options are available for depression, including medications and psychological therapies, but a considerable proportion of people with major depression fail to obtain adequate benefit and therefore may be candidates for DBS.

The brain targets selected for DBS to treat major depression are comparable with those used to treat OCD (Blomstedt *et al.* 2011; Schlaepfer and Bewernick 2013). This is partly because some early interventions involving either lesioning or electrical stimulation in patients with OCD showed an antidepressive effect. DBS targets that have been studied include the anterior cingulate gyrus, the ALIC and the NAc. Although the numbers of subjects in most studies reported to date are small (up to ~20 patients per cohort), outcomes have been generally positive. Assessments are typically based on the Hamilton Rating Scale for Depression (Malone *et al.* 2009), on which responders can be defined as patients who experience a symptom reduction of at least 50%. Using this criterion, approximately half of the patients who have received DBS were responders, and the response rate appeared to be relatively stable over follow-up periods of up to 6 years. In common with other studies of DBS, a number of adverse side effects were also reported, including failure or migration of the implanted devices, infections and mood changes. Current developments involve selection and evaluation of alternative DBS targets to improve the response rate and reduce the incidence of side effects (Lujan *et al.* 2012).

8.3 MOVEMENT DISORDERS

8.3.1 Essential Tremor

Essential Tremor (ET) is the most common movement disorder, with a prevalence of at least 0.3% of the general population. Among people aged 65 years or more, the prevalence is over 4%, and it is frequently observed that tremor symptoms tend to worsen over time. Medications such as beta-adrenergic antagonists are often effective in reducing tremor, but in severe cases when drugs are inadequate the effects of tremor can be disabling. DBS received approval from the FDA as a treatment option for ET in 1997. In general, the same type of DBS system is used as described above for the treatment of OCD (Kuncel *et al.* 2006). The most common target for electrical stimulation to alleviate ET is the ventral intermediate nucleus of the thalamus. It is now widely accepted that DBS of the ventral intermediate nucleus is an effective treatment option for many patients who have severe ET that have not responded to other therapies. Several studies have reported average tremor reduction of 50–80%, with concomitant improvements in patients' quality of life (Nazzaro *et al.* 2013).

8.3.2 Parkinson's disease

PD is a common movement disorder affecting approximately 0.3% of the population. It is a neurodegenerative condition more prevalent among elderly people; for instance, about 1% of people aged over 60 years have PD. The signs and symptoms of PD include tremor, rigidity, akinesia and postural instability. The characteristics of the tremor observed in PD differ from those of ET. In particular, people with PD typically have a rest tremor which reduces with movement, whereas ET is characterized by a postural or kinetic tremor which becomes more problematic during fine motor tasks such as writing. As PD involves loss of dopaminergic neurons in the substantia nigra, the usual medical treatment includes drugs such as levodopa, which increase the level of dopamine in the brain. However, it is frequently observed that such medications lose efficacy over long periods of use. No currently available therapies prevent or reverse the progression of PD, and consequently the life expectancy of PD patients is lower than that of unaffected people.

DBS is an established treatment option for many people with PD (Foltynie and Hariz 2010). It is used mainly to treat motor symptoms. The FDA approved DBS therapy for use in PD in 2002, and by 2010 about 80 000 people worldwide had been implanted with a DBS system to alleviate movement disorders. A frequently-used neuroanatomical target for electric stimulation is the STN (Figure 8.3). Other targets include the globus pallidus interna (GPi) and occasionally the pedunculopontine nucleus (PPN). Considerable evidence has been published confirming that DBS improves motor symptoms leading to a better quality of life for PD patients. For example, a review of numerous studies of STN stimulation (Kleiner-Fisman *et al.* 2006; Deuschl *et al.* 2013) reported that mean scores on the Unified Parkinson's Disease Rating Scale decreased significantly for DBS in comparison with the preoperative condition. In addition, dyskinesia was reduced on average by 69.1%, and quality

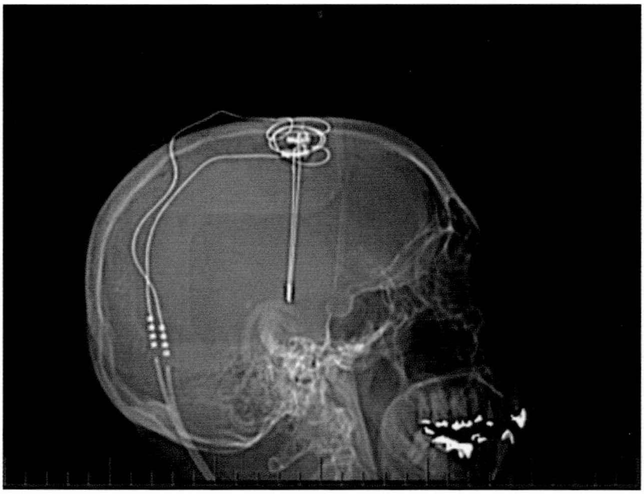

Figure 8.3 An image from a computed tomography scan showing DBS electrode arrays (center), leadwires, and connectors attached to extension cables (left).

of life as evaluated using a standard instrument improved substantially. However, side effects and complications were found to be relatively frequent; for example, intracranial hemorrhage occurred during surgery in 3.9% of patients. Similar outcomes have been reported for DBS of the GPi (Collins *et al.* 2010), although one study suggested that STN stimulation may be preferable to GPi stimulation, for example because it might enable a greater reduction in anti-Parkinsonian medication for patients (Anderson *et al.* 2005).

8.3.3 Dystonia

The characteristics of dystonia include abnormal body posture and movements, including involuntary sustained muscle contractions and tremor. Severe cases may involve pain, impaired motor skills, and injuries arising from repetitive movements. Dystonia can be greatly disabling and result in a substantial reduction in quality of life. As with other chronic movement disorders, most types of treatment are aimed at alleviating symptoms rather than directly affecting the underlying cause of the condition. DBS is one therapeutic option that has received increasing attention over the past 3–4 decades (Collins *et al.* 2010). Most reported studies have applied DBS to the GPi, with stimulation parameter settings usually selected individually for each patient (Vidailhet *et al.* 2013). The largest studies to date have involved mainly patients with primary dystonia, and outcomes have typically been assessed using standard rating scales. Although considerable variability among patients has been reported, the findings based on the rating scales show, in general, that DBS results in average improvements of about 50%. There is some evidence that positive outcomes are more likely to occur in younger patients and when the duration of the symptoms has been relatively short. In addition to alleviation of movement and posture abnormalities, significant reduction in disability and improvements in quality of life have been observed in many patients.

8.3.4 Tourette's syndrome

Tourette's syndrome is a condition that typically has an onset in childhood or early adolescence, and is characterized by frequently occurring tics that involve both movements and vocal productions. It affects about 0.05% of the population, and is more common in males than females. The frequency, location and severity of the tics generally change over time, but they most often include repetitive movements of the head or limbs combined with stereotyped utterances such as grunts, coughs and occasionally inappropriate words. There is evidence that Tourette's syndrome is related to OCD, and in some people, a diagnosis includes both disorders (Sachdev *et al.* 2012). In many instances, Tourette's syndrome is a lifelong condition that can be disabling, particularly when the tics result in physical injury or social isolation. For a small proportion of patients, DBS may be an effective treatment (Hariz and Robertson 2010). Although the number of Tourette's syndrome patients who have received a DBS implant so far is small, the reported results have been generally positive. The targets selected have mostly been similar to those used in DBS for dystonia or OCD. For example, across several studies evaluating DBS of the GPi in a total of 9 patients, the degree of tic reduction ranged from 34 to 88% (Visser-Vandewalle and Kuhn 2013). Given the limited amount of clinical experience with DBS for Tourette's syndrome, more research is required to identify the most effective brain targets, minimize the risks associated with the treatment, and assist in selecting which patients are most likely to benefit.

8.4 EPILEPSY

Epilepsy is a relatively common condition characterized by unpredictable seizures that may occur frequently. In severe cases, people with epilepsy encounter difficulties with many everyday tasks, and their quality of life is often greatly impaired. At least 3% of the general population experience a seizure within their lifetime, while about 1% have recurring seizures that require medical investigation and intervention. When epilepsy is diagnosed, the usual treatment involves one or a combination of specific pharmaceutical agents. Although these drugs are effective in controlling seizures for approximately 70% of patients, the remaining 30% obtain little benefit. In some cases, surgery may be carried out to resect a restricted volume of brain tissue. However, such an approach is feasible for only a limited number of patients. Even when resection is practical there may be serious side effects including impairment of brain function, and it is not possible to predict with certainty for each patient whether the surgery will completely abolish the seizures. Consequently, there has been considerable ongoing interest in devices which either suppress seizures by applying electric stimuli to the brain, or monitor electrical brain activity in order to predict when seizures are likely to occur (Fisher 2012). The latter type of device may be used either by itself to warn patients or carers of an impending seizure, or as part of a closed-loop system that automatically delivers stimuli or drugs to terminate seizures before they cause debilitating symptoms.

Vagus Nerve Stimulation (VNS) has been applied to treat epilepsy for many years; it was approved by the FDA as an adjunctive therapy for partial-onset epilepsy in 1997. In particular, the US company Cyberonics has developed a series of VNS systems that have been implanted in over 65 000 patients with epilepsy worldwide. The implant typically includes two helical electrodes that are placed around the vagus nerve in the neck and connected via leads to a stimulator which is usually implanted subcutaneously in the upper chest. An external programmer enables clinicians to set the parameters of stimulation, and patients are also able to use a magnetic device to enable or disable the stimulator without using the programmer. The usual programmed settings of the device provide a cyclic pattern of stimulation such that pulses are delivered to the electrodes for brief periods of time separated by longer periods of inactivity. Reports of the efficacy of VNS suggest that the number of patients experiencing a reduction in frequency of seizures tends to increase over time since device activation. For example, in one major study (DeGiorgio *et al.* 2000), it was found that out of 195 patients, 35% had at least a 50% reduction in seizures at 12 months. Side effects of VNS include disordered breathing during sleep, and laryngeal problems, such as hoarseness and sore throat. Recent versions of the device vary the stimulation parameters in accordance with a cardiac-based seizure detection algorithm, but it is not yet clear whether this technique improves the therapeutic efficacy of VNS.

DBS has also been applied as a therapy for epilepsy, particularly in the clinical trial known as SANTE (Stimulation of the Anterior Nucleus of the Thalamus in Epilepsy). The device being used in this trial, manufactured by Medtronic, is very similar to the DBS systems described earlier. Although the trial was not completed until 2015, preliminary results were promising. After two years of participation in the trial, about half of the 110 patients had a reduction in seizures of at least 50%, and 14 patients were seizure-free for at least 6 months (Fisher *et al.* 2010). The median reduction in seizure frequency was 56%, and the rate of complications or adverse events was described as modest.

Another relatively recent development is the NeuroPace RNS System, which delivers electric stimulation in response to abnormal electrical activity detected in specific brain regions (Morrell 2011). The RNS System includes a number of discrete electrodes which may be implanted either on the surface of the brain or close to deep targets, and which may be used either to monitor electrical activity or to provide local electrical stimulation. The electrodes are connected to a cranially implanted electronics package which functions autonomously once programming has been completed to suit the particular characteristics of the patient's epilepsy. The system is designed to detect epileptiform brain activity, and in response to such abnormalities, to deliver brief stimuli in an attempt to terminate seizures or prevent them from occurring. A report of outcomes from a multicentre trial of the RNS System with 191 patients showed a significant reduction in seizure occurrence of about 50% at the end of a 2-year follow-up period (Heck *et al.* 2014). However, there appears to be no published evidence at this time supporting the presumed benefits of "responsive" stimulation (i.e. stimulation controlled by detection of abnormal electrical activity in the brain) relative to open-loop stimulation. It is noteworthy that the overall reduction of seizure frequency reported from the recent RNS System clinical trial is similar to that of the SANTE trial, in which only open-loop stimulation was applied.

8.5 PAIN

The treatment of chronic, severe pain by electrical stimulation has a very long history (Fodstad and Hariz 2007). Numerous targets have been selected for stimulation, particularly the spinal cord and various peripheral nerves as well as diverse brain locations. The latter include periaqueductal and periventricular gray regions, thalamic nuclei and anterior cingulate cortex. Similarly, the types and origins of pain that have been addressed with DBS therapy have varied widely. Pain etiologies have included stroke, amputation, malignancy, headache and injuries. Partly because of such variability across the published clinical studies, and the lack of high-quality randomized, controlled trials, it is difficult to extract reliable and consistent evidence concerning efficacy of DBS as a pain treatment. However, a criterion for efficacy similar to that mentioned above in relation to treatment of other disorders can also be applied in the case of pain alleviation. With this criterion, a positive response is defined as a reduction of at least 50% on a subjective pain-rating scale when comparing between conditions with stimulation enabled and disabled. Further criteria for success include improved outcomes from instruments assessing quality of life, and reduction or cessation of pain-relieving medications. A relatively large study of DBS for pain alleviation included 178 patients (Pereira *et al.* 2013), all of whom completed a brief initial trial to assess stimulation efficacy. Subsequently, 89 patients were implanted permanently with a neurostimulator; the remaining patients failed to obtain adequate pain relief during the trial. Based on the above criteria, 62% of patients implanted with the permanent DBS system obtained levels of pain alleviation that were deemed successful. Although these findings appear typical of the published reports on this topic, further research is needed to clarify the potential benefits and side effects of DBS for pain relief. A thorough discussion of the complex methodological issues related to this topic is provided in a recent review publication (Coffey *et al.* 2011).

8.6 FUTURE DIRECTIONS

The generally promising results summarized briefly above showing the benefits of electrical stimulation to treat a range of disorders of the CNS imply that in the near future there will be a substantial increase in the number of patients receiving implants. Furthermore, different neuroanatomical structures may be targeted, and the range of clinical indications will most likely be expanded. Already, DBS has been used in trials to treat conditions such as drug addiction, anorexia nervosa and cognitive disorders (Fodstad and Hariz 2007; Kuhn *et al.* 2013; Laxton *et al.* 2013; Lipsman *et al.* 2011, 2013a). Although patients who are currently receiving DBS systems and other invasive devices have typically failed all other available forms of therapy, it is probable that electric stimulation will become more widely accepted and applied. In future, DBS is likely to be used more commonly as a conventional treatment alongside drugs and other types of therapy. However, considerable further research and technical developments are required to improve the clinical uptake and success rates of DBS treatment. Some of the issues needing particular attention at present are noted below.

Fitting of DBS systems to individual patients is typically a time-consuming and imprecise process that may often result in less than optimal outcomes (Kuncel and Grill 2004). Assessment of the benefits and side effects of the stimulation is usually based mainly on subjective observations by clinicians and patient reports. Improved programming could be achieved by greater use of objective measures, such as monitoring of tremor and rigidity in patients with PD by means of sensors attached to the patient's limbs. Suitable sensors can be used either temporarily during a programming session or for longer periods, so that clinicians can observe any changes in objective signs that may have a time course of hours, days or weeks. Analyses of such measures can provide information to enable optimization of the DBS parameter settings, either directly by the clinician or autonomously (i.e. through "closed-loop" stimulation, discussed below). The range of stimulus parameters that are available for adjustment is too large for efficient optimization using procedures that are currently conventional. Selectable parameters include the mode of stimulation (constant-voltage or constant-current), the electrode configuration (monopolar or bipolar, etc), the level (voltage or current), the pulse width and the pulse rate. Many of these parameters have a large number of possible settings. As DBS stimulators become more sophisticated, the parameter space will expand even further; for example, leads with many more than four discrete electrodes available for stimulation are likely to become commonplace in the near future.

Closed-loop stimulation requires the implanted device to monitor one or more biomarkers associated with the condition being treated, and to adaptively adjust the stimulation parameters so that the biomarker signals indicate a reduction in the disease symptoms, or a reduction in undesirable stimulation side effects, or both. For example, electrical activity recorded in the beta frequency band from electrodes on the DBS leads has been used successfully to enable and disable stimulation of the STN in patients with PD (Little *et al.* 2013). The biomarker was obtained by monitoring local field potentials (LFPs) on selected electrodes. After appropriate signal processing to extract the most relevant components of the LFPs, stimulation was controlled by comparing the level of biomarker activity with a threshold. This relatively simple algorithm resulted in a 56% reduction in stimulation time relative to the continuous DBS, which otherwise would have been used with these patients. Furthermore, there were significant improvements in the patients' clinical signs. Closed-loop stimulation schemes such as this have potential not only to improve benefit for each patient, but also to increase battery lifetime, as the stimulation is delivered with better energy efficiency. Providing longer battery lifetime is important, both in terms of reducing the total cost of DBS treatment and providing greater convenience and comfort for patients. Similar stimulation schemes are feasible, in principle, for controlling DBS when applied to treat other disorders (Stanslaski *et al.* 2012). Although at present it may be easier to identify and monitor suitable biomarkers in movement disorders (Little and Brown 2012), there is potential to extend these techniques to a range of other conditions, possibly including some neuropsychiatric disorders.

One clear advantage of controlling stimulation with a closed-loop strategy is that changes in the clinical or physiological state of an individual patient can be monitored continuously, so that stimulation levels can be adjusted automatically in response to those changes. This is important for many conditions that may be treated using DBS, because such changes are common, frequent, and in many instances unpredictable (Schiff 2010). A further enhancement of this type of closed-loop scheme would be to adjust one or more of the stimulation parameters continuously rather than in discrete steps (e.g. by simply switching the pulse generator on and off, as in the above example). Continuous and automatic adjustment of multiple parameters may reduce the present need for considerable time to be spent with each patient postoperatively to program the DBS device, and would likely lead to better clinical outcomes with minimal side effects. However, much additional research is required to devise stimulation strategies of this type and to ensure that they are safe, efficacious and efficient. In particular, biomarkers need to be identified for each DBS patient, which correlate reliably with that patient's clinical state, even as it varies over extended periods of time. In addition, appropriate stimulation parameter values need to be selected from within the multidimensional space of parameters available in implantable devices, which is extensive, as noted above. It is probable that application of computational models will assist with this task. For example, if a general model can be devised that accurately predicts the effect of a change in stimulation settings on a given patient's clinical state, then only a relatively small number of free parameters in the model may need to be adjusted for each patient. This would be more practicable than optimization of the large number of parameters involved in extracting useful features from recordings of biomarker signals and controlling the numerous variables of the stimuli.

Although published reports generally confirm that most patients obtain benefit from existing DBS devices that do not contain the advanced techniques outlined above, there remain an unacceptably large number of patients who either obtain insufficient benefit or experience serious complications. In addition to providing closed-loop stimulation, other innovative techniques could help to address this problem. In many cases, suboptimal outcomes seem to be related to inaccurate positioning of the DBS electrodes. Improvements in imaging and in guidance systems that assist surgeons to optimize the electrode location in each patient may partially solve this problem (Hemm and Wardell 2010), but once electrodes have been implanted there are very few options for adjusting the locus of stimulation. Typically these options include selecting the active electrode when stimulating in monopolar mode (i.e. with the current-return electrode located far from the stimulation site) or choosing a different electrode configuration, such as the bipolar mode (i.e. with the active and return electrodes located nearby on the same electrode lead). However, as noted earlier, today's DBS devices often have only four discrete electrodes on each lead, and those electrodes are relatively large and spaced widely apart. New electrode arrays are under development to make available many more sites of current delivery that can be selected under program control post-operatively. Furthermore, it has been established that the locus of tissue excitation can be controlled by adjusting currents delivered simultaneously via a number of closely spaced electrodes (Hariz 2014; van den Honert and Kelsall 2007). This technique enables stimulation patterns to be focused spatially and to be "steered" with much greater precision than is possible with existing implants. By these means, it will become easier to activate one or more carefully selected regions of neural tissue while avoiding unintended excitation of other nearby areas. Additional improvements to electrode design may enhance the ability to record biomarker signals from the same leads as are used for stimulation, facilitating the development of sophisticated closed-loop stimulators.

Medical complications (such as hemorrhage and infection associated with DBS implantation) have been reported relatively frequently, as have malfunctions of devices (Chan *et al.* 2009; Miller and Gross 2009; Oh *et al.* 2002). These problems may be addressed in part by improved accuracy of electrode insertion and potentially also by placement of the implantable stimulator closer to the electrode arrays. For example, implantation of the stimulator in the temporal bone of the skull would obviate the requirement for long extension cables to connect a stimulator placed conventionally at a remote location (such as in the subclavicular region) with the intracranial DBS electrodes. Typically, extension cables are implanted subcutaneously along the neck where they are subjected to movement and occasionally placed under tension, sometimes resulting in fractures or unintended disconnections. Failures of this kind would be less likely if the stimulator was fixed securely in close proximity to the implanted electrodes, with relatively short connecting cables that would not be expected to move in normal use of the device.

Finally, it is worthwhile to note that general improvements in the outcomes obtained from application of DBS will result from better selection of candidates for the treatment. This requires that each potential patient receives an accurate diagnosis of their condition and any relevant comorbidities. Furthermore, well-designed clinical trials of DBS need to be expanded, particularly as the scope of indications that seem to be appropriate for DBS treatment continues to broaden. Where feasible, randomized controlled trials with sufficiently large numbers of subjects should be conducted, so that the effects of DBS can be evaluated more objectively and more precisely. As information is accumulated from such trials, it will become easier to specify which patients are most likely to obtain benefit that would adequately outweigh any likely adverse side effects. At present, reliable evidence concerning the efficacy of DBS and its possible side effects is limited, particularly for conditions other than movement disorders. Nevertheless, it appears highly probable that the benefits and safety of DBS will be demonstrated for a growing range of indications and an increasing number of patients. Ongoing clinical research in this area will help to optimize patient selection and ensure that DBS is applied appropriately to address some of the most challenging medical conditions.

ACKNOWLEDGEMENTS

The Bionics Institute acknowledges the support it receives for this work from the Colonial Foundation and from the Victorian Government through its Operational Infrastructure Support Program. Important contributions to the figures and text have been made by Amanda Walmsley, Peter McNeill, Wesley Thevathasan and Aimee Clague. Rob Shepherd provided valuable comments on a draft of this chapter.

REFERENCES

Anderson, V.C., Burchiel, K.J., Hogarth, P., Favre, J. and Hammerstad, J.P. (2005) Pallidal vs. subthalamic nucleus deep brain stimulation in Parkinson's disease. *Arch. Neurol.,* **62**(4): 554–560.

Benabid, A.L., Chabardes, S,, Torres, N., Piallat, B., Krack, P. *et al.* (2009) Functional neurosurgery for movement disorders: A historical perspective. *Prog. Brain Res.,* **175**: 379–391.

Blomstedt, P., Sjoberg, R.L., Hansson, M., Bodlund O. and Hariz, M.I.. (2011) Deep brain stimulation in the treatment of depression. *Acta Psychiatr. Scand.,* **123**(1): 4–11.

Burdick, A., Goodman, W.K. and Foote, K.D. (2009) Deep brain stimulation for refractory obsessive-compulsive disorder. *Front Biosci.,* **14**: 1880–1890.

Chan, D.T., Zhu, X.L., Yeung, J.H., Mok, V.C. Wong, E. *et al.* (2009) Complications of deep brain stimulation: a collective review. *Asian J. Surg.,* **32**(4): 258–263.

Coffey, R.J., Hamani, C. and Lozano, A.M. (2011) Evidence base: neuro-stimulation for pain, in *Youmans Neurological Surgery*, **2**, Winn, H.R. (ed.), Elsevier Saunders, Philadelphia, pp 1809–1820.

Collins, K.L., Lehmann, E.M. and Patil, P.G. (2010) Deep brain stimulation for movement disorders. *Neurobiol. Dis.*, **38**(3): 338–345.

de Koning, P.P., Figee, M, van den Munckhof, P., Schuurman, P.R. and Denys, D. (2011) Current status of deep brain stimulation for obsessive-compulsive disorder: a clinical review of different targets. *Curr. Psychiatry Rep.*, **13**(4): 274–282.

DeGiorgio, C.M., Schachter, S.C., Handforth, A., Salinsky, M., Thompson, J. *et al.* (2000) Prospective long-term study of vagus nerve stimulation for the treatment of refractory seizures. *Epilepsia*, **41**(9): 1195–1200.

Denys, D., Mantione, M., Figee, M., van den Munckhof, P., Koerselman, F. *et al.* (2010) Deep brain stimulation of the nucleus accumbens for treatment-refractory obsessive-compulsive disorder. *Arch. Gen. Psychiatry,* **67**(10): 1061–1068.

Deuschl, G., Paschen, S. and Witt, K. (2013) Clinical outcome of deep brain stimulation for Parkinson's disease, in *Handbook of Clinical Neurology*, vol **116**, Lozano, A.M. and Hallett, M. (eds), Elsevier, Amsterdam, pp 107–128.

Fisher, R.S. (2012) Therapeutic devices for epilepsy. *Ann. Neurol.,* **71**(2): 157–168.

Fisher, R.S, Salanova, V., Witt, T., Worth, R., Henry, T. *et al.* (2010) Electrical stimulation of the anterior nucleus of thalamus for treatment of refractory epilepsy. *Epilepsia* **51**(5): 899–908.

Fodstad, H. and Hariz, M.I. (2007) Electricity in the treatment of nervous system disease. *Acta Neurochir. Suppl.*, **97**(Pt 1): 11–19.

Foltynie, T. and Hariz, M.I. (2010.) Surgical management of Parkinson's disease. *Expert Rev. Neurother.*, **10**(6): 903–914.

Greenberg, B.D., Malone, D.A., Friehs, G.M., Rezai, A.R., Kubu, C.S. *et al.* (2006) Three-year outcomes in deep brain stimulation for highly resistant obsessive-compulsive disorder. *Neuropsychopharmacology,* **31**(11): 2384–2393.

Greenberg, B.D., Rauch, S.L. and Haber, D.N. (2010) Invasive circuitry-based neurotherapeutics: stereotactic ablation and deep brain stimulation for OCD. *Neuropsychopharmacology,* **35**(1): 317–336.

Hariz, M. (2014) Deep brain stimulation: new techniques. *Parkinsonism Relat. Disord.,* **20** Suppl. 1: S192–S196.

Hariz, M. and Robertson, M. (2010) Gilles de la Tourette's syndrome and deep brain stimulation. *Eur. J. Neurosci.,* **32**(7): 1128–1134.

Heck, C.N., King-Stephens, D., Massey, A.D., Nair, D.R., Jobst, B.C. *et al.* (2014.) Two-year seizure reduction in adults with medically intractable partial onset epilepsy treated with responsive neurostimulation: final results of the RNS System Pivotal trial. *Epilepsia,* **55**(3): 432–441.

Hemm, S. and Wardell, K. (2010) Stereotactic implantation of deep brain stimulation electrodes: a review of technical systems, methods and emerging tools. *Med. Biol. Eng. Comput.*, **48**(7): 611–624.

Hoy, K.E. and Fitzgerald, P.B. (2010) Brain stimulation in psychiatry and its effects on cognition. *Nat. Rev. Neurol.,* **6**(5): 267–275.

Kleiner-Fisman, G., Herzog, J., Fisman, D.N., Tamma, F., Lyons, K.E. *et al.* (2006) Subthalamic nucleus deep brain stimulation: summary and meta-analysis of outcomes. *Mov. Disord.,* **21** Suppl. 14: S290–3S04.

Kuhn, J., Bührle, C.P., Lenartz, D. and Sturm, V. (2013) Deep brain stimulation in addiction due to psychoactive substance use, in *Handbook of Clinical Neurology*, vol **116**, Lozano, A.M. and Hallett, M. (eds), Elsevier, pp 259–269.

Kuncel, A.M. and Grill, W.M. (2004) Selection of stimulus parameters for deep brain stimulation. *Clin. Neurophys.,* **115**(11): 2431–2441.

Kuncel, A.M., Cooper, S.E., Wolgamuth, B.R., Clyde, M.A., Snyder, S.A. *et al.* (2006) Clinical response to varying the stimulus parameters in deep brain stimulation for essential tremor. *Movement Disorders,* **21**(11): 1920–1928.

Laxton, A.W., Lipsman, N. and Lozano, S.M. (2013) Deep brain stimulation for cognitive disorders, in *Handbook of Clinical Neurology*, vol **116**, Lozano, A.M. and Hallett, M. (eds), Elsevier, pp 307–311.

Lefaucheur, J.P. 2009. Methods of therapeutic cortical stimulation. *Neurophysiol. Clin.,* **39**(1): 1–14.

Lipsman, N., Mendelsohn, D., Taira, T. and Bernstein, M. (2011) The contemporary practice of psychiatric surgery: results from a survey of North American functional neurosurgeons. *Stereotact. Funct. Neurosurg.,* **89**(2): 103–110.

Lipsman, N., Giacobbe, P. and Lozano, A.M. (2013a) Deep brain stimulation in obsessive–compulsive disorder: neurocircuitry and clinical experience, in *Handbook of Clinical Neurology*, vol **116**, Lozano, A.M. and Hallett, M. (eds), Elsevier, pp 245–250.

Lipsman, N., Woodside, D.B., Giacobbe, P., Hamani, C., Carter, J.C. *et al.* (2013b) Subcallosal cingulate deep brain stimulation for treatment-refractory anorexia nervosa: A phase 1 pilot trial. *Lancet,* **381**(9875): 1361–1370.

Little, S. and Brown, P. (2012) What brain signals are suitable for feedback control of deep brain stimulation in Parkinson's disease? *Ann. New York Acad. Sci.,* **1265**(1): 9–24.

Little, S., Pogosyan, A., Neal, S., Zavala, B., Zrinzo, L. *et al.* (2013) Adaptive deep brain stimulation in advanced Parkinson's disease. *Ann. Neurol.,***74**(3): 449–457.

Lujan, J.L., Chaturvedi, A., Malone, D.A, Rezai, A.R, Machado, A.G. and McIntyre, C.C. (2012) Axonal pathways linked to therapeutic and nontherapeutic outcomes during psychiatric deep brain stimulation. *Hum Brain Mapp.*, **33**(4): 958–968.

Mallet, L., Polosan, M., Jaafari, N., Baup, N., Welter, M.L. *et al.* (2008) Subthalamic nucleus stimulation in severe obsessive-compulsive disorder. *N. Engl. J. Med.,* **359**(20): 2121–2134.

Malone, D.A. Jr., Dougherty, D.D., Rezai, A.R., Carpenter, L.L., Friehs, G.M. *et al.* (2009) Deep brain stimulation of the ventral capsule/ventral striatum for treatment-resistant depression. *Biol. Psychiatry,* **65**(4): 267–275.

Mian, M.K., Campos, M., Sheth, S.A. and Eskandar, E.N. (2010) Deep brain stimulation for obsessive-compulsive disorder: past, present, and future. *Neurosurg. Focus,* **29**(2): E10.

Miller, P.M. and Gross, R.E. (2009) Wire tethering or "bowstringing" as a long-term hardware-related complication of deep brain stimulation. *Stereotact. Funct. Neurosurg.,* **87**(6): 353–359.

Morrell, M.J. (2011) Responsive cortical stimulation for the treatment of medically intractable partial epilepsy. *Neurology,* **77**(13): 1295–1304.

Nazzaro, J.M., Lyons, K.E. and Pahwa, R. (2013) Deep brain stimulation for essential tremor, in *Handbook of Clinical Neurology*, vol **116**, Lozano, A.M. and Hallett, M. (eds), Elsevier, pp 155–166.

Nitsche, M.A., Boggio, P.S., Fregni, F. and Pascual-Leone, A. (2009) Treatment of depression with transcranial direct current stimulation (tDCS): a review. *Exp. Neurol.,* **219**(1): 14–19.

Oh, M.Y., Abosch, A., Kim, S.H., Lang, A.E. and Lozano, A.M. (2002) Long-term hardware-related complications of deep brain stimulation. *Neurosurgery,* **50**(6): 1268–1274.

Patel, S.R. and Simpson, H.B. (2010) Patient preferences for obsessive-compulsive disorder treatment. *J. Clin. Psychiatry*, **71**(11): 1434–1439.

Pereira, E.A.C., Green, A.L. and Aziz, T.Z. (2013) Deep brain stimulation for pain, in *Handbook of Clinical Neurology*, vol **116**, Lozano, A.M. and Hallett, M. (eds) Elsevier, pp 277–294.

Read, C.N. and Greenberg, B.D. (2009) Psychiatric neurosurgery 2009: review and perspective. *Semin. Neurol.,* **29**(3): 256–265.

Sachdev, P.S., Cannon, E., Coyne, T.J. and Silburn, P. (2012) Bilateral deep brain stimulation of the nucleus accumbens for comorbid obsessive compulsive disorder and Tourette's syndrome. *BMJ Case Report.*

Schiff, S.J. (2010) Towards model-based control of Parkinson's disease. *Phil. Trans. Royal Soc. A – Math. Phys. Eng.Sci.,* **368**(1918): 2269–2308.

Schlaepfer, T.E. and Bewernick, B.H. (2013) Deep brain stimulation for major depression, in *Handbook of Clinical Neurology*, vol **116**, Lozano, A.M. and Hallett, M. (eds), Elsevier, pp 235–243.

Stanslaski, S., Afshar, Cong, P., Giftakis, J., Stypulkowski, P. *et al.* (2012) Design and validation of a fully implantable, chronic, closed-loop neuromodulation device with concurrent sensing and stimulation. *IEEE Trans. Neur. Syst. Rehab. Eng.,* **20**(4): 410–421.

van den Honert, C. and Kelsall, D.C. (2007) Focused intracochlear electric stimulation with phased array channels. *J. Acousti. Soc. Am.,* **121**(6): 3703–3716.

Vidailhet, M., Jutras, M.-F., Roze, E. and Grabli, D. (2013) Deep brain stimulation for dystonia, in *Handbook of Clinical Neurology*, vol **116**, Lozano, A.M. and Hallett, M. (eds), Elsevier, pp 167–187.

Visser-Vandewalle, V. and Kuhn, J. (2013) Deep brain stimulation for Tourette's syndrome, in *Handbook of Clinical Neurology*, vol. **116**, A Lozano, A.M. and Hallett, M. (eds), Elsevier, pp 251–258.

9

BRAIN COMPUTER INTERFACES

David M. Brandman and Leigh R. Hochberg

Department of Neuroscience, Brown University, Providence, RI, USA

9.1 INTRODUCTION

Brain Computer Interfaces (BCIs) use neural information recorded from the brain for voluntary control of external devices. BCI systems are often targeted towards improving functional independence for individuals with severe motor limitations. Early pilot clinical trials are currently investigating the role of BCIs for the purpose of restoring communication (i.e. typing on a computer screen) or as an effector for physical responses (i.e. controlling a robotic arm).

There are many types of BCIs, but they all have three major components: a *sensor* to record neural activity or its proxy, a *decoder* that converts neural activity into a command signal, and an *effector* such as a computer cursor or robotic arm. Sensor choice is informed primarily by considering the signal we wish to record, such as action potentials from individual neurons, field potentials from more spatially disparate neurons, or blood flow reflecting underlying cortical activity. Similarly, sensor choice will impact the technology's portability, its requirement for expert assistance or caregiver set-up, its need for surgical placement, the availability for use, the cost of acquiring or maintaining the system, and other key characteristics. Indeed, even the choice of sensor reflects an earlier decision about which of the available neural signals (e.g. action potentials, field potentials, neurochemical concentrations, etc.) or biological signals (e.g. blood flow reflecting underlying metabolic activity) we wish to use in a BCI, and each of those signals is associated with its own benefits and risks.

In this chapter, we provide an introduction for understanding intracortical BCI (iBCI) devices, one of the many types listed above. iBCIs record from ensembles of

Neurobionics: The Biomedical Engineering of Neural Prostheses, First Edition.
Edited by Robert K. Shepherd.

individual neurons, which are directly engaged in generating voluntary limb movement – or, in the case of a person with paralysis, neurons directly engaged in the intention to move the arm or hand. We begin with an overview of motor anatomy and physiology from an engineering standpoint. We then discuss populations of individuals who could potentially benefit from iBCIs. We provide a survey of various BCI designs, focusing on the technical aspects of iBCI decoding. Finally, we discuss directions for future developments for the field.

9.2 MOTOR PHYSIOLOGY

Let us consider the highly rewarding act of reaching for a cup of coffee on a table. From a control-system perspective, we can conceptualize this task by subdividing it into several components. First, we will determine where the cup is relative to the body, using incoming visual cues; second, we will coordinate the appropriate muscle activity to reach the cup; and third, we will need to grip the cup correctly. During this process, real-time visual and somatosensory feedback would improve our performance and allow for error correction. A brief outline of the organization of the nervous system from a motor systems perspective, beginning with a discussion of neurons, helps to frame the problem from the perspective of designing a restorative neurotechnology.

9.2.1 Neurons are the fundamental unit of the brain

The functional unit of the nervous system is the neuron. Neurons are remarkably complex cells that specialize in synthesizing inputs from and conveying outputs to other neurons (or, in the peripheral nervous system, to muscles and other organs). Information is conveyed within and between neurons using electrical signals called action potentials, which are localized changes in voltage potentials controlled by sophisticated interplay between ion channels (Chapter 4). When action potentials reach the distal terminal of a neuron, chemical neurotransmitters are released at a synapse that evokes voltage changes in the closely approximated dendrite of the receiving neuron. Action potentials effectively act as an all-or-nothing information-encoding scheme over the timescale of a millisecond, and their presence can be identified by extracellular electrodes placed within approximately 100 microns of the neuron.

The brain and spinal cord are collectively called the central nervous system (CNS). There are roughly 10^9 neurons in the brain, with immense heterogeneity in terms of their structure, behavior and size. For instance, while neurons have cell bodies ranging within one order of magnitude, their lengths range over six orders of magnitude. On average, neurons form 10^5 connections (synapses) with other neurons. Neurons often have a single output pathway, referred to as an axon, which leaves the cell body and connects to (projects to) to other neurons. Groups of axons that travel together in the CNS are called tracts.

Understanding the process of how information is encoded in neurons has been one of the most important research pursuits in neuroscience. Generally speaking, neurons encode information as some combination of rate and recruitment. Rate refers to the number of action potentials of a particular neuron per unit time, with or without the details for the specific timing of the events. Neuronal recruitment refers to which of the many neurons change their firing behavior according to a given stimulus. The parameters of information encoded by neurons generally emerge in part from their physical location in the CNS.

As noted above, performing a complex task requires multiple steps. Reaching for a cup of coffee requires many brain regions to work simultaneously, with each region having specialized functions.

9.2.2 Movement occurs through coordinated activity between multiple regions of the nervous system

Historically, neural control has been conceptually divided into two components: lower motor and upper motor neurons. Lower-motor neurons are analogous to electrical wires: they receive inputs from other neurons located either within the spinal cord or the brain, and then project to specific muscle fibers. Action potentials from large numbers of neurons cause muscle fibers to contract in a highly coordinated fashion, resulting in limb movement. In general, lower-motor neurons with similar muscle projections tend to aggregate together in the spinal cord.

Upper motor neurons coordinate the activity of lower-motor neurons. In humans, the representative upper motor neuron pathway is the corticospinal (CS) tract, which is a critical motor system for controlling planned, coordinated movement (especially in the arms and hands). This tract originates from various regions of the cortex and projects onto specific collections of neurons in the spinal cord.

Real-time feedback regulating motor control comes from somatosensory information. Specialized sensory neurons project from within skin, muscles and tendons to the spinal cord and the brain, providing information about joint angles, muscle tension and joint velocity changes. Other sensory modalities that influence motor control include tactile information from the object, and visual information about the arm and object position. Thus, real-time input to the system is synthesized, allowing for continuous movement adjustments.

Beyond the upper and lower motor neuron systems, several other brain regions are involved in movement. In the cortex, the CS tract works in concert with several other motor pathways to orchestrate movement. Error detection and motor learning involves regions of the brain, including the anterior cingulate cortex and the ventral tegmentum. The pre-frontal cortex is involved in planning movements based on context. The thalamus is a subcortical structure that can be thought of as a relay station, wherein incoming and outgoing information is synthesized and modulated. The basal ganglia are a collection of subcortical regions that, among other functions, serve to modulate a "stop/go" signal for purposeful movement. Located directly adjacent to the brainstem, the cerebellum helps to smooth and coordinate movements.

With this overview in mind, we can now consider sources of information in the brain that could serve as inputs for an iBCI system. Given the number of brain regions involved in movement (described above), it is useful to start with sites that provide rich and reliable information. The primary motor cortex, which includes neurons that synapse directly with lower motor neurons, fits the bill nicely.

9.2.3 Motor cortex: a first source for iBCI signals

The primary motor cortex (referred to as M1) plays a key role in directing voluntary, dexterous movement (Cheney *et al.* 2000; Hatsopoulos and Donoghue 2009; Kaas *et al.* 2012; Rizzolatti *et al.* 1998). Decades of research have revealed that firing rates of M1 neurons correlate with various kinematic parameters. Pioneering work by Evarts (1968) in non-human primates (NHPs) demonstrated that the magnitude of wrist torque in a one-dimensional movement correlated with increasing firing rates in CS neurons (Figure 9.1A). The reverse of this was also found to be true: the firing rate for M1 neurons could be used to predict the torque exerted (Humphrey *et al.* 1970).

In now classic experiments, a rhesus monkey was trained to move a manipulandum to one of eight different targets on a two-dimensional surface (this is called a center-out task) (Georgopoulos and Kalaska 1982). The CS neurons were found to correlate with the movement direction: movement in a particular direction resulted in some neurons increasing their firing rate, whereas other neurons decreased their firing rate (Figure 9.1B). The relationship between firing rate and movement direction could be fit to a sinusoidal function; hence this description of M1 CS neurons is referred to as a cosine-tuning model. The direction of movement corresponding to the maximum firing rate is referred to as the preferred direction. Preferred directions generalize to three dimensions (Georgopoulos *et al.* 1986).

In the Georgopoulos model, a neuron in M1 can be thought of as a vector in space: the direction and magnitude of the vector reflects the preferred three-dimensional direction and the firing rate of the neuron, respectively. Thus, over a population of neurons, adding all vectors together generates a net vector, which may correlate with reaching to a point in space (Georgopoulos *et al.* 1986). Beyond describing limb position, M1 neurons have been found to reliably encode velocity (Schwartz 1994), acceleration (Paninski *et al.* 2004), muscle activation (Pohlmeyer *et al.* 2007; Todorov 2003), and several other variables related to movement.

With the development of tools that allowed for simultaneous recordings of a population of neurons in M1, it became clear that a neural population encodes more movement information than could be extrapolated from simply recording many individual neurons (Hatsopoulos *et al.* 1998). For instance, on a population level, as many as 20% of M1 neurons carry information not reflected in kinematic variables (Paninski *et al.* 2004), potentially reflecting information concerning joint sensation (Shaikhouni *et al.* 2013) or visual information related to the task (Eisenberg *et al.* 2011). Thus, the aggregate neural activity from neurons in M1 reflects multiple movement-related properties simultaneously.

Although muscle activation correlates with neural activity in M1, neural activity need not result in muscle activation (Hochberg *et al.* 2006; Schieber 2011; Shenoy *et al.* 2013). For instance, even though no movement activity is recorded in a subject, neurons in M1 modulate their activity based on observed movements of other individuals (Eisenberg *et al.* 2011; Kilner and Lemon 2013; Vigneswaran *et al.* 2013).

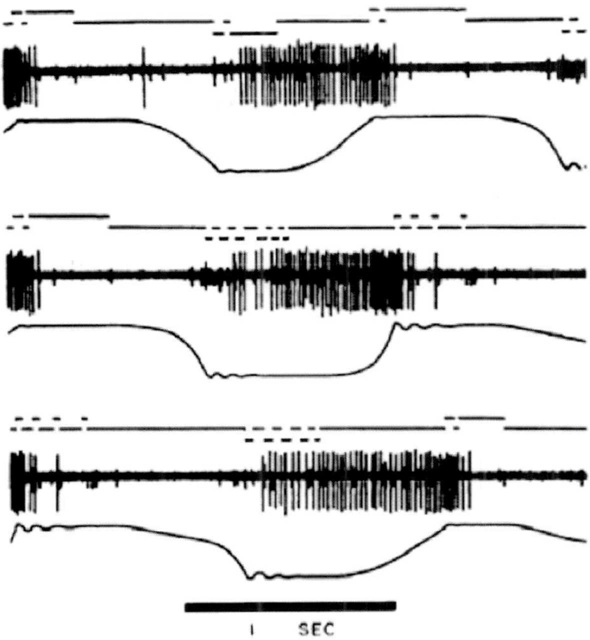

(a)

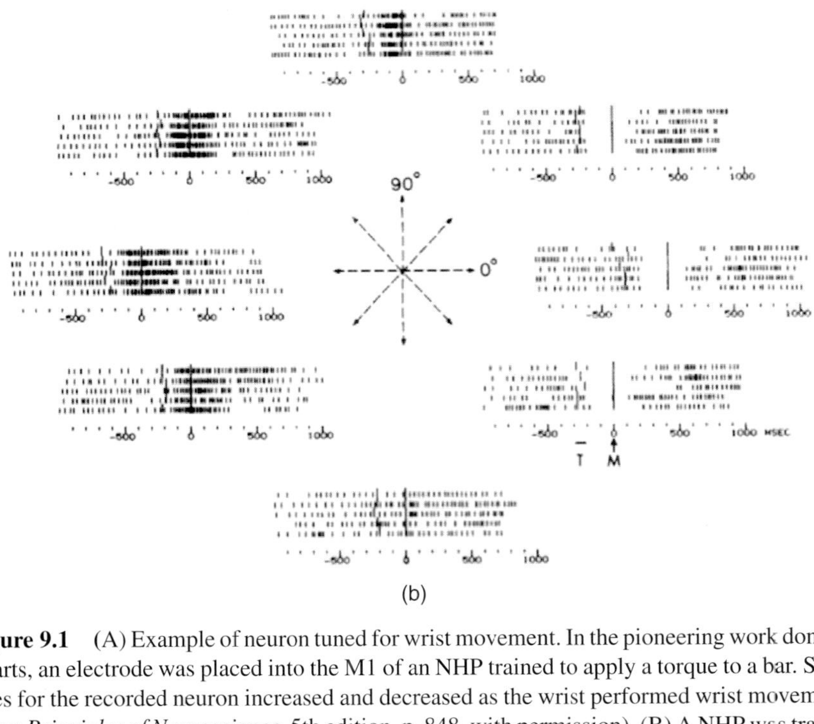

(b)

Figure 9.1 (A) Example of neuron tuned for wrist movement. In the pioneering work done by Evarts, an electrode was placed into the M1 of an NHP trained to apply a torque to a bar. Spike rates for the recorded neuron increased and decreased as the wrist performed wrist movements (from *Principles of Neuroscience*, 5th edition, p. 848, with permission). (B) A NHP was trained to move a manipulandum to one of eight different positions. The firing rate of the neuron was modeled using a cosine function, relating the angular arm displacement to the firing rate (from *Principles of Neuroscience*, 5th edition, p. 849, with permission).

Neural activity also occurs during the delay period between an instruction cue and the onset of movement (Rao and Donoghue 2014), potentially reflecting a preparatory activity (Shenoy *et al.* 2013). One hypothesis posits that although individual neurons may reliably reflect kinematic variables, on a population level group firing-rate configurations need not produce movements (Kaufman *et al.* 2014).

There are some characteristics that make M1 an especially attractive target for iBCI devices (Cheney *et al.* 2000; Humphrey and Hochberg 1995). First, neurons in M1 yield a rich variety of signals from a small area. A single CS neuron can project to multiple pools of motor neurons; conversely, pools of motor neurons often have CS neurons from multiple cortical regions that coordinate their activity. Neurons separated micrometers apart may thus display highly variable behavior (Kaas *et al.* 2012), although neurons with similar preferred directions have a periodic spatial arrangement (Georgopoulos *et al.* 2007). Taken together, a small iBCI recording array is likely to yield information from neurons that modulate their behavior during the generation of a wide repertoire of intended movements.

Second, the hand area of the motor cortex is an accessible surgical target, and is relatively easy to identify pre-operatively using known anatomical landmarks (Ahdab *et al.* 2013; Yousry *et al.* 1997), as part of pre-operative functional MRI experiments (Shoham *et al.* 2001), or intra-operatively with electrical stimulation (Penfield and Boldrey 1937), although intraoperative stimulation may not be relevant in individuals who are paralyzed. This point is non-trivial: many cortical regions of potential interest to iBCI technology (e.g. areas of motor planning) lack either clear structural or functionally derivable landmarks; thus, at the time of implantation, a neurosurgeon would be much less confident as to whether electrodes are being implanted in the correct brain region. In the future, it is possible that surgical placement of iBCI electrodes could be directed by "awake" mapping of cortical or subcortical areas, as is commonplace for the placement of deep brain stimulating electrodes in people with movements disorders such as Parkinson's disease (PD).

Third, evidence suggests that sufficient neural modulation is present in M1 despite downstream destruction of axonal projections (Section 9.2.6). Destruction of a neuron's axonal projection results in retrograde and anterograde degeneration of the axon – often resulting in cell death – in a process called Wallerian degeneration. A presence of a measurable loss of CS cells resulting from downstream destruction of axons is controversial in the non-human literature (for an excellent review, see the background in Wannier *et al.* 2005); regardless of whether or not this occurs, evidence suggests that primary processing of motor control in humans is intact. First, human functional magnetic resonance imaging (fMRI) (Sabbah *et al.* 2002; Shoham *et al.* 2001) experiments suggest robust signal changes to the M1 in individuals with longstanding paralysis. Second, pilot clinical trials (Collinger *et al.* 2013; Hochberg *et al.* 2006; Truccolo *et al.* 2008) have successfully demonstrated robust neural signal decoding in individuals with paralysis. Thus, it appears as though M1 retains relevant physiological properties, despite longstanding damage to the motor system. It is worth noting that this would not necessarily have been predicted by prior cortical plasticity research which had led, perhaps too broadly, to the conclusion that gross reorganizations of motor somatotopy always follows downstream central or peripheral nervous system injury (Sanes and Donoghue 2000).

Many experimental paradigms have taken advantage of these physiological characteristics of M1 to create brain machine interface devices in both NHPs and clinical investigations. There are, however, other cortical regions involved in movement that have physiologically relevant activity and are surgically tenable. We will briefly review the parietal, premotor and supplementary motor areas (SMAs), highlighting important features that allow them to serve as potential targets for iBCI devices.

9.2.4 The parietal cortex is implicated in spatial coordination

Let us return to our coffee cup example. In the process of reaching out for the cup, multiple lines of sensory information are computed: visual cues as to where the cup is in space are integrated with proprioceptive information from the head, neck, limbs and eyes. In this way, spatial decisions are integrated in real time to update movement trajectories.

The posterior parietal cortex (PPC) is involved in synthesizing information from various sensory modalities. The PPC has many functions, but germane to our discussion is its role in generating a reference map of the world based on incoming information (Buneo and Andersen 2006). Individual neurons are involved in integrating sensory information, such as visual and somatosensory information pertaining to a region of space (Batista *et al.* 1999; Buneo *et al.* 2002). The parietal cortex then projects to the frontal motor areas to influence motor control (Filimon 2010; Hwang *et al.* 2014; Mulliken *et al.* 2008; Rizzolatti *et al.* 1998).

In addition, regions of the parietal cortex are directly implicated in limb movement; parietal neural recordings correlate with NHP movements (Mulliken *et al.* 2008; Musallam *et al.* 2004; Philip *et al.* 2012). With easy surgical accessibility, and its direct role in motor control, the PPC may serve as a useful recording target for iBCI technology.

9.2.5 The premotor and supplementary motor cortices are engaged in movement goals

Once an object's spatial information is synthesized, it is then projected to various motor regions that coordinate the movement (Rizzolatti *et al.* 1998). The dorsal pre-motor cortex (PMd) receives projections from the PPC, and is thought to be involved in the transformation of sensory and coordinate information into the appropriate motor commands that are required for reaching (Cohen *et al.* 2009; Luppino and Rizzolatti 2000).

While reach is coordinated by PPC projections to PMd, grasping is believed to be coordinated largely by the ventral premotor cortex (PMv). Neuronal activity in the PMv correlates with NHP representations of power-grip, precision grip or swiping movements (Hendrix *et al.* 2009). Conceptually, we can think about the PPC as providing visual cues about spatial characteristics of an object, while the PMv generates the corresponding grip (Davare *et al.* 2011).

The SMA is involved in motor planning (Fried *et al.* 2011). Activity in SMA occurs prior to spontaneous volitional movements (Brass and Haggard 2008; Romo and Schultz 1987): this activity stems from a series of instructions and a previously learned movement (Mushiake *et al.* 1991).

All three of these cortical regions have CS tract neurons (He *et al.* 1993, 1995), and also project to the M1 (Dum and Strick 2005). An advantage of using M1 is its receipt of informative, multi-modality signals from other motor-planning regions, thus making it an attractive target for iBCI recording devices.

9.2.6 Functional brain organization is constantly changing

The fact that people are able to learn and remember new facts, pick up a hobby or play a new sport, all hint at the fact that the nervous system and its underlying neuronal interactions are dynamic. The umbrella term used to describe this phenomenon is neural plasticity. Broadly speaking, neural plasticity involves the creation, modification, and (most importantly) dissolution of synapses between neurons.

One robust finding of the CNS is that regions of the brain responsible for mediating the behavior of damaged components can begin to take on new functions. In now classic experiments, Donoghue and colleagues demonstrated that transection of the facial motor nerves in rats results in functional loss of the whisker area in the rat motor cortex (Sanes *et al.* 1992). Said another way, neurons that previously were found to modulate their activity to control whisker movement, began to modulate neuronal activity to control another action once the previously controlled activity becomes obsolete. Of note, reorganization emerges within hours of nerve transection (Nudo 2013; Nudo and McNeal 2013; Sanes *et al.* 1992; Sanes and Donoghue 2000).

In the context of iBCIs, it is reasonable to think of the iBCI as a "tool" or "instrument" for which a motor skill can be developed, and that skill acquisition is accompanied by the expected plasticity of motor areas of the brain. In one important study, NHPs were implanted with intracortical electrodes and learned to move a cursor in a center-out task in three dimensions under neural control using a population-vector algorithm (Jarosiewicz *et al.* 2008) (see Section 9.5 in this chapter for a discussion of decoding). Once the population-vector algorithm was calibrated, a subset of identified neurons had their preferred directions perturbed. As NHPs learned the new settings to establish neural control, neurons that were perturbed developed larger compensatory changes than those that were not, suggesting that the functional properties of neurons – used as part of decoding algorithms for iBCI devices – are in fact affected by the choice of decoding algorithms.

9.2.7 Section summary

Neural activity in the motor cortex results from a complex series of inputs from multiple cortical and subcortical regions. Some regions of the cortex generate spatial maps that continuously update themselves based on internal and external cues, while others are involved in learning new or recalling previously stored motor movements. The entire system is continuously updating itself with feedback from several cortical and subcortical regions, working together to direct reaching, grasping, and many other dexterous movements.

Having established an anatomical and physiological framework of motor control, we now shift our attention to discussing individuals who have motor impairment.

9.3 THE CLINICAL POPULATION FOR BRAIN COMPUTER INTERFACES

Motor control is essential for everyday life. The way in which people interact with the surrounding world depends almost exclusively on controlled, coordinated movement. Reaching for a cup of coffee is difficult, if not impossible, when the systems responsible for movement are damaged. Unfortunately, the diseases and injuries that interfere with movement are relatively common, and hundreds of thousands of individuals develop functional motor impairments every year. BCI research is promising, given its potential benefits for individuals with severe motor control and communication impairments. In this section, we describe some of the individuals who may benefit from these emerging technologies.

9.3.1 Paralysis may result from damage to the motor system

The CS system is required for strong, coordinated and fluid voluntary movement; damage occurring at any point in the system may affect motor control. Some of the causes of motor weakness stem from damage to: the CS neurons within the various cortical regions; the axon fibers projecting from these cortical centers through the various portions of the brain and the spinal cord; the basal ganglia and cerebellum; the alpha-motor neurons and peripheral projections onwards; the neuromuscular junction; or, even the muscles themselves.

Motor limb weakness may be partial (-paresis) or complete, as in paralysis (-plegia). One instance of complete motor limb weakness is tetraplegia – paralysis of all four limbs, with normal facial muscle strength. An even more extensive form of paralysis is referred to as Locked-In Syndrome (LIS), wherein individuals lack the ability for most voluntary movements (which usually includes limb and facial movements, speaking), though retaining cognition. Classical LIS is characterized by preserved vertical eye movements and blinking, whereas individuals who are completely locked-in lack all movements, including volitional eye movements (Laureys *et al.* 2005; Lulé *et al.* 2009; Plum and Posner 1969).

Individuals can develop tetraplegia or LIS from a variety of causes. While an exhaustive list is beyond the scope of this chapter, the most common cause of tetraplegia is due to traumatic spinal cord injuries (SCIs) in the cervical neck region (such as motor vehicle collisions, dive injuries or penetrating injuries). Common forms of LIS are due to brainstem (commonly, pontine) strokes and neurodegenerative diseases such amyotrophic lateral sclerosis (ALS). In addition, reduced motor function may result from traumatic limb amputation, damage to nerves innervating muscles or from diseases that interfere with the neuromuscular junction (such as myasthenia gravis).

In the sections below, we emphasize diseases where brain information from the cortex appears viable despite paralysis: tetraplegia from traumatic SCIs and LIS from ALS. The following sections outline the pathophysiology, epidemiology and functional limitations for individuals with these diseases.

9.3.2 Individuals with spinal cord injuries develop motor impairments that may impact hand function

There are roughly 275,000 people in the USA (NSCISC 2012) and millions worldwide (Oudega and Perez 2012) who have sustained a traumatic SCI. These injuries tend to result from high-impact traumatic events (i.e. motor vehicle collisions and falls) – young men between the ages of 16 and 30 represent the largest demographic. Of those who suffer a traumatic SCI, approximately half will develop some degree of tetraparesis or paralysis (NSCISC 2012).

At the time of acute injury, common clinical interventions include surgery (for the purpose of aligning and stabilizing the spine and decompressing the spinal cord) (Fehlings *et al.* 2012) and altering physiological parameters (putatively increasing blood flow to the spinal cord) (Vale *et al.* 1997). While there have been hundreds of clinical studies investigating interventions for improving outcomes for patients with SCI, unfortunately, at present, there are only limited interventions available for clinicians to improve the functional independence of individuals with SCI once injury has occurred (Anderson 2009). Rehabilitation techniques remain the best option for improving the quality of life of people with tetraplegia (Oudega and Perez 2012), although spinal cord stimulation is a very active and exciting area of research with tremendous potential (Angeli *et al.* 2014; Harkema *et al.* 2011).

It is through greater awareness, improved medical care, and recognition of the unique challenges faced by people with tetraplegia that life expectancy has shown significant improvement over the past decades. With modern medical management, individuals with tetraplegia live 35–40 years after sustaining their injury (NSCISC 2012). Tetraplegia affects people in myriad ways (Anderson 2004; Anderson *et al.* 2011; Dijkers 2005). Physical health impairments include difficulties with locomotion or arm control, bladder/bowel care, sexual dysfunction and skin care (Anderson 2004). Complications also include muscle contractures and bone fragility, chronic neurological pain, autonomic instability and chronic fatigue. Beyond these physical impairments, however, individuals with SCI can be affected in the social realm as well, including challenges with employment, relationships and social inclusion (Simpson *et al.* 2012).

Clearly, given the wide physical, social and economic impacts of paralyzing conditions, it is important to recognize that the role of interventions must be focused towards the needs of the population. Accordingly, as a research and clinical community, our interventions should prioritize the needs and desires of the individuals living with the condition, rather than the needs perceived in the laboratory as being most important. To identify specific interventions that would substantially improve the quality of life of individuals living with a SCI, a survey of 347 people with quadriplegia was conducted by Anderson (2004). When individuals with tetraplegia were asked, "What gain of function would dramatically improve your life?" the ability to regain arm/hand function was the most common outcome selected from seven different physiological outcomes. A more recent systematic literature review including 3187 patients over 11 studies also supports this finding (Simpson *et al.* 2012).

iBCI technologies are well suited to realize the needs identified in the aforementioned surveys. The goal is to utilize the neural information related to intended movement available from neurons in the motor cortex for the purpose of restoring reliable arm and hand function.

9.3.3 Individuals with LIS develop motor impairment that impacts communication

ALS is a progressive neurodegenerative disease in which both upper motor and lower motor neurons are affected. There are as many as 30,000 individuals with ALS in the United States (ALS Association 2014), yielding a prevalence of roughly 4 cases per 100,000 people year (Mehta *et al.* 2014). The incidence of ALS in the US has been estimated as approximately 5500 new cases diagnosed each year, with an estimated 1.5–2 people diagnosed per 100,000 per year worldwide (Mehta *et al.* 2014). Most people are diagnosed in their 50s or 60s, although some are diagnosed much earlier.

While the disease may begin with weakness in the arms, legs or facial muscles, eventually all motor groups affected throughout the body. As the disease progresses, cranial nerves are affected resulting in difficulty swallowing, and weakness of the diaphragm and chest wall muscles impair respiratory function. Individuals with ALS will eventually develop tetraplegia and difficulty breathing; if they receive mechanical ventilation, some will eventually develop LIS. Median survival time from onset ranges between two and four years, though life can continue for many years after mechanical ventilation is initiated. The only US FDA-approved medication for ALS is riluzole; unfortunately, while it has been demonstrated to prolong survival, its benefits are modest (Miller *et al.* 2012).

Beyond ALS, other common causes of LIS include stroke, specifically ischemic and hemorrhagic infarcts of the brainstem. In these cases, vertical eye movements are usually spared because the responsible neuronal cell bodies are located in the uppermost portion of the midbrain.

Individuals with LIS may survive for years after their injury (Lulé *et al.* 2009). Like individuals living with tetraplegia, the treatment of individuals with LIS is complex and requires multidisciplinary coordination from various healthcare professionals. Individuals who lack the ability to communicate their needs face challenges even beyond those with other forms of severe physical disability. Our direct experience in clinical neuroscience suggests that tools to enhance communication are a priority for people with LIS. It is likely that any device allowing for greater ability to communicate – beyond blinking or performing vertical eye movements – would provide considerable improvements to the quality of life of people with LIS.

9.4 BCI MODALITIES

A complete BCI system includes at least three components (Figure 9.2) (Hatsopoulos and Donoghue 2009; Hochberg *et al.* 2006; Homer *et al.* 2013; Schwartz *et al.* 2006). First, the system uses an electrode array to record neural signals from the brain. Second, those signals are amplified and filtered to extract useable information, and then classified or decoded into command signals for an external device. The third component is an effector, which could be a computer cursor or a robotic arm, which is controlled through the modulated neural signals recorded and decoded by the first two components. A system is said to be open-loop when the processed neural information is not displayed back to the user (or the user is unaware of any output); a closed-loop system incorporates all three steps (Cunningham *et al.* 2011; Jarosiewicz *et al.* 2013). For each step listed above, we enumerate various techniques and focus on the approaches that are currently being used for intracortical recording investigations (Collinger *et al.* 2013; Hochberg *et al.* 2012).

One way to classify BCI systems involves identifying the spatial resolution of their source of brain information. The spatial resolution refers to the spatial volume of neurons that are averaged together to generate a signal. BCI systems may be classified as macroscopic, mesoscopic or microscopic, reflecting the amount of spatial and temporal resolution of the usable signal (NIH 2014). As outlined by Table 9.1, we will now list several BCI paradigms according to the three scales.

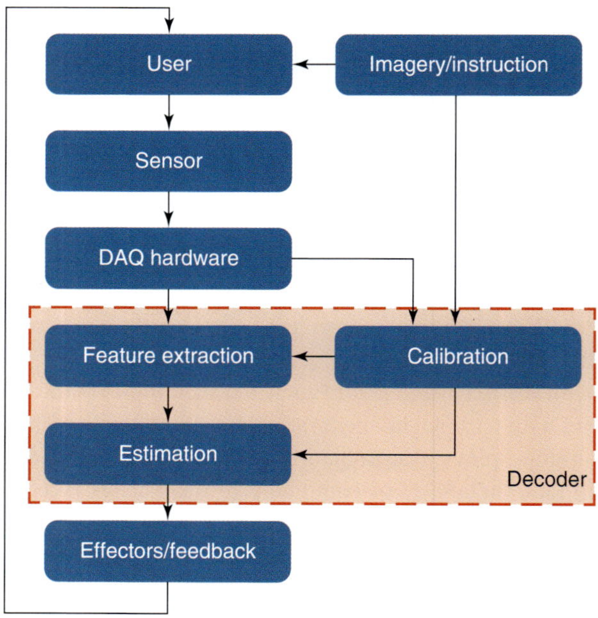

Figure 9.2 Block diagram of iBCI architecture (from Homer *et al.* 2013, with permission).

Table 9.1 Sources of neural information for brain computer interfaces

Scope	Device Name	Recording Device	Type of Brain Information	Hardware	Surgical Requirements	Timescale of signal resolution	Scale of spatial averaging
Macroscopic	Electroencephalogram	Scalp Electrodes	Spatially averaged neural action potentials	Bedside computer	No operations required	Milliseconds	Centimeters
	Functional Magnetic Resonance Imaging	MRI machine	Cerebral changes in ratio of oxy/de-oxy hemoglobin	Specialized facility	No operations required	Milliseconds to Seconds	Centimeters to millimeters
	Near Infrared Spectroscopy	Infrared light detectors	Cerebral changes in ratio of oxy/ de-oxy hemoglobin	Bedside computer	No operations required	Seconds	Centimeters
Mesoscopic	Magnetoencephalogram	MEG machine	Localized Magnetic fields	Specialized facility	No operations required	Milliseconds	Centimeters to millimeters
	Electrocorticogram	Cortical strip electrodes	Spatially averaged neural action potentials	Bedside computer	Craniotomy, meninges intact or opened	Milliseconds	Millimeters
Microscopic	Local Field Potentials/ multi-unit activity	Intracortical electrodes	Spatially averaged neural action potentials	Bedside computer	Meninges opened, cortex penetrated	Milliseconds	Millimeters to micrometers
	Single cell action potentials	Intracortical electrodes	Single neuron action potentials	Bedside computer	Meninges opened, cortex penetrated	Milliseconds	Micrometers
	and photon microscopy	Intracellular molecular dyes	Calcium/voltage concentration	Specialized microscope	Meninges opened, cortex penetrated	Milliseconds	Micrometers

9.4.1 Other neural activity-based signals for BCI devices

The process of generating action potentials for neurons is metabolically demanding; the energy is derived from the oxidation of glucose, as cells take oxygen and release carbon dioxide. Changes in local oxy- and deoxyhemoglobin concentrations putatively reflect aggregate neural activity. It is the ratio of these forms of hemoglobin that are the source of brain information in fMRI and near-infrared spectroscopy (fNIRS) devices.

Both devices thus use surrogates of neural activity rather than directly reflecting electric field changes from action potentials. fMRI (Lee *et al.* 2009) has been used as part of a BCI paradigm; however, the infrastructure requirements for fMRI are substantial and devices are restricted to academic and clinical research institutions. Within the foreseeable future, the large size, infrastructure requirements and spatio-temporal resolutions preclude it as a practical BCI intervention for daily use (Hochberg and Donoghue 2006). By contrast, fNIRS does not have the requirement for large, immobile equipment. In its current instantiation, fNIRS requires a substantial time-window to generate reliable binary responses, which makes it viable as a slow but potentially useful BCI (Gallegos-Ayala *et al.* 2014).

Since action potentials are changing electric fields, magnetic fields are simultaneously generated; localizing the sources of the magnetic field to specific regions of the brain is the underlying principle behind magnetoencephalography (MEG). MEG has been used as a BCI (Mellinger *et al.* 2007); however, much like fMRI, the overhead for utilizing these devices in its current form makes them restricted to academic settings.

Developed in the early part of the 20th century, the electroencephalogram (EEG) is routinely used as a tool in neuroscience, a diagnostic tool in clinical medicine, and more recently, as a source of neural signals for BCI. Recordings provided by EEGs are obtained by placing electrodes on the scalp. Electrical signals reflect volume averaging of huge numbers of changes in neural electrical fields oriented perpendicular to the skull. In general, inputs for EEG-BCI devices rely on changes in specific frequencies of bands (Niedermeyer *et al.* 2011), whether generated reflecting voluntary or non-voluntary signals (Daly and Wolpaw 2008; Hochberg and Donoghue 2006).

This technique does not require surgical intervention and may be performed without specialized institutional equipment. However, many EEG devices require multiple individual electrodes to be placed on the participant's scalp every time the device is to be used (often with conducting gel). Not only does this process require a technician or a caretaker, but it also requires a substantial time investment. Devices have been developed which allow for the electrodes to be collected within a hat, although in its current instantiation the signal-to-noise ratio (SNR) is generally worse than the traditionally used electrode setups. As a source of neural signals, recordings are made roughly 1–3 cm off the surface of the cortex (Cuffin 1993; Stokes *et al.* 2005); consequently, the signals are filtered, attenuated and spatially averaged by the skull as well as the multiple soft tissue layers.

Subdural electrodes have been used for decades as part of the pre-surgical evaluation for individuals potentially undergoing surgical resection for epilepsy. In an operating room, neurosurgeons place flexible strips of electrodes onto the cortical surface. The electrodes are then used to identify specific seizure foci based on aggregate signals of the thousands of neurons around the electrodes. More recently, they have been investigated as sensors for BCI devices (Schalk and Leuthardt 2011). In able-bodied participants undergoing surgical evaluation, they can be used for decoding movements (Bundy *et al.* 2012; Milekovic *et al.* 2012), imagined words (Pei *et al.* 2011) and for closed-loop control of prosthetic limbs (Fifer *et al.* 2014).

There are several advantages of using cortical surface electrodes. First, compared to surface EEGs, the recorded brain information has substantially higher SNR, due to substantially less signal attenuation and filtering. Second, surface electrodes are a standard neurosurgical procedure (i.e. are placed at most academic neurosurgical centers).

By contrast, in their current form, electrodes are designed for short-term use only: patients are generally admitted to hospital and are continuously watched for days to weeks, as multiple wires directly connect the brain-surface electrodes to the outside world. Patients are often placed on continuous antibiotics (which has its own set of deleterious consequences), and prolonged use of the electrodes is known to increase infection risk (Voorhies and Cohen-Gadol 2013). We would expect, however, that as surface electrodes become smaller, the silicone substrate between electrodes decreases in mass, and wireless systems for transmitting their signals become available, such electrocorticography (ECoG) approaches to BCI will prove valuable.

Motor control is ultimately controlled at the level of individual neurons *viz.* their complex spiking behavior. The highest possible resolution for decoding motor control is therefore reflected in recording action potentials directly from neurons. In the next sections, we detail how recordings are obtained and subsequently processed into useful signals.

9.4.2 Electrodes placed in the cortex record action potentials from neurons

Single-cell recordings are obtained using small conducting electrodes placed directly into the cortex. Essentially, electrode tips record voltage changes in the local extracellular environment within a sphere with a radius on the order of 50–100 micrometers (Quiroga 2009). The dynamics of how electrodes interact with brain tissue are reviewed in greater detail in Chapter 4, as well as by Buzsáki and colleagues (Buzsáki *et al.* 2012; Gold 2006). In brief, electrode voltage signals depend on: (i) the type and orientation of neurons being recorded; (ii) the size, shape and conductance properties of the electrodes; and (iii) the position of the electrode with respect to neurons. It should be noted that foreign objects inserted into the cortex can induce an inflammatory reaction. The extent to which the inflammatory response affects the recording device itself, or the local neural activity available to be recorded, is an active field of investigation (Chapter 2).

Multiple kinds of recording devices exist to record from single cells in the cortex (Homer *et al.* 2013). The device currently used as part of ongoing clinical trials is the Utah microarray (formally, Blackrock/NeuroPort array) (Barrese *et al.* 2013; Maynard *et al.* 1997). The device is a 4.2 × 4.2 mm platform upon which 100 tapered, boron-doped silicon shanks are placed. The shanks are 1.0 and 1.5 mm in length, separated from each other by 400 μm. The tip of each shank exposes a platinum (or iridium oxide)-tipped microelectrode 50–150 m long, coated with vapor-deposited platinum. Voltage recordings from the microelectrodes are transferred through 25 μm diameter gold/platinum alloy cables potted with silicone elastomer, which connect the array to a skull-mounted 96-pin titanium pedestal. Information is then transferred via cable to signal processing hardware and subsequently to computers for decoding.

Utah arrays are typically inserted up to 1.5 mm into the cortex using a pneumatic insertion device (Rousche and Normann 1992). These devices displace a piston by a precise amount, ensuring that an inserted Utah array is implanted in a consistent manner. Pneumatic insertion may optimize signal quality (in part, by placing the electrodes at the intended cortical depth) and minimize trauma to the cortex (Rennaker *et al.* 2005).

The electrical signals from each electrode provide raw voltage recordings sampled at 20–40 KHz per channel. These signals may be processed in a variety of ways, but are generally band-pass filtered to yield information in one of two categories: local field potentials (LFPs) or spikes. LFPs reflect spatially averaged electrical fields generated from groups of neurons (Buzsáki 2004). A spike refers to the recorded time of the maximum voltage change of a neuron's action potentials. Note that this can cause some confusion since EEGs also have transient amplitude fluctuations referred to as "spikes", though these are the result of synchronous activity in a large number of neurons as described above.

Both LFPs and single-unit spikes correlate with physiological behavior and have been investigated as sources for decoding (Ajiboye *et al.* 2012; Bansal *et al.* 2012; Perge *et al.* 2014). Spikes have historically been used as the primary input for decoders in BCI devices, and have been most commonly used in the ongoing clinical trials. In the following section, we describe the process of extracting spikes from raw voltages.

9.4.3 Raw voltage signals are processed into spikes

The process in taking voltage information to spike times involves several steps (Quiroga 2012), and are outlined in Figure 9.3. The electrical signal from each electrode is first band-pass filtered (i.e. using a Butterworth filter) to isolate a particular type of signal – frequencies less than 200 Hz are generally used for LFP signals, whereas the 300 and 10 000 Hz range are used for spikes. Next, a researcher-specified SNR threshold is assigned, and any voltage signal that surpasses the cutoff is designated as a (putative) spike event. Intuitively, the researcher-defined SNR threshold sets the cutoff in a receiver-operator curve: decreasing the recording threshold increases the rate of false positives and raising the threshold increases the rate of false negatives.

As described above, the electrode position relative to the neuron affects the characteristic waveform recorded during an action potential. When spikes are identified using threshold crossing from a single electrode, several different waveform patterns may sometimes be found: these are thought to correspond to spikes from multiple

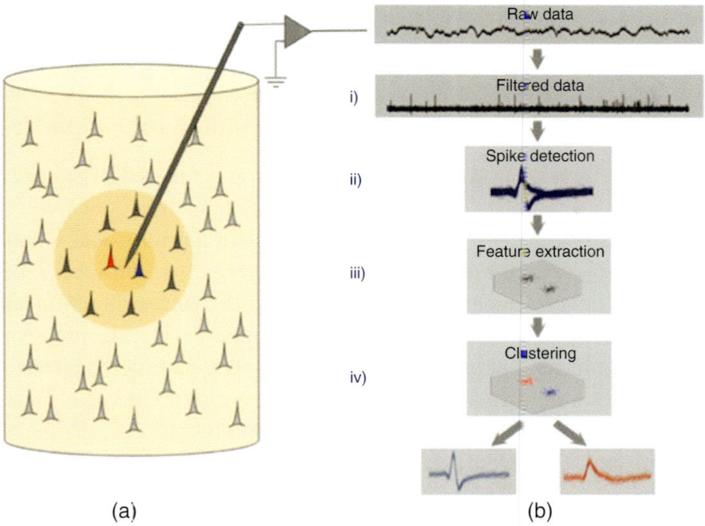

Figure 9.3 Schematic describing the conceptual steps involved in obtaining neural information. An electrode is placed into the cortex, which records voltage at a high sample rate. The voltage signals are band-pass filtered, and the spikes are identified where the filtered voltage exceeds some threshold. Multiple algorithms exist for spike sorting, that is, identifying specific waveforms of voltage signals corresponding to neuronal depolarizations (from Quiroga 2012, with permission).

neurons (Figure 9.3). To distinguish between these neurons, a fourth step called spike sorting may be performed, which identifies features of the spiking event voltage waveforms and clusters them based on shapes (Lewicki 1998; Pedreira *et al.* 2012; Wild *et al.* 2012). Multiple algorithms exist for spike sorting, based on alternative methods of waveform identification and clustering methods (Hulata *et al.* 2002; Quiroga *et al.* 2004; Vargas-Irwin and Donoghue 2007). The process of performing spike sorting into "single units" is computationally intensive, and often requires researcher input to perform optimally (and is therefore subject to human error) (Wood *et al.* 2004). Recent studies suggest that spike sorting may not be necessary to obtain reliable spike signals for BCI decoding (Hochberg *et al.* 2012; Perge *et al.* 2014), although see Todorova *et al.* (2014).

Having discussed the physiological basis of spikes, and how spikes are extracted from the underlying voltage fluctuations, we now turn our attention to exploring how BCI decoding algorithms use spikes to control external devices.

9.5 BCI DECODING AND APPLICATIONS

iBCIs often use spike information to control external devices, such as a computer cursor, wheelchair or robotic arm. This process is called signal decoding. In this section, we begin with a discussion of some common decoding algorithms, and then describe how the algorithms have been applied in NHPs as well as the recent pilot clinical investigations.

9.5.1 BCI decoders convert neural information into control of devices

As described in the first section of this chapter, neurons in the NHP motor cortex can have preferred directions: they increase or decrease their firing rates when the animal reaches towards or away from this direction (Georgopoulos and Kalaska 1982; Georgopoulos *et al.* 1986). By determining the preferred direction of each neuron, we can then construct a three-dimensional vector, where the vector direction and magnitude are representative of the neuron's preferred direction and spike rate respectively. Summing all vectors over a population of simultaneously recorded neurons generates a direction in three-dimensional space which may be interpreted as a decoded location of reach endpoint. An extension of this principle is the basis of linear position decoding algorithm. This idea may extend to other kinematic variables, including velocity.

In previous experiments, Utah arrays recorded from CS neurons in the motor cortex to allow NHPs to control the velocity of a cursor on a computer screen (Serruya *et al.* 2002; Taylor *et al.* 2002). After spike sorting, the array provides spikes from N neurons simultaneously (note that the Utah array provides 96 channels, but after spike sorting, each electrode may reflecting spiking behavior from zero, one or a few neurons). We can bin the spikes into time intervals (e.g. 50 milliseconds long). We will call $\mathbf{z}_t$ an $N \times 1$ vector describing the firing rate of the neuronal population at time t. We want to translate this recorded firing rate to control a cursor in D dimensions, and we can represent a parameter such as velocity of this cursor as $\mathbf{v} = [\mathrm{v}_x\ \mathrm{v}_y]$. A linear decoder transforms $\mathbf{z}_t$ into $\mathbf{x}_t$ according to a contribution of the firing rates for the N neurons. That is:

$$\mathrm{v}_t = \mathrm{b} + \mathrm{w}\,\mathrm{z}_t \tag{9.1}$$

where b is a constant offset vector and w is a weighting matrix of dimensions $D \times N$, specifying the preferred direction of each neuron. The values of b and w are determined experimentally. This decoder takes advantage of what is understood from cortical physiology: namely, that the brain encodes information in rate and recruitment, and that neurons in M1 have directional tuning properties. This example linear decoder is excellent for understanding the underlying principles of decoding; however, it has some important limitations that have motivated algorithmic advancements in the BCI field (Homer *et al.* 2013; Koyama *et al.* 2009).

First, the signals are noisy. Neuronal firing is an inherently stochastic process (Kostal *et al.* 2007), and there is tremendous variability in the timing of spike trains, even within a single neuron between trials of the same movement (Cunningham *et al.* 2009). Furthermore, the voltage recordings are sensitive to electrical noise in the environment and/or positional effects of the electrodes and tissue response around the electrode array (Chapter 2). One approach to data smoothing is to bin spikes into time windows, as was done in the preceding example. In establishing a time window over which to determine firing rates, BCI decoding algorithms must therefore balance two conflicting properties: a time window that is too small is more sensitive to noise (Koyama *et al.* 2009), whereas one that is too large averages too much time-varying information and functions poorly. One approach is to include several previously binned time steps as the input into the decoder, thereby integrating neural signals over time (Kim *et al.* 2008). An alternative approach is to perform a convolution of spike times represented with multiple Kronecker delta functions with some sort of finite-integral function (i.e. a Gaussian distribution). This process reduces noise and may facilitate some decoding algorithms (Cunningham *et al.* 2009).

Second, linear decoders are sensitive to experimental sampling. When only a small sub-population of neurons is used as the input for the decoder, the most basic linear decoder may fail to provide optimal control of an external device. Thus, many alternative decoding algorithms have been developed over the last few decades that have provided improvements on the basic design.

One of the most successful algorithms is the Kalman filter (Wu *et al.* 2006). Briefly, the Kalman filter is a hidden Markov model in which the known variable is the current estimation of the firing rate, and the hidden variables depend on the firing rate and the previous estimate of the hidden variable and Gaussian noise. To say it another way, the Kalman filter generates a "best guess" for intended device control combining (i) the system's previous state with (ii) the current readout of the neuron firing rates. Many improvements have been made to the implementation of the Kalman filter in BCI devices (Gilja *et al.* 2012; Homer *et al.* 2014; Jarosiewicz *et al.* 2013; Kao *et al.* 2013; Malik *et al.* 2011).

Another approach to decoding disregards data binning altogether, focusing instead on developing models to predict future-spiking behavior based on previous behavior (Brown *et al.* 2002; Truccolo *et al.* 2004, 2009). These so-called point-process methods do not apply convolution or binning approaches to data, but instead use generalized linear models to generate distributions describing the timing of individual spikes. Point-process methods can predict CS neuron behavior in human participants remarkably well, and have tremendous promise as a real-time decoding system (Truccolo *et al.* 2009).

9.5.2 BCI decoders allow for the control of prosthetic devices

Insights into the underlying neurophysiology, engineering advances allowing for simultaneous recording from populations of neurons, and the development in decoding algorithms all set the stage for position decoding in NHPs. Multiple cortical electrodes were demonstrated to predict actual performed movements in one dimension (Burrow *et al.* 1997) and subsequently three-dimensional tasks (Isaacs *et al.* 2000; Wessberg *et al.* 2000). Subsequent work demonstrated that small, localized ensembles of neurons in M1 could decode 25 joint angles, thereby containing sufficient information to recreate naturalistic grasping (Vargas-Irwin *et al.* 2010). It should be emphasized here that these experiments are "open loop", that is, there was no feedback from the decoded neural signal to influence the primates' behavior. This open-loop recording paradigm is fundamentally different from closed-loop systems. As defined in Section 9.3, closed-loop decoding allows for users to have direct, real-time control over external devices using decoded signals.

An early example of closed-loop control was performed by Fetz (1969), who demonstrated that NHPs could modulate firing activity of individual neurons based on auditory feedback. Complex closed-loop control from motor cortex neurons was demonstrated by having NHPs control computer cursors in real time (Carmena *et al.* 2003; Musallam *et al.* 2004; Santhanam *et al.* 2006; Serruya *et al.* 2002; Taylor *et al.* 2002). Later experiments have used closed-loop control of visual targets on a computer monitor as a mechanism of control for robotic prosthetic arms (Carmena *et al.* 2003; Taylor *et al.* 2003). A demonstration of closed-loop control of a robotic arm (without this intermediate cursor-control step) subsequently allowed a Rhesus monkey to feed itself marshmallows (Velliste *et al.* 2008).

Intracortical recordings from single neurons in humans for external device control was first performed in the 1990s by Kennedy and Bakay (1998). A specialized electrode with two recording wires and a neutrophic factor (Kennedy 1989) was inserted into the M1 region of the cortex of a participant with ALS. Within weeks of implantation, neural signals could be reliably recorded from the subject. Real-time auditory and visual feedback allowed the participant to modulate motor cortex neurons in a closed-loop BCI paradigm. A follow-up publication showed that a participant with an implant was able to drive the unidirectional movement of a computer cursor, at a rate proportional to the firing rate (Kennedy *et al.* 2000).

While Kennedy and colleagues were able to demonstrate recordings from small numbers neurons, previous NHP work suggested that populations of neurons were required as input for decoders to provide robust control of devices with multiple degrees of freedom. The first example of simultaneous recording from a population of neurons in a closed-loop BCI system in humans was demonstrated by Hochberg *et al.* (2006) – a conceptual drawing of the set-up is shown in Figure 9.4. Two individuals with SCIs resulting in tetraplegia were implanted with intracortical Utah microarrays in M1. In both instances, neurons in M1 were isolated that were directionally tuned to imagined and/or intended movements, and were found to modulate their activity to the onset of movement cues (Truccolo *et al.* 2008).

Spikes were manually discriminated by a technician using visual features, after a manually adjusted signal threshold. A linear position decoder was used with 20 consecutive 50-ms time bins in outputting a cursor. With this set-up, a subject demonstrated control of a two-dimensional neural cursor; to play "Neural Pong" (similar to the classic type but controlled entirely through neuronal M1 spike trains); manipulate television controls (see supplementary videos from this publication); and even control a rudimentary robotic device. Importantly, the subject quickly learned how to achieve control of external devices and could perform these tasks while conversing with researchers, suggesting that control of a prosthetic device could become as intuitive as voluntary limb movement is for able-bodied individuals.

Since the publication in 2006, numerous advances have been made in neural decoding. For instance, velocity-based decoding (as outlined in the example in Section 9.3) has shown superior neural control over position, due to both greater correlations between cursor kinematics and neural activity, as well as ease of system

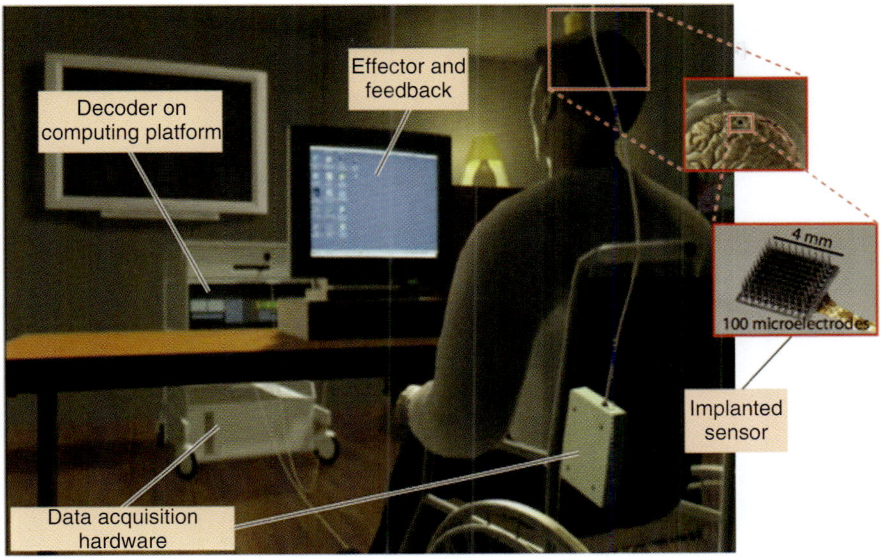

Figure 9.4 Conceptual drawing of a iBCI set-up. An implanted sensor (in this figure, Utah array) is implanted into the M1 of an individual with motor impairment. The neural signal is transmitted from the electrode to a pedestal fixated to the skull. Wires from the pedestal then descend to the data acquisition hardware located on the wheelchair, and subsequently to an external computer for additional processing. In a closed-loop setting, information from the decoded signal is interpreted and the real-time results are provided to the participant as feedback (from Homer *et al.* 2013, with permission).

calibration (Kim *et al.* 2008). Kalman filters have been shown to provide superior decoding performance over linear filters (Kim *et al.* 2008). The tuning properties of neurons may be dramatically different between off-line and on-line cursor control (Jarosiewicz *et al.* 2013; Kim *et al.* 2008; Perge *et al.* 2013), and Kalman decoding incorporating real-time detection of these changes has suggested the potential for improved cursor control (Homer *et al.* 2014). Decoders have also been developed to provide multi-state decoding to allow for a click-signal in cursor control (Kim *et al.* 2011).

The feasibility of long-term intracortical recording was realized when participants demonstrated successful decoding in an implanted array over 1000 days (Simeral *et al.* 2011) and subsequently 5 years after implantation (Hochberg *et al.* 2012). The same participant described in this report was able to use a robotic prosthetic arm to grasp a bottle, bring it towards her mouth and drink from it (Hochberg *et al.* 2012) – this was the first time in nearly 15 years that she was able to take a drink solely of her own volition (Figure 9.5). This research was independently replicated and expanded with the participation of a 52-year-old woman diagnosed with spinocerebellar degeneration (Collinger *et al.* 2013). The participant was able to show gradual improvement with control of a complex robotic arm (Bridges *et al.* 2011), until she was able to control seven independent dimensions of a motor prosthetic arm and hand.

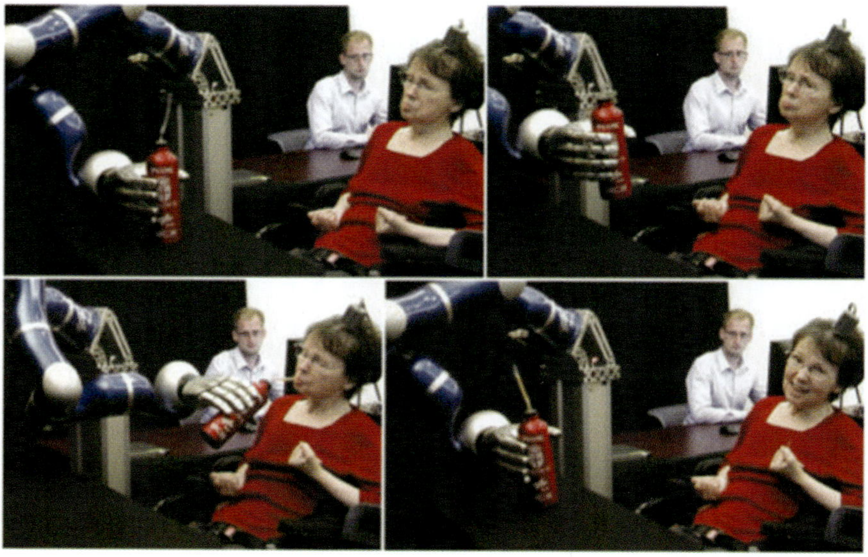

Figure 9.5 iBCI-based neural control of a robotic arm in a person with LIS. In this photograph, the iBCI user is bringing herself a cup of coffee. This was the first time in nearly 15 years that she was able to take a drink solely of her own volition (from Hochberg *et al.* 2012, with permission).

For both of the above studies, leads running from the cortical recording device were attached to a titanium pedestal affixed to the skull and the pedestal protruded through a gap in the scalp. The pedestal was then connected by a cable to a cart of desktop computers for processing the necessary calculations for the control of an external device.

The current generation of iBCI devices have shortcomings, and are not yet ready for widespread clinical use. The field, however, is rapidly evolving as a result of multidisciplinary collaborations between healthcare professionals, neuroscientists, biomedical engineers, computer scientists and mathematicians. In the following section, we will outline some future directions for this technology.

9.6 FUTURE DIRECTIONS

Health-related medical devices undergo iterative improvements. Cardiac pacemakers for arrythmias and deep brain stimulators for PD began as big, bulky, experimental devices requiring wires protruding through the skin. Through collaborations between academia and industry, these devices have become miniaturized and fully implantable (Chapter 1). They provide safe and effective treatment and have become the standard of care for individuals worldwide.

The ultimate goal of an intracortical recording for people with paralysis and whether it is paralysis of a limb, amputations, tetraplegia or LIS, is to circumvent the dysfunctional motor system and to restore functional independence. An ideal device would provide an effective, safe and practical route to this goal (Hochberg and Donoghue 2006).

9.6.1 Scientific and engineering directions for developing BCI technology

Any break in the surface of the skin is susceptible to bacterial infection. Ideally, therefore, intracortical recording devices will be fully implantable. In recent publications by engineers at Brown University, intracortical recordings were wirelessly transmitted through a hermetically sealed and fully implanted titanium system (Borton and Nurmikko 2013; Yin *et al.* 2013) (Chapter 5). The device has been successfully used in NHPs and is being prepared for potential investigational use in people.

While the results of the ongoing pilot clinical trials are promising, the current level of control achieved by participants controlling prosthetic devices does not resemble able-bodied human movements. The Utah arrays used for these trials record dozens of neurons in M1. Although NHP work suggests that this order of magnitude is sufficient for reasonable external device control (Vargas-Irwin *et al.* 2010), it may be that additional devices (or even more closely spaced devices) could provide superior input signals. Furthermore, for reasons outlined in Section 9.2, Utah arrays have been primarily focused in iBCI work on neurons in the motor cortex. Sampling from other cortical layers implicated in local neural circuits in M1, or recording other brain regions involved in movement (i.e. SMA, PMv, PMd, parietal cortex as well as subcortical areas) could optimize decoding. Superior and lateral structures of the brain are easily accessible by neurosurgical techniques and could use the Utah array; deeper brain and medial structures would require different recording probes.

In addition to selecting other cortical regions, one obstacle towards providing a level of control comparable to able-bodied human movements may be a lack of proper feedback (Bensmaia and Miller 2014). Currently, real-time feedback from decoded iBCI signals are presented visually back to the participant. This approach does not take advantage of the usual somatosensory feedback that occurs in normal human movements.

Providing sensory feedback could occur in several ways. For instance, for people in whom spinal cord sensory pathways are intact, the sensation of proprioception can be generated by stimulating muscle tendons (De Nunzio *et al.* 2008; Pettorossi and Schieppati 2014). This approach is used as part of physical rehabilitation, and has been demonstrated recently in a medical trial, which reported improvement in gait with tendon vibration in individuals with PD – a population with an intact somatosensory system but aberrant posture and gait (Volpe *et al.* 2014). In future, iBCI research, tendon vibration could be used to provide the sensation of proprioception for individuals with iBCIs and intact somatosensory inputs – such as in the case of a pontine stroke or ALS.

However, not all populations that stand to benefit from iBCI have intact somatosensory pathways. Direct stimulation of the sensory cortex may be a viable approach to providing feedback and can emulate somatosensory precepts (Tabot *et al.* 2013). Experiments in NHPs suggest congruent proprioceptive information improves motor control (Suminski *et al.* 2010). Furthermore, initial experiments in NHPs have demonstrated that direct stimulation of area 3A (primary sensory cortex involved in proprioception) can decrease reaction time in a two-way decision task (London *et al.* 2008). Thus, while sensory feedback has already been demonstrated not to be required for initial, useful BCI performance, there is tremendous potential for supplementing visual feedback with sensory feedback in improving iBCI control in the future.

The current implementation of iBCI technology requires a technician to oversee device use. A goal is to move away from intensive technician supervision – for instance, changing from technician-supervised spike-sorting to using computer-controlled threshold crossing (Hochberg *et al.* 2012; Quiroga and Panzeri 2009). However, the set-up process (i.e. connection of cables) and start-up calibration process still requires human intervention. With ongoing improvements in algorithms and development of fully-implantable systems, the requirement for technician intervention will gradually decrease. Thus, a future goal for the field is to permit system function with only minimal intervention from trained technicians or caretakers, who would only be involved in the initial device training and periodic maintenance (such as the current model of deep brain stimulators and cardiac pacemakers).

A wireless intracortical recording device must not only be safe and effective, but also durable. Current deep-brain stimulators use either rechargeable or one-time use batteries. The former are charged by induction by placing a coil above the surface of the battery, while the latter must be replaced every few years (depending on the current requirements for an individual patient). A fully implantable iBCI technology that is powerful enough to record from multiple electrodes simultaneously (at a high sampling rate) and transmit the information wirelessly will require substantial power supply. Thus, an ongoing challenge in the field is matching power consumption needs to battery longevity (whether replaceable or rechargeable), while minimizing heat loss into surrounding tissues – an issue that is especially important for implanted brain devices that contact the brain.

Implanting electrodes into the cortex results in a local inflammatory response (Chapter 2). While this may impact recording quality, a recent study investigating modes of failure with implanted electrodes in NHPs identified other causative factors for signal degradation (Barrese *et al.* 2013). While reliable decoding in humans has been demonstrated years after implantation (Hochberg *et al.* 2012), any device must be developed with the intention of being effective over decades. Further investigations are required to ensure the devices remain effective and safe on this timescale.

9.6.2 Clinical directions for development of BCI technology

Given the rapid evolution of BCI technology over the last decade, its multidisciplinary nature and the substantial patient need, the development of BCI technology will require input from experts from a variety of fields. Such a multifaceted approach, in the context of specialized rehabilitation requirements for people with LIS or tetraplegia, will be important.

The use of clinical outcome scales allows clinicians to quantify intrinsically qualitative physiological processes. One example is the utilization of a modified clinical measure for arm control (commonly used as a metric for stroke rehabilitation), used to provide a framework from which to quantify the participant's ability to use a prosthetic arm (Collinger *et al.* 2013; Lang *et al.* 2006). A challenge for the field is to develop clinical metrics responsive to BCI usage that quantifies improvements in clinically relevant outcomes. In doing so, implementation of these devices can be systematically studied, and hopefully will be shown to provide tangible improvements for people with paralysis.

Multiple other modalities are being explored to improve functional independence with individuals with severe motor deficits. For instance, functional electrical stimulation is a technique that uses electrical current to result in direct coordinated stimulation of muscles, generating controlled contractions (Collinger *et al.* 2013; Lang *et al.* 2009; Memberg *et al.* 2014a, 2014b). Potential future work would integrate neural signals derived from M1 to result in direct stimulation of users' muscles, with the intention of reanimating fluid, intuitive movement (Chadwick *et al.* 2009).

9.7 CONCLUSION

The field of iBCIs is a multidisciplinary endeavor. Building on observations made in neuroscience laboratories from the second half of the 20th century, the field has grown from proof-of-concept animal models to initial clinical trials in humans. As the field continues to evolve, new innovations will yield clinically useful iBCI devices to help people with paralysis, and the early experience with those devices will undoubtedly lead to an evolution of evermore flexible and powerful devices for restoring lost communication, mobility and independence.

REFERENCES

Ahdab, R. *et al.* (2013) Reappraisal of the anatomical landmarks of motor and premotor cortical regions for image-guided brain navigation in TMS practice. *Hum. Brain Mapp.*, **35**(5): 2435–2447.

Ajiboye, A.B., Simeral, J.D. and Donoghue, J.P. (2012) Prediction of imagined single-joint movements in a person with high-level tetraplegia. *IEEE Trans. Biomed. Eng.*, **59**(10): 2755–2765

Anderson, K.D. (2004) Targeting recovery: priorities of the spinal cord-injured population. *J. Neurotrauma*, **21**(10): 1371–1383.

Anderson, K.D. (2009) Consideration of user priorities when developing neural prosthetics. *J. Neural. Eng.*, **6**(5): 055003.

Anderson, K.D. *et al.* (2011) United States (US) multi-center study to assess the validity and reliability of the Spinal Cord Independence Measure (SCIM III). *Spinal Cord*, **49**(8): 880–885.

Angeli, C.A. *et al.* (2014) Altering spinal cord excitability enables voluntary movements after chronic complete paralysis in humans. *Brain*, **137**(5): 1394–1409.

ALS Association (2014) About ALS. Available at: http://www.alsa.org/about-als/ [Accessed December 1, 2014].

Bansal, A.K. *et al.* (2012) Decoding 3D reach and grasp from hybrid signals in motor and premotor cortices: spikes, multiunit activity, and local field potentials. *J. Neurophys.*, **107**(5): 1337–1355.

Barrese, J.C. *et al.* (2013) Failure mode analysis of silicon-based intracortical microelectrode arrays in non-human primates. *J. Neural Eng.*, **10**(6): 066014.

Batista, A.P. *et al.* (1999) Reach plans in eye-centered coordinates. *Science*, **285**(5425): 257–260.

Bensmaia, S.J. and Miller, L.E. (2014) Restoring sensorimotor function through intracortical interfaces: progress and looming challenges. *Nature Rev. Neurosci.*, **15**(5): 313–325.

Borton, D.A. and Nurmikko, A.V. (2013) *Future Trends in Microelectronics.* S. Luryi, J. Xu and A. Zaslavsky (Eds), John Wiley & Sons, Inc., Hoboken NJ.

Brass, M. and Haggard, P. (2008) The what, when, whether model of intentional action. *The Neuroscientist*, **14**(4): 319–325.

Bridges, M.M., Para, M.P. and Mashner, M.J. (2011) Control system architecture for the modular prosthetic limb. *Johns Hopkins APL Technical Digest*, **30**(3): 217–222.

Brown, E. *et al.* (2002) The time-rescaling theorem and its application to neural spike train data analysis. *Neural Computation*, **14**(2): 325–346.

Bundy, D.T. *et al.* (2012) Using ipsilateral motor signals in the unaffected cerebral hemisphere as a signal platform for brain–computer interfaces in hemiplegic stroke survivors. *J. Neural Eng.*, **9**(3): 036011.

Buneo, C.A. and Andersen, R.A. (2006) The posterior parietal cortex: sensorimotor interface for the planning and online control of visually guided movements. *Neuropsychologia*, **44**(13): 2594–2606.

Buneo, C.A. *et al.* (2002) Direct visuomotor transformations for reaching. *Nature*, **416**(6881): 632–636.

Burrow, M. *et al.* (1997) Cortical control of a robot using a time-delay neural network, in *Proceedings of the Fifth International Conference on Rehabilitation Robotics*.

Buzsáki, G. (2004) Large-scale recording of neuronal ensembles. *Nature Neurosci.*, **7**(5): 446–451.

Buzsáki, G., Anastassiou, C.A. and Koch, C. (2012) The origin of extracellular fields and currents – EEG, ECoG, LFP and spikes. *Nature Publishing Group*, **13**(6): 407–420.

Carmena, J.M. *et al.* (2003) Learning to control a brain–machine interface for reaching and grasping by primates. *PLoS Biology*, **1**(2): e42.

Chadwick, E.K. *et al.* (2009) A real-time, 3-D musculoskeletal model for dynamic simulation of arm movements. *IEEE Trans. Biomed. Eng.*, **56**(4): 941–948.

Cheney, P.D. *et al.* (2000) Cortical motor areas and their properties: implications for neuroprosthetics. *Progress in brain research*, **128**: 135–160.

Cohen, N.R. *et al.* (2009) Ventral and dorsal stream contributions to the online control of immediate and delayed grasping: a TMS approach. *Neuropsychologia*, **47**(6): 1553–1562.

Collinger, J.L. *et al.* (2013) High-performance neuroprosthetic control by an individual with tetraplegia. *The Lancet*, **381**(9866): 557–564.

Cuffin, B.N. (1993) Effects of local variations in skull and scalp thickness on EEGs and MEGs. *IEEE Trans. Biomed. Eng.*, **40**(1): 42–48.

Cunningham, J.P. *et al.* (2009) Methods for estimating neural firing rates, and their application to brain–machine interfaces. *Neural Networks*, **22**(9): 1235–1246.

Cunningham, J.P. *et al.* (2011) A closed-loop human simulator for investigating the role of feedback control in brain-machine interfaces. *J. Neurophys.*, **105**(4), 1932–1949.

Daly, J.J. and Wolpaw, J.R. (2008) Brain–computer interfaces in neurological rehabilitation. *The Lancet Neurology*, **7:** 1032–1043.

Davare, M. *et al.* (2011) Interactions between areas of the cortical grasping network. *Curr. Op. Neurobiol.,* **21**(4): 565–570.

De Nunzio, A.M. *et al.* (2008) Alternate trains of postural muscle vibration promote cyclic body displacement in standing Parkinsonian patients. *Movement Disorders*, **23**(15): 2186–2193.

Dijkers, M.P.J.M. (2005) Quality of life of individuals with spinal cord injury: a review of conceptualization, measurement, and research findings. *J. Rehab Res. Develop.*, **42**(3, Suppl. 1): 87–110.

Dum, R.P. and Strick, P.L. (2005) Frontal lobe inputs to the digit representations of the motor areas on the lateral surface of the hemisphere. *J. Neurosci.*, **25**(6): 1375–1386.

Eisenberg, M. *et al.* (2011) The representation of visual and motor aspects of reaching movements in the human motor cortex. *J. Neurosci.,* **31**(34): 12377–12384.

Evarts, E.V. (1968) Relation of pyramidal tract activity to force exerted during voluntary movement. *J. Neurophys.,* **31**(1): 14–27.

Fehlings, M.G. *et al.* (2012) Early versus delayed decompression for traumatic cervical spinal cord injury: results of the surgical timing in acute spinal cord injury study (STASCIS). *PLoS ONE*, **7**(2): e32037.

Fetz, E.E. (1969) Operant conditioning of cortical unit activity. *Science*, **163**(3870): 955–958.

Fifer, M.S. *et al.* (2014) Simultaneous neural control of simple reaching and grasping with the modular prosthetic limb using intracranial EEG. *IEEE Trans. Neural Syst. Rehabil. Eng.* **22**: 695–705.

Filimon, F. (2010) Human cortical control of hand movements: parietofrontal networks for reaching, grasping, and pointing. *The Neuroscientist*, **16**(4): 388–407.

Fried, I., Mukamel, R. and Kreiman, G. (2011) Internally generated preactivation of single neurons in human medial frontal cortex predicts volition. *Neuron*, **69**(3): 548–562.

Gallegos-Ayala, G. *et al.* (2014) Brain communication in a completely locked-in patient using bedside near-infrared spectroscopy. *Neurology*, **82**(21): 1930–1932.

Georgopoulos, A.P. and Kalaska, J.F. (1982) On the relations between the direction of two-dimensional arm movements and cell discharge in primate motor cortex. *J. Neurosci.*, **2**(11): 1527–1537.

Georgopoulos, A.P., Schwartz, A.B. and Kettner, R.E. (1986) Neuronal population coding of movement direction. *Science*, **233**(4771): 1416–1419.

Georgopoulos, A.P. *et al.* (2007) Mapping of the preferred direction in the motor cortex. *Proc. Nat. Acad. Sci. USA*, **104**(26): 11068–11072.

Gilja, V. *et al.* (2012) A brain machine interface control algorithm designed from a feedback control perspective. *Ann. Int. Conf. IEEE Eng. Med. Biol. Soc.*, pp. 1318–1322.

Gold, C. (2006) On the origin of the extracellular action potential waveform: a modeling study. *J. Neurophys.*, **95**(5): 3113–3128.

Harkema, S. *et al.* (2011) Effect of epidural stimulation of the lumbosacral spinal cord on voluntary movement, standing, and assisted stepping after motor complete paraplegia: a case study. *The Lancet*, **377**(9781): 1938–1947.

Hatsopoulos, N.G. and Donoghue, J.P. (2009) The science of neural interface systems. *Ann. Rev. Neurosci.*, **32**(1): 249–266.

Hatsopoulos, N.G. *et al.* (1998) Information about movement direction obtained from synchronous activity of motor cortical neurons. *Proc. Nat. Acad. Sci. USA*, **95**(26): 15706–15711.

He, S.Q., Dum, R.P. and Strick, P.L. (1993) Topographic organization of corticospinal projections from the frontal lobe: motor areas on the lateral surface of the hemisphere. *J. Neurosci.*, **13**(3):.952–980.

He, S.Q., Dum, R.P. and Strick, P.L. (1995) Topographic organization of corticospinal projections from the frontal lobe: motor areas on the medial surface of the hemisphere. *J. Neurosci.*, **15**(5): 3284–3306.

Hendrix, C.M., Mason, C.R. and Ebner, T.J. (2009) Signaling of grasp dimension and grasp force in dorsal premotor cortex and primary motor cortex neurons during reach to grasp in the monkey. *J. Neurophys.*, **102**(1): 132–145.

Hochberg, L.R. and Donoghue, J.P. (2006) Sensors for brain-computer interfaces. *IEEE Eng. Med. Biol.*, **25**(5): 32–38.

Hochberg, L.R. *et al.* (2006) Neuronal ensemble control of prosthetic devices by a human with tetraplegia. *Nature*, **442**(7099): 164–171.

Hochberg, L.R. *et al.* (2012) Reach and grasp by people with tetraplegia using a neurally controlled robotic arm. *Nature*, **485**(7398): 372–375.

Homer, M.L. *et al.* (2013) Sensors and decoding for intracortical brain computer interfaces. *Ann. Rev. Biomed. Eng.*, **15**(1): 383–405.

Homer, M.L. *et al.* (2014) Adaptive offset correction for intracortical brain computer interfaces. *IEEE Trans. Neur. Syst. Rehab. Eng.*, **22**(2): 239–248.

Hulata, E., Segev, R. and Ben-Jacob, E. (2002) A method for spike sorting and detection based on wavelet packets and Shannon's mutual information. *J. Neurosci. Meth.*, **117**(1): 1–12.

Humphrey, D.R. and Hochberg, L.R. (1995) Intracortical recording of brain activity for control of limb prostheses. *Proc. Rehab. Eng. Soc. N. Am.*, pp 650–658.

Humphrey, D.R., Schmidt, E.M. and Thompson, W.D. (1970) Predicting measures of motor performance from multiple cortical spike trains. *Science*, **170**(3959): 758–762.

Hwang, E.J. *et al.* (2014) Spatial and temporal eye-hand coordination relies on the parietal reach region. *J. Neurosci.*, **34**(38): 12884–12892.

Isaacs, R.E., Weber, D.J. and Schwartz, A.B. (2000) Work toward real-time control of a cortical neural prothesis. *IEEE Trans. Rehab. Eng.: IEEE Eng. Med. Biol. Soc.*, **8**(2): 196–198.

Jarosiewicz, B. *et al.* (2008) Functional network reorganization during learning in a brain–computer interface paradigm. *Proc. Nat. Acad. Sci. USA.*, **105**(49): 19486–19491.

Jarosiewicz, B. *et al.* (2013) Advantages of closed-loop calibration in intracortical brain–computer interfaces for people with tetraplegia. *J. Neural. Eng.*, **10**(4): 046012.

Kaas, J.H., Stepniewska, I. and Gharbawie, O. (2012) Cortical networks subserving upper limb movements in primates. *Eur. J. Phys. Rehab. Med.*, **48**(2): 299–306.

Kao, J. *et al.* (2013) Investigating the role of firing-rate normalization and dimensionality reduction in brain-machine interface robustness. *IEEE Eng. Med. Biol. Soc. Ann. Conf.*, pp 293–298.

Kaufman, M.T. *et al.* (2014) Cortical activity in the null space: permitting preparation without movement. *Nature*, **17**(3): 440–448.

Kennedy, P.R. (1989) The cone electrode: a long-term electrode that records from neurites grown onto its recording surface. *J. Neurosci. Meth.*, **29**(3): 181–193.

Kennedy, P.R. and Bakay, R. (1998) Restoration of neural output from a paralyzed patient by a direct brain connection. *Neuroreport*, **9**(8): 1707–1711.

Kennedy, P.R. *et al.* (2000) Direct control of a computer from the human central nervous system. *IEEE Trans. Rehab. Eng.*, **8**(2): 198–202.

Kilner, J.M. and Lemon, R.N. (2013) What we know currently about mirror neurons. *CURBIO*, **23**(23): R1057–R1062.

Kim, S. P. *et al.* (2008) Neural control of computer cursor velocity by decoding motor cortical spiking activity in humans with tetraplegia. *J. Neural Eng.*, **5**(4): 455–476.

Kim, S.P., Simeral, J.D. and Hochberg, L.R. (2011) Point-and-click cursor control with an intracortical neural interface system by humans with tetraplegia. *IEEE Tran. Neural Syst. Rehab. Eng.*, **19**(2): 193–203.

Kostal, L., Lansky, P. and Rospars, J.-P. (2007) Neuronal coding and spiking randomness. *Eur. J. Neurosci.*, **26**(10): 2693–2701.

Koyama, S. *et al.* (2009) Comparison of brain–computer interface decoding algorithms in open-loop and closed-loop control. *J. Comp. Neurosci.*, **29**(1–2): 73–87.

Lang, C.E. *et al.* (2006) Measurement of upper-extremity function early after stroke: properties of the action research arm test. *Arch. Phys. Med. Rehab.*, **87**(12): 1605–1610.

Laureys, S. *et al.* (2005) The locked-in syndrome: what is it like to be conscious but paralyzed and voiceless? in The Boundaries of Consciousness: Neurobiology and Neuropathology. Progress in Brain Research. Elsevier, pp 495–611.

Lee, J.H. *et al.* (2009) Brain–machine interface via real-time fMRI: preliminary study on thought-controlled robotic arm. *Neurosci. Lett.*, **450**(1): 1–6.

Lewicki, M.S. (1998) A review of methods for spike sorting: the detection and classification of neural action potentials. *Network (Bristol UK)*, **9**(4): 53–78.

London, B.M. *et al.* (2008) Electrical stimulation of the proprioceptive cortex (area 3a) used to instruct a behaving monkey. *IEEE Trans. Neural Syst. Rehab. Eng.*, **16**(1): 32–36.

Lulé, D. *et al.* (2009) Life can be worth living in locked-in syndrome. *Prog. Brain Res.*, **177**: 339–351.

Luppino, G. and Rizzolatti, G. (2000) The organization of the frontal motor cortex. *News Physiol. Sci.*, **15**: 219–224.

Malik, W.Q. *et al.* (2011) Efficient decoding with steady-state Kalman filter in neural interface systems. *IEEE Trans. Neur. Syst. Rehab. Eng.*, **19**(1): 25–34.

Maynard, E.M., Nordhausen, C.T. and Normann, R.A. (1997) The Utah intracortical electrode array: a recording structure for potential brain–computer interfaces. *Electroencep. Clin. Neurophys.*, **102**(3): 228–239.

Mehta, P. *et al.* (2014) Prevalence of amyotrophic lateral sclerosis and the United States, 2010–2011. *Morbidity and mortality weekly report. Surveillance summaries*, **63** (Suppl. 7): 1–14.

Mellinger, J. *et al.* (2007) An MEG-based brain–computer interface (BCI). *NeuroImage*, **36**(3): 581–593.

Memberg, W.D., Polasek, K.H. *et al.* (2014a) Implanted neuroprosthesis for restoring arm and hand function in people with high level tetraplegia. *Arch. Phys. Med. Rehab.*, **95**(6): 1201–1211.e1.

Memberg, W.D., Stage, T.G. and Kirsch, R.F. (2014b) A fully implanted intramuscular bipolar myoelectric signal recording electrode. *Neuromodulation: Tech. Neural Interface*, **17**(8): 794–799

Milekovic, T. *et al.* (2012) An online brain–machine interface using decoding of movement direction from the human electrocorticogram. *J. Neural Eng.*, **9**(4): 046003.

Miller, R.G., Mitchell, J.D. and Moore, D.H. (2012) Riluzole for amyotrophic lateral sclerosis (ALS)/motor neuron disease (MND). *The Cochrane Database of Systematic Reviews*, **3**: CD001447.

Mitchell, J.D. and Borasio, G.D. (2007) Amyotrophic lateral sclerosis. *The Lancet*, **369**(9578): 2031–2041.

Mulliken, G.H., Musallam, S. and Andersen, R.A. (2008) Decoding trajectories from posterior parietal cortex ensembles. *J. Neurosci.,* **28**(48): 12913–12926.

Musallam, S. *et al.* (2004) Cognitive control signals for neural prosthetics. *Science*, **305**(5681): 258–262.

Mushiake, H., Inase, M. and Tanji, J. (1991) Neuronal activity in the primate premotor, supplementary, and precentral motor cortex during visually guided and internally determined sequential movements. *J. Neurophysiol.* **66**(3): 705–18.

National Institute of Health (2014) *BRAIN 2025: A Scientific Vision*. pp 1–146.

Niedermeyer, E., Schomer, D.L. and Lopes da Silva, F.H. (2011) *Niedermeyer's electroencephalography Basic Principles, Clinical Applications, and Related Fields*, Wolters Kluwer Health/Lippincott Williams & Wilkins, Philadelphia.

NSCISC (2012) *The 2012 Annual Statistical Report for the Spinal Cord Injury Model Systems*, National Spinal Cord Injury Statistical Center.

Nudo, R.J. (2013) Recovery after brain injury: mechanisms and principles. *Frontiers Hum. Neurosci.*, **7**: 887.

Nudo, R.J. and McNeal, D. (2013) Plasticity of cerebral functions. *Handbook Clin. Neurol.*, **110**: 13–21.

Oudega, M. and Perez, M.A. (2012) Corticospinal reorganization after spinal cord injury. *J. Physiol.*, **590**(16): 3647–3663.

Paninski, L. *et al.* (2004) Superlinear population encoding of dynamic hand trajectory in primary motor cortex. *J. Neurosci.*, **24**(39): 8551–8561.

Pedreira, C. *et al.* (2012) How many neurons can we see with current spike sorting algorithms? *J. Neurosci. Meth.*, **211**(1): 58–65.

Pei, X. *et al.* (2011) Decoding vowels and consonants in spoken and imagined words using electrocorticographic signals in humans. *J. Neural Eng.*, **8**(4): 046028.

Penfield, W. and Boldrey, E. (1937) Somatic motor and sensory representation in the cerebral cortex of man as studied by electrical stimulation. *Brain: A Journal of Neurology*, **60**(4): 389–443.

Perge, J.A. *et al.* (2013) Intra-day signal instabilities affect decoding performance in an intracortical neural interface system. *J. Neural Eng.*, **10**(3): 036004.

Perge, J.A. *et al.* (2014) Reliability of directional information in unsorted spikes and local field potentials recorded in human motor cortex. *J. Neural Eng.* **11**(4): 046007.

Pettorossi, V.E. and Schieppati, M. (2014) Neck proprioception shapes body orientation and perception of motion. *Frontiers Hum. Neurosci.,* **4**: 1–13

Philip, B.A., Rao, N. and Donoghue, J.P. (2012) Simultaneous reconstruction of continuous hand movements from primary motor and posterior parietal cortex. *Experim. Brain Res.*, **225**(3): 361–375.

Plum, F. and Posner, J.B. (1969) *The Diagnosis of Stupor and Coma*. Davis Co., Philadelphia PA.

Pohlmeyer, E.A. *et al.* (2007) Prediction of upper limb muscle activity from motor cortical discharge during reaching. *J. Neural Eng.*, **4**(4): 369–379.

Quiroga, R.Q. (2009) What is the real shape of extracellular spikes? *J. Neurosci. Meth.*, **177**(1): 194–198.

Quiroga, R.Q. (2012) Spike sorting. *CURBIO*, **22**(2): R45–R46.

Quiroga, R.Q. and Panzeri, S. (2009) Extracting information from neuronal populations: information theory and decoding approaches. *Nat. Rev. Neurosci.*, **10**(3): 173–185.

Quiroga, R.Q., Nadasdy, Z. and Ben-Shaul, Y. (2004) Unsupervised spike detection and sorting with wavelets and superparamagnetic clustering. *Neural Computation*, **16**(8): 1661–1687.

Rao, N.G. and Donoghue, J.P. (2014) Cue to action processing in motor cortex populations. *J. Neurophys.*, **111**(2); 441–453.

Rennaker, R.L. *et al.* (2005) A comparison of chronic multi-channel cortical implantation techniques: manual versus mechanical insertion. *J. Neurosci. Meth.*, **142**(2): 169–176.

Rizzolatti, G., Luppino, G. and Matelli, M. (1998) The organization of the cortical motor system: new concepts. *Electroenceph. Clin. Neurophys.*, **106**(4): 283–296.

Romo, R. and Schultz, W. (1987) Neuronal activity preceding self-initiated or externally timed arm movements in area 6 of monkey cortex. *Experim. Brain Res.*, **67**: 656–662.

Rousche, P.J. and Normann, R.A. (1992) A method for pneumatically inserting an array of penetrating electrodes into cortical tissue. *Ann. Biomed. Eng.*, **20**(4): 413–422.

Sabbah, P. *et al.* (2002) Sensorimotor cortical activity in patients with complete spinal cord injury: a functional magnetic resonance imaging study. *J. Neurotrauma*, **19**(1): 53–60.

Sanes, J.N. and Donoghue, J.P. (2000) Plasticity and primary motor cortex. *Ann. Rev. Neurosci.,* **23**(1): 393–415.

Sanes, J.N., Wang, J. and Donoghue, J.P. (1992) Immediate and delayed changes of rat motor cortical output representation with new forelimb configurations. *Cerebral Cortex*, **2**(2): 141–152.

Santhanam, G. *et al.* (2006) A high-performance brain–computer interface. *Nature*, **442**(7099): 195–198.

Schalk, G. and Leuthardt, E.C. (2011) Brain-computer interfaces using electrocorticographic signals. *IEEE Rev. Biomed. Eng.*, **4**: 140–154.

Schieber, M.H. (2011) Dissociating motor cortex from the motor. *J. Physiol.* **589**: 5613–5624.

Schwartz, A.B. (1994) Direct cortical representation of drawing. *Science*, **265**(5171): 540–542.

Schwartz, A.B. *et al.* (2006) Brain-controlled interfaces: movement restoration with neural prosthetics. *Neuron*, **52**(1): 205–220.

Serruya, M.D. *et al.* (2002) Instant neural control of a movement signal. *Nature*, **416**(6877): 141–142.

Shaikhouni, A., Donoghue, J.P. and Hochberg, L.R. (2013) Somatosensory responses in a human motor cortex. *J. Neurophys.*, **109**(8): 2192–2204.

Shenoy, K.V., Sahani, M. and Churchland, M.M. (2013) Cortical control of arm movements: a dynamical systems perspective. *Ann. Rev. Neurosci.*, **36**(1): 337–359.

Shoham, S. *et al.* (2001) Motor-cortical activity in tetraplegics. *Nature*, **413**(6858): 793–793.

Simeral, J.D. *et al.* (2011) Neural control of cursor trajectory and click by a human with tetraplegia 1000 days after implant of an intracortical microelectrode array. *J. Neural Eng.*, **8**(2): 025027.

Simpson, L.A. *et al.* (2012) The health and life priorities of individuals with spinal cord injury: a systematic review. *J. Neurotrauma*, **29**(8): 1548–1555.

Stokes, M.G. *et al.* (2005) Simple metric for scaling motor threshold based on scalp-cortex distance: application to studies using transcranial magnetic stimulation. *J. Neurophys.*, **94**(6): 4520–4527.

Suminski, A.J. *et al.* (2010) Incorporating feedback from multiple sensory modalities enhances brain-machine interface control. *J. Neurosci.*, **30**(50): 16777–16787.

Tabot, G.A., Dammann, J.F. and Berg, J.A. (2013) Restoring the sense of touch with a prosthetic hand through a brain interface. *PNAS*, **10**(45): 18279–18284.

Taylor, D.M., Tillery, S.I.H. and Schwartz, A.B. (2002) Direct cortical control of 3D neuroprosthetic devices. *Science*, **296**(5574): 1829–1832.

Taylor, D.M., Helms Tillery, S.I. and Schwartz, A.B. (2003) Information conveyed through brain-control: cursor versus robot. *IEEE Trans. Neural Syst. Rehab. Eng.*, **11**(2): 195–199.

Todorov E. (2003) On the role of primary motor cortex in arm movement control, in *Progress in Motor Control III*, Latash, M.L. and Levin, M.F. (eds), pp 125–166,

Todorova, S., Sadtler, P., Batista, A., Chase, S. and Ventura, V. (2014) To sort or not to sort: the impact of spike-sorting on neural decoding performance. 1–14.

Truccolo, W., *et al.* (2004) A point process framework for relating neural spiking activity to spiking history, neural ensemble, and extrinsic covariate effects. *J. Neurophys.* **93**(2): 1074–1089.

Truccolo, W. *et al.* (2008) Primary motor cortex tuning to intended movement kinematics in humans with tetraplegia. *J. Neurosci.*, **28**(5): 1163–1178.

Truccolo, W., Hochberg, L.R. and Donoghue, J.P. (2009) Collective dynamics in human and monkey sensorimotor cortex: predicting single neuron spikes. *Nature Neurosci.*, **13**(1): 105–111.

Vale, F.L. *et al.* (1997) Combined medical and surgical treatment after acute spinal cord injury: results of a prospective pilot study to assess the merits of aggressive medical resuscitation and blood pressure management. *J. Neurosurg.*, **87**(2): 239–246.

Vargas-Irwin, C. and Donoghue, J.P. (2007) Automated spike sorting using density grid contour clustering and subtractive waveform decomposition. *J. Neurosci., Meth.*, **164**(1): 1–18.

Vargas-Irwin, C.E. *et al.* (2010) Decoding complete reach and grasp actions from local primary motor cortex populations. *J. Neurosci.*, **30**(29): 9659–9669.

Velliste, M. *et al.* (2008) Cortical control of a prosthetic arm for self-feeding. *Nature*, **453**(7198): 1098–1101.

Vigneswaran, G. *et al.* (2013) M1 Corticospinal mirror neurons and their role in movement suppression during action observation. *CURBIO*, **23**(3): 236–243.

Volpe, D., Giantin, M.G. and Fasano, A. (2014) A wearable proprioceptive stabilizer (Equistasi®) for rehabilitation of postural instability in Parkinson's Disease: a phase II randomized double-blind, double-dummy, controlled study. Quinn, T.J. (ed.), *PLoS ONE*, **9**(11): e112065.

Voorhies, J.M. and Cohen-Gadol, A. (2013) Techniques for placement of grid and strip electrodes for intracranial epilepsy surgery monitoring: pearls and pitfalls. *Surg. Neurol. Int.*, **4**(1): 98.

Wannier, T., Schmidlin, E., Bloch J. and Rouller, E.M. (2005) A unilateral section of the corticospinal tract at cervical level does not lead to measurable cell loss in motor cortex. *J. Neurotrauma*, **22**(6): 703–717

Wessberg, J. *et al.* (2000) Real-time prediction of hand trajectory by ensembles of cortical neurons in primates. *Nature*, **408**(6810): 361–365.

Wild, J. *et al.* (2012) Performance comparison of extracellular spike sorting algorithms for single-channel recordings. *J. Neurosci. Meth.*, **203**(2): 369–376.

Wood, F. *et al.* (2004) On the variability of manual spike sorting *IEEE Trans. Biomed. Eng.*, **51**(6): 912–918.

Wu, W. *et al.* (2006) Bayesian population decoding of motor cortical activity using a Kalman filter. *Neural Comput.*, **18**(1): 80–118.

Yin, M. *et al.* (2013) A 100-channel hermetically sealed implantable device for chronic wireless neurosensing applications. *IEEE Trans. Biomed. Circ. Syst.*, **7**(2): 115–128.

Yousry, T.A. *et al.* (1997) Localization of the motor hand area to a knob on the precentral gyrus. A new landmark. *Brain*, **120**(1): 141–157.

PART IV

COMMERCIAL AND ETHICAL CONSIDERATIONS

10

TAKING A DEVICE TO MARKET: REGULATORY AND COMMERCIAL ISSUES

John L. Parker
Saluda Medical Pty Ltd, Artarmon, NSW, Australia

10.1 INTRODUCTION

The initial underestimation of how difficult and expensive the journey from concept to making a profit is a consistent theme in success stories of commercialization of medical devices. This is perhaps why few teams embark on the journey in the first place; full knowledge of the coming trials would deter most. Why is it so difficult? Why are costs and efforts underestimated? What could we do to simplify the process? These will be the central themes of this chapter.

In one sense, all of the activity from basic research through feasibility and prototype development finally leading to sales could be considered as part of the commercialization process. However, there is a point where the nature of the funding for the activity changes from purely speculative (not necessarily looking for financial returns) to sources of funding that expect to see a return and this is often considered the point where commercialization begins. Commercialization can be considered a process in reducing risk. At the beginning of the project there are a large number of unquantifiable risks but as the project proceeds the risk is gradually reduced. Matching the risk in the project to the risk an investor is willing to take is a difficult but essential part of funding the new venture.

We can define the commencement of commercialization as the point at which the project team seeks investment from sources which expect a financial return. At this

Neurobionics: The Biomedical Engineering of Neural Prostheses, First Edition.
Edited by Robert K. Shepherd.

point there should be proof that the device works, that there is a market for the product and that there is a reasonable expectation that payers or individuals would be willing to pay for it. Understanding who pays for the technology and how they judge efficacy and if they should pay is very complex. It varies from country to country, depending on the nature of the health care system, public or private, and even varies between insurance companies. In some countries the absence of either public or private health care systems requires individuals to fund the therapies themselves. At this point there should also exist a reasonable plan to make the device and procedure practical, that there is sufficient freedom to operate and finally – and perhaps most importantly to an investor – that the people in the new organisation can make the project happen.

The commercialization pathway is illustrated in Figure 10.1. The point at which professional investment is sought and good answers are available to the issues noted above depends on the nature of the opportunity. For some projects, commercialization readiness can be achieved after the basic research phase, but for others this may take until clinical trials have commenced or even later. Choosing the time to begin this phase is critical for the long-term success of the venture. Too early, when the risk is high, will limit the number of investors that are willing to take up the opportunity. Leave it too late and the project starves for capital and progresses slowly, allowing others to follow, execute more quickly and overtake.

Timing for seeking funding depends on being able to answer the basic questions of performance, market, payment and execution. All the phases of the development should be directed at providing the basic data to answer these questions and create a story which describes the business. In business parlance, this is called a "strategy" or business plan.

This pathway consists of four phases: (i) basic research; (ii) preclinical development; (iii) clinical trials; and finally (iv) regulatory approval and sales (Figure 10.1). Commercialization starts when professional investment is brought into the company; however, there are a number of areas where early focus during the basic research phase can greatly facilitate the later commercialization process.

10.2 BASIC RESEARCH

The basic research phase focuses on establishing the performance and feasibility of the therapy. In many instances basic research begins without commercialization objectives; there has been many cases of lost value and missed opportunities as a result. For example, the development of spinal cord stimulation (SCS) for treatment of chronic back pain began in the 1960s and the original investigators failed to patent the technology (Shealy *et al.* 1967: Mortimer *et al.* 1970). SCS was commercialised by Medtronic and is now the most prevalent neuromodulation device in the world. Perhaps the most important commercialization decision in the entire lifecycle of a new neurobionic technology is the choice of what to pursue. The evolution of technology and the pace of technological advance provide a space where the choices of what therapy and application to pursue are seemingly limitless.

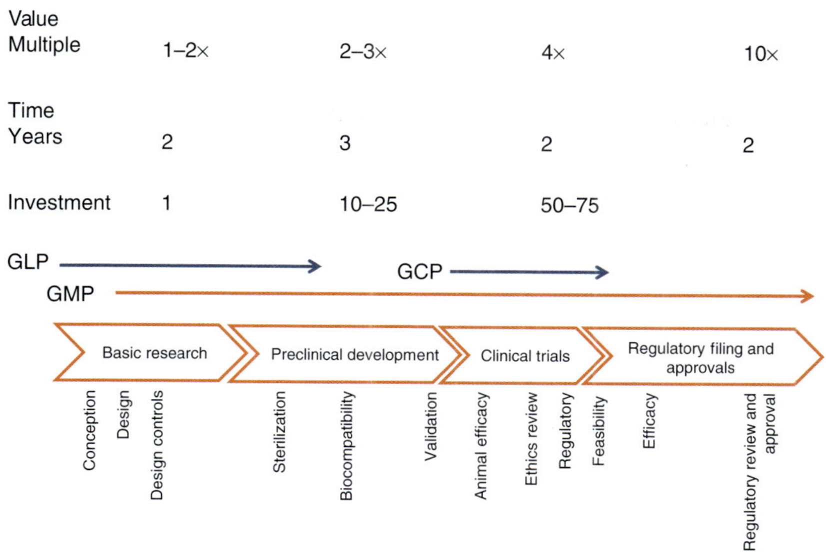

Figure 10.1 Overview of the commercialization process for medical devices. The four phases: (i) basic research, (ii) preclinical development, (iii) clinical trials, and (iv) regulatory approvals, are shown along with the major milestones which need to be accomplished in the phases. The value multiple is how much commercialization effort may be worth as a multiple of the investment and the time taken and a typical invested amount. The quality standards which operate over the three phases are also indicated (good laboratory practice (GLP), good clinical practice (GCP) and good manufacturing practice (GMP)).

There are a large number of developed and approved therapies which are based on neurobionic applications There is an even longer list of applications currently in development (Figure 10.2). Choosing a project requires an understanding of this landscape. Re-inventing an existing therapy or competing in a hotly contested area will make it difficult to generate a competitive advantage.

Work, thought and planning directed at ensuring that technology will develop into a commercial opportunity can be a powerful guide in the conduct of the basic research phase of this pathway and can save time, millions of dollars and increase the likelihood of success. One way to think about this is to understand the specific targets that a technology and therapy must achieve in order to be a viable competitor to non-treatment or the best available alternative therapy. Understanding the value proposition to the patient who is offered the therapy is essential. It may seem difficult to judge this from the outside prior to the existence of the technology, but a useful framework is to consider the therapy in terms of the performance dimensions critical to success.

One framework for assessing the performance of a therapy consists of six dimensions. Any payer or purchaser considers the following, to different degrees, in making a decision to proceed with a therapy. The dimensions are performance, safety, habilitation, ergonomics, total cost of ownership (TCO) and cosmetics. These criteria can

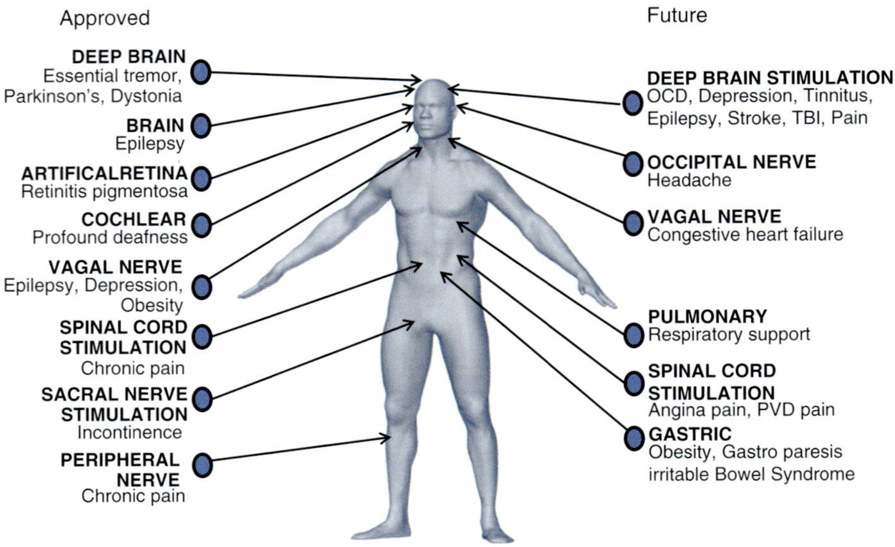

Figure 10.2 The neuromodulation/neurobionic industry landscape. To the left of the figure are the currently approved therapies and on the right the therapeutic targets currently being developed.

be used for any type of product, but dimensions can be tailored for specific neurobionic applications.

Often the early phase research is focused on the performance of the therapy, and this dimension captures the actual primary benefit that the technology delivers. For cochlear implants, it is the measurement of speech perception; for retinal prosthesis, it is the level of visual acuity which can be achieved; and for deep brain stimulation (DBS) for Parkinson's disease, it is the level of improvement in motor tremor scores post-operatively. For every disease state there is a method to score the impact of the disease.

Assessing performance can be complex and there are several layers of performance to consider. Ultimately the intervention will either prolong life or improve the quality of life. The therapy achieves these improvements by function changes, for example improvement in motor scores for DBS. Almost all neurobionic therapies have some side effects and are subject to variable outcomes, some patients can receive tremendous benefit while others do not do so well, and for some the surgery and side effects can outweigh the potential benefit.

One of the most successful therapies is DBS for Parkinson's disease, which improves motor scores by 5090 or more in 8090 of patients (Kleiner 2006). The subthalamic nucleus, a target for DBS in Parkinson's disease, has three functional subdivisions with only one as a target. The therapeutic window for electrical stimulation is defined by the limits at which stimulation causes side effects which may include behavioural (Temel 2006) and speech and language (Rousseax 2004) changes. New developments in DBS are directed to improving control of the stimulation and increasing the therapeutic window (Pollo 2014) by developing higher density electrode arrays and the use of current focusing techniques. Two companies, Aleva Therapeutics and Sapiens, have recently been established and have raised funds to pursue this objective. Sapiens have raised 35.5 M Euros to peruse its high resolution DBS electrodes. The prospect of providing current steering in DBS applications offers the potential to reduce some of the side effects related to DBS and improve the motor scores.

The basic research phase is directed at ensuring the therapy has a promising level of performance, but attention must be paid to the cost of achieving the performance and any side effects of the therapy and how all these contribute to the quality of life.

There are standard measures for quality of life and as we will see, getting reimbursement for a therapy can depend on the overall improvement in quality of life. This is measured by Quality Adjusted Life Year (QALY) (Sassi 2006), which is a measure of disease burden and factors in the number of years lived and the quality of life enjoyed. The SF-36 survey instrument is the most common measure of health of a patient. It scores well-being across eight factors including vitality, physical function, body pain and general health.

Sapiens and Aleva therapeutics have recognised that although the motor performance scores for the larger majority of patients with existing DBS technology and quality of life is improved substantially, there is room for improvement of the therapy by extending the therapeutic window and stimulation range available without inducing side effects.

The safety of the product or therapy, attempts to capture the risk associated with the therapy; despite the huge therapeutic success of DBS for Parkinson's disease, the fact that it involves brain surgery with a potential for death, even though vanishingly small, has an impact on decision-making by both clinicians and recipients. The risk and benefit from the therapy have to be aligned. The extent of risk that a patient is prepared to bare to have a procedure depends on the benefit from the therapy. An individual is likely to accept a very dangerous procedure if its life saving, whereas procedural risk for treatment of non-threatening chronic conditions must be relatively low.

A technology may have great performance and be essentially safe, but if the technology takes decades of intensive habilitation or effort to learn how to use it then it may have little or no adoption. The history of assistive technology developed for the blind is full of examples of innovations which work for a small group of people prepared to learn and adapt to it, but not generally applicable across the wider population. The use of echo location, although not a technological intervention in its own right, is an extreme example of this. Habilitation can be measured in the time and efficiency with which the product or therapy reaches maximum performance for the user.

The addition of extra electrodes and steering capacity for DBS offered in Sapiens and Aleva technologies may in the end improve the quality-of-life outcomes and reduce adverse side effects; however, the expansion of the number of programming options places additional burden on the clinicians and may require longer time frames to reach the performance levels of simpler systems. Both companies have strategies to reduce this risk. In general, the addition of features even though technologically feasible may not result in effective products. The basis of competition in the cochlear implant market for a time was the length of time required for individuals to reach maximum performance.

The technology therapy may provide good performance and take little time to learn to use; however, it may not be very easy to use on an ongoing basis. SCS for pain relief suffers from the problem of needing constant adjustment of stimulation parameters with variation in posture. To overcome this problem, devices have been developed which attempt automatic adjustment of the stimulation intensity based on the accelerometer readings of the position of the subject (Shultz 2012).

The TCO is defined by the full life cycle cost of the product or therapy. Reimbursement is discussed later in the chapter; however, failure of technology to meet health economics criteria can lead to failure of the therapy.

The last performance criterion is cosmetics, but often this is ignored in the development of medical products but it can be a potent inhibitor for the uptake of new technology. Cosmetics is not just the look and feel of the product but the impact of the perception of others on the user of the product. The first attempt at creating a cosmetically appealing Cochlear implant sound processor, the Esprit 3g, resulted in a dramatic shift in the cosmetics of these devices and established a new benchmark which all competitors in the cochlear implant field eventually followed.

Technology can be developed to advance performance along any or all of the above dimensions. All these categories of performance can be defined, measured and compared to existing therapies, and in the case where no competing therapy exists, the comparisons are made to non-treatment. Understanding the targets for the performance of the technology and therapy at the earliest possible stage of the project is the most valuable contribution that can be made to the commercialization process.

A visual way to represent this information is with a spider diagram. Figure 10.3 illustrates a diagram of the estimated performance characteristics required for the development of a successful commercial bionic eye. Every therapy and technology has minimal acceptable level of performance and for some therapies, a significant number of people who are considered for a therapy have an experience which results in them not accepting the therapy or non-use. For instance, patient selection for implantation of permanent SCS systems is made on the basis of a trial with an externalised lead. Approximately 65% of people who undergo this trial proceed to a full implant. Some patients undertaking the trial decline the therapy because of the level of movement induced over stimulation. Others who accept the therapy will only use it during periods when they can maintain a stable position.

Each of the performance dimensions has a rational basis for choice. In order to obtain reimbursement for a product in managed health care environments, the benefit to cost ratio needs to be established. This is done by determining the cost of QLAY. The cost utility ratio for a cochlear implant has been estimated at between 12 500 and 30 000$/QLAY (Wyatt 1996). A chronic disability such as blindness results in a reduction of quality of life, which can be measured with disability adjusted life year (DALY).

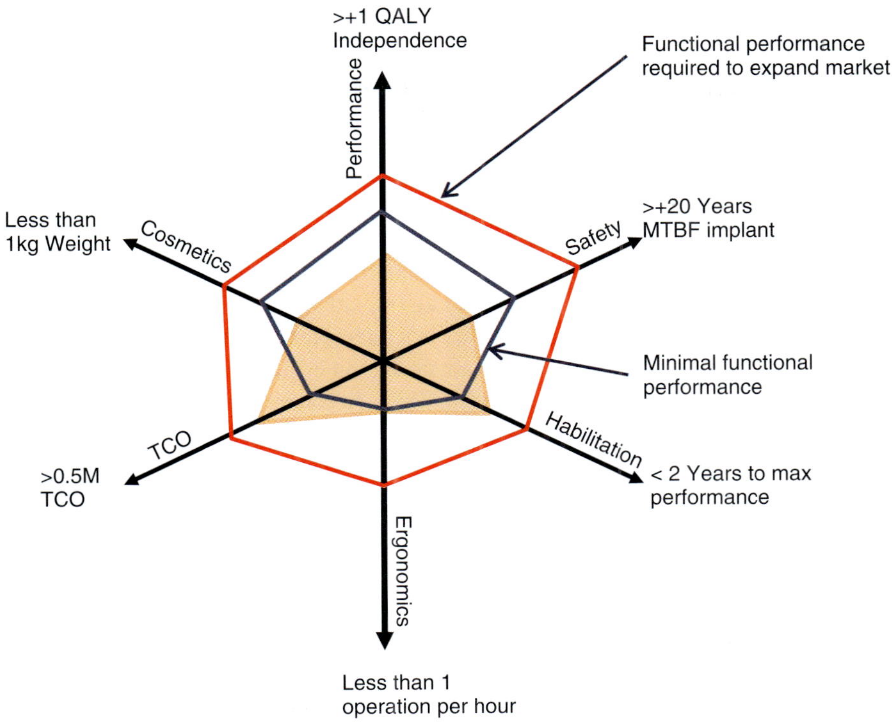

Figure 10.3 An example assessment of performance dimensions for a bionic eye. The dimensions (see text for explanation) are drawn in a spider diagram. Lines indicating the differences between non-treatment (shaded area), minimal performance required and the performance criteria required to expand the market. QALY, quality-adjusted life year; MTBF, Mean Time Before Failure (a measure of the reliability of the device); TCO (i.e. the sum of all costs for the therapy, including running and maintenance costs).

The safety criteria can be difficult to judge and interacts strongly with the level of performance. The promised better performance will result in a willingness of individuals to take more risks. An implanted device is safe if it causes no harm and whatever is done surgically to implant the device can be undone if the device needs to be removed. There have been rare examples where devices fail over time. For example, cardiac pacing leads have failed, resulting in the death of the patient. In other instances where the perceived risk is low, for example breast implants, failure, although not life-threatening, can be catastrophic for the individual. Although devices can fail, the science behind reliability engineering is well established and so device design and manufacture for high reliability is essential. In the bionic eye, this translates to an estimated MTBF of at least 20 years.

The time it takes to reach optimal clinical performance for many neurobionic applications can depend a great deal on the neuroplastic changes that occur with device use. The typical time for a cochlear implant patient to reach optimal performance is approximately one year. Pantev *et al.* demonstrated almost normal stimulus evoked brain activity in post-lingual deaf individuals implanted with cochlear implants due to neuroplastic recovery of auditory processing (Pantev 2006). For a patient using a visual prosthesis this may be longer, but there is an upper limit to how long this should take. Ultimately, every year without reaching optimal performance reduces the cost utility benefit, and for an intervention such as a bionic eye the cost could be significant. The average age of the population also has an impact, as in older populations there is less time to amortise this cost.

Ergonomics for a visual prosthesis and cosmetics are probably the least important factors. The device has to be a wearable (<1 kg) including all external hardware; it has to be readily useable by the patient. If the device requires too much user interaction, such as constant adjustment or a significant change in behaviour from the individual, this may impact the acceptance of the technology. At some point in the development, defining an acceptable amount of user interaction is important.

The TCO defines the limit of cost for the intervention and it is this number which is divided by the benefit in QALY to estimate the cost per QALY. The opposite of QALY is the DALY. The cost for being blind is estimated at £1325–£16 800 per year (Meads and Hyde 2003), and so the economics allows at least this amount to be spent on an intervention. It is likely that a vision prosthesis TCO of ownership is going to be higher compared with a cochlear implant. The TCO of a Second Sight epiretinal prosthesis is approximately $120 000 USD. Use of performance metrics in this way allows a targeted approach to the development of the technology and provides direction for the conduct of basic research.

The basic research phase begins with an academic interest in the subject matter and often proceeds along the steps outlined in Figure 10.4. Commercialization is often not even considered by the key people involved in the project at the outset.

In a traditional academic setting, basic research is aimed at producing publications. The initial idea generates research questions which are then worked on and commercialization thoughts are left to the end of the process. Although most academic groups understand the need to patent before publication, often it is not a priority and can be overwhelmed by the need to demonstrate success in peer reviewed journals in order

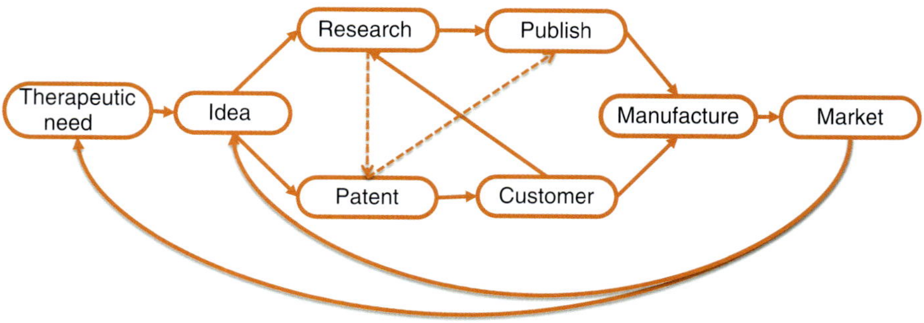

Figure 10.4 Basic research flow diagram illustrating the stages of development from identification of the therapeutic need through idea generation and then to market.

to attract the next round of research funding. Publication before considering patents can limit or completely exclude the grant of a patent. The development of technology and the basic idea may have tremendous value as a therapy, but failure to engage with customers and users of the technology can result in a device or a therapy, which is unacceptable in the market for reasons other than the core technology or its efficacy. A commercial organization will often follow the process by taking the lower path illustrated in Figure 10.4. Patenting early in the process allows the organization to engage with suppliers and partners to help realise the technology without the concern that the idea may be adopted elsewhere. It also allows the idea to be tested on the market and customer feedback obtained, which in turn allows the development of devices that are much closer to useable forms. Following this path is difficult in academic organizations. As publication is the major metric of success and this is delayed by patent activity and customer engagement.

Early engagement with customers and the market ensures that the technology is useful and useable as its being developed. Done well, this process ensures that there are willing recipients to trial the technology as soon as it becomes available. It may seem strange to talk to customers and identify key opinion leaders so early in the process, and it is dangerous without adequate patent protection, but often individuals are more willing to collaborate and share ideas and thoughts before everything is set in stone. They can be more willing when the project is young and forming, as they feel they can influence the direction.

The other consistent feature of successful medical device companies, other than under estimation of effort, is that successful companies develop as early as possible the techniques to trial their ideas on real patients in a clinical setting. Identification of clinical partners early in the project life cycle is essential for success.

This requires a special relationship to form between clinicians who must place the interest of their patients above all else and the engineers and technologists who develop the therapy. Building a relationship of trust and collaboration can be a significant effort on both sides. The selection of a clinical partner(s) and how to grow the interest in the technology at the early stages can be critical. Often the support from such individuals is required to obtain funding. Many venture capitalists will require opinions from leaders in the field as one of the funding hurdles. In the end, these clinicians will be the first customers of the product and its greatest supporters.

Clinicians are often the originator of the ideas for many new medical devices. They are acutely aware of the needs of patients and often frustrated by the lack of appropriate and effective therapies. For neurobionic applications, clinicians will be responsible for implantation of the hardware and long-term management of the patient. The design of devices is often a highly iterative process requiring a great deal of trial and error, which requires the involvement of clinicians with each iteration. Engaging with clinicians and the market so early in the process can only be done when there is clarity on the ownership of the core IP. The best way to ensure clarity on IP is to obtain patents.

A granted patent can be very expensive and there is a reluctance sometimes to file patents, as the cost to a small project can be prohibitive. However, provisional patents can be filed relatively cheaply and then later abandoned should further work and investigation show that the idea has little value. Patents are granted on the basis that the invention described has the properties of being novel, inventive and having industrial utility. Prior to being granted, the examiner will assess if the disclosure and importantly the claims that define the scope of the invention fulfill those essential requirements. Novelty requires the patent be new and not described previously in patent literature or in the public domain. It is insufficient for the invention to be new; it also must be inventive or non-obvious to "someone skilled in the art". The simple combination of two previously described inventions, if obvious to someone skilled in the art, is not inventive. The last criteria is industrial utility, so the invention must be useful and provide some benefit as described in the disclosure. There is little point in filing patents which will simply be rejected by the patent office during prosecution and so patent searching and understanding the IP landscape early, even before the work begins, can be a very useful exercise. The patent literature provides a rich source of information, which can be used to help define areas to work on and can be used to generate deep insight into the competition in the market.

Granstrand (2000) asks "Patents – the most potent source of competitive intelligence?" In the medical device domain the answer is definitely yes. The published patents are a terminal or lag indicator of R&D activity. They appear as publications several years after the conception of the idea and its reduction to practice. There is a great deal that can be learned by careful analysis of a patent portfolio or an industry IP landscape, including:

- The individual entities strengths and sometimes weakness and their position in technologies and markets;
- Company innovation rate benchmarking;
- Identify new or potential entrants, or strategic groupings of firms;

- Identify technological conditions and trends in general;
  - Fluctuations, interdependencies, ageing of the technology base
  - Fertile R&D directions and directions for research
  - Revival of old technological ideas
  - Infringement risks
  - Strategic patent positions
- Identification of suitable targets for technology acquisition;
  - Mergers, acquisitions, joint ventures, alliances, licensing opportunities
  - Recruitment of key inventors
- How R&D and technological investment strategies are being shaped;
- Identification of international marketing strategies of rivals;
- Valuation of a company's or research groups technological assets;
- Identify opportunities for cross-licensing.

Simply visualising an IP landscape from a strategic perspective can be complex. A number of models have been developed to map the landscapes and a number of commercial tools and services are available to allow this to happen. A simple process in understanding the IP landscape is to classify patents according to criteria and use these to judge the movements in the industry. For example, one method of classification is to consider how the IP contributes to the performance of a therapy. It is a relatively simple matter, when reading patents to assess them in terms of which of the previously described performance dimensions the product or therapy the patent is primarily aimed at improving. The patent offices require that only one invention is described in a patent – a term referred to as unity of application – and so it is rare that a patent actually covers more than one therapy performance dimension.

Figure 10.5 illustrates data from an analysis of a subset of more than 1200 patents across the neuromodulation sector, where the patents were assigned a performance dimension based on the criteria in Figure 10.3. This was done by assessing the impact of the invention on the device. It is no surprise that most of the inventions are directed at improving the performance of the systems, followed next in number by inventions aimed at producing improvements in safety of the devices. The data demonstrates a change in emphasis of activity over time from performance to safety. Of particular interest is that there is little focus on total cost of ownership and cosmetics. The fact that all these devices require reimbursement and in general there has been little pressure on costs in the health care systems of developed countries, results in little effort by manufactures to drive their costs down and hence there is little IP development in this area. The advent of Obama Care in the USA and the application of rational health economics may see this picture change over the coming decade.

Although performance of the therapy remains the focus of R&D activity leading to patents, there has been a trend towards more activity in safety and habilitation.

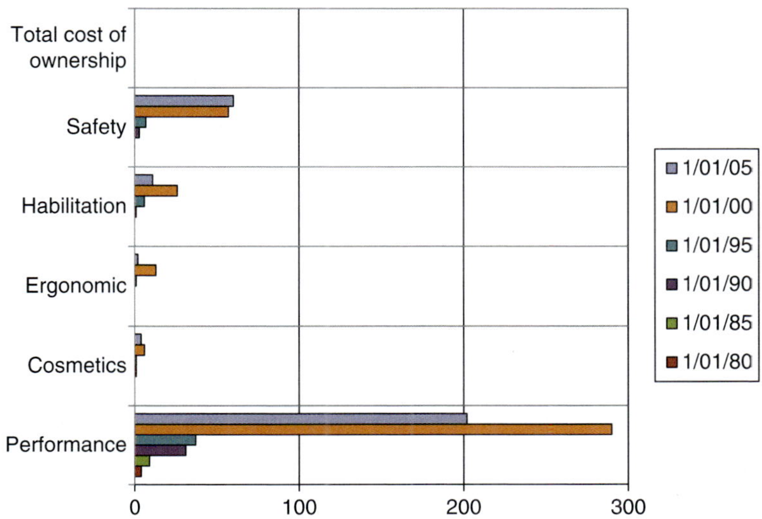

Figure 10.5 The number of patents in the neuromodulation industry over five year epochs from 1980 to 2010 categorised into product performance dimensions.

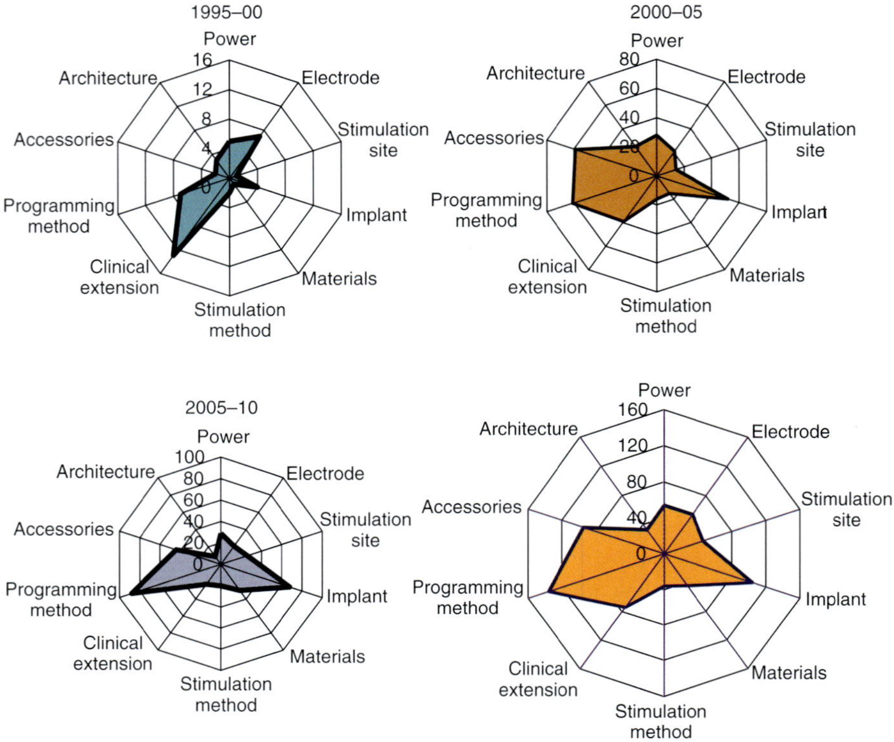

Figure 10.6 Spider diagram showing the number of patents in each technology area over five-year epochs from 1995 to 2010 and the total number of patents for each technology area from 1980 to 2010.

This may be a result of the maturing of some of the technology which is represented in the patent database and companies and researchers are working on technology to differentiate themselves from competitors in applications where the core performance of the products is similar.

Another useful analysis to perform is to identify the areas of technology focus in the industry. Figure 10.6 shows a series of spider diagrams illustrating the changes in technology focus for the neuromodulation sector over a 15-year period.

The patents that formed the data in Figure 10.6 were sorted into technological categories. This is a simple task of looking at the patent and determining what technological feature or characteristic of the product is improved by the patent disclosure/invention.

The categories chosen for neuromodulation were:

- *Architecture*: The arrangement and connection between sub-systems of the product, including the clinician and patient interfaces;
- *Power*: Technology aimed at reducing power consumption;

- *Electrode*: The electrode tissue interface and stimulating electrode itself;
- *Stimulation Site*: Variations in place to stimulate;
- *Implant*: Captures technology for manufacture of implants and mechanical design;
- *Materials*: New materials for fabricating implant components;
- *Stimulation Method*: Pulse shapes and combinations, including stimulation algorithms;
- *Clinical Extensions*: Application of the same technology but for a different therapy;
- *Programming Method*: Captures everything to do with device programming;
- *Accessories*: Parts of the device to use as an adjunct to make use or implantation simpler.

Categorising patents into their different technological contributions provides an insight into the areas that participants in the field regard as important drivers of the performance of the therapy.

The three cartwheels in Figure 10.6 illustrate the number of patents devoted to a particular area over different epochs. Patents in the 1995–2000 period were dominated by clinical extensions, which is defined as the use of an existing technology but for a different therapeutic application. In the neuromodulation market, the technology was established by 1995 with Medtronic as the dominant player. Medtronic were largely uncontested in the period prior to 1995 and as a result most of the patents filed were devoted to clinical extensions. The company was busy seeing how many markets and diseases that it could apply its technology to and so it is not surprising that DBS devices remain very similar to SCS devices. Advanced Neuromodulation Systems (ANS) was established in 1995, and by 2001 had obtained a market share of about 14%.

The rapid growth of ANS triggered a change in focus of innovation and the period from 2000–2005 was devoted to programming methods and accessories. Innovation was focused on simplifying the delivery and control of the therapy. There was renewed focus on implant technology and this trend remained for the next five year period (2005–2010).

This sort of analysis and knowledge of what has been occurring in the industry graphically illustrates the technological battleground and helps determine the underlying strategies of new entrants. Patents are published years before the technology they describe ends up being trialled and in fact the vast majority of patents never end up being employed in products. The fact that they are written down in the first place provides a direct indication of what the inventors think is important and this is visible long before a product appears. In turn, this allows new ventures to plan research and development accordingly.

The chances of a successful outcome from a patent dispute in the medical device sector are higher than any other sector and the settlements tend to be higher. This combined with the long timescales to market and the overall investment means that the medical device sector is highly litigious. The long development cycles expose the technology to interested parties a long time before it is possible to generate a commercial result from their efforts.

Most professional investors will want some assurance that you have freedom to operate. This means the ability to conduct business without being sued for patent infringement. A formal freedom to operate a search is an expensive exercise and requires detailed searching of the patent literature. The fact that a patent may have been granted does not mean the assignee cannot be sued because use of the IP may infringe a third party's rights. The use of someone else's invention may be required in order to realise the product or therapy being developed. Understanding how the pieces fit together and who owns which piece is an essential part of building a validated story for the business to attract funding. If third-party IP rights are required to enable the technology or therapy, then investors will want to see the plans to obtain licences for this IP.

The basic research phase should demonstrate the science behind the idea and provide sufficient proof that the idea works to obtain funding for the next phase of development. Often some good work, which is conducted in the basic science phase, ends up being repeated during the next phases of commercialization due to a lack of compliance with regulatory expectations. Commercialization of medical devices requires compliance to regulatory authorities. The Food and Drug Administration (FDA) regulates medical devices in the USA, and in Europe market entry is gained via the CE mark process. ISO 13485 is an International Organization for Standardization (ISO) standard that represents the requirements for a comprehensive quality management system for the design and manufacture of medical devices and compliance to this standard is essential in order to obtain CE mark.

The FDA regulations are described in Good Manufacturing Practices (GMP) – 21 CFR Part 820. All potential neurobionic applications will target global markets and that means compliance to both these regulations will be required. A critical decision which needs to be made on the commercialization pathway is to choose when the effort to move to compliance with these regulations should take place. The first area which requires focus is design control as development of the products and trial devices requires the application of this part of the standard. Achieving compliance with design control regulations as early as possible in the project brings significant advantages.

The introduction of design control for the development of new devices and compliance with GLP will allow any data that is collected during the basic research phase to be used in the support of later regulatory approvals. Design control is a mandatory requirement for regulatory approval for a medical device. When first encountered, it appears like a substantial bureaucratic overhead, but it is not when implemented well. It actually speeds up the development process and reduces errors.

The principles of design control can be implemented with a simple process that research teams are often reluctant to implement, as they are seen as a bureaucratic overhead. Good design control is based on the idea that the user's needs are translated through to end design under a control and through a number of phases of development as illustrated in Figure 10.7. User needs are simple statements about the needs of all the stakeholders in the use of therapy. They are the input to design and compliance requirement that must be traceable through the design cycle. The other source of input

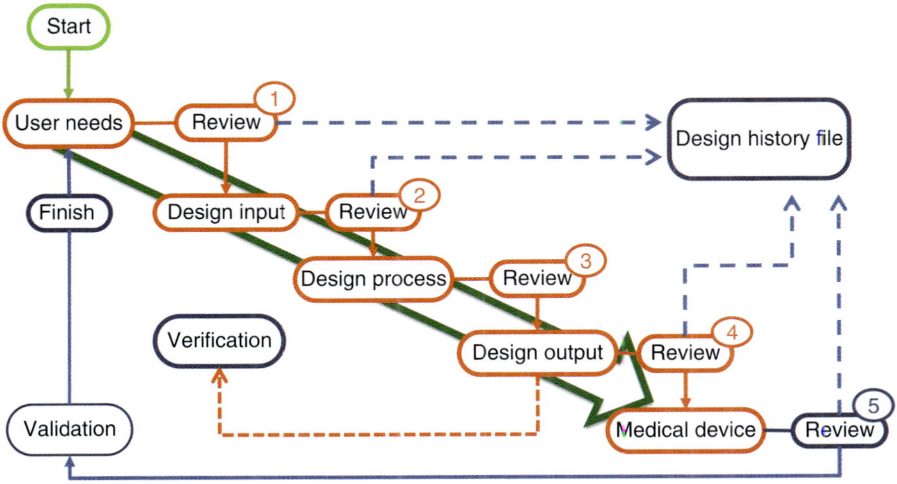

Figure 10.7 Application of Design Controls to Waterfall Design Process adapted from Figure 1 in http://www.fda.gov/medicaldevices/deviceregulationandguidance/guidancedocuments/ucm070627.htm

is through the estimation and understanding of risks and hazards associated with the device. Needs and risk form the input to design and each requirement captured in this part of the design needs to have a corresponding design element which addresses the requirement. The performance targets, which were discussed previously, form a great starting point to build a list of user requirements.

The process consists of a definition of user needs which are then used to create a system specification which forms the input to the design control process. The design process creates the product design and it and all of the processes used to manufacture are defined in the design output. At each stage reviews are conducted to ensure that what is being designed and constructed meets the requirements defined in the previous stage. Verification is the activity of checking if the design output meets the specification and validation is the activity that checks if the device meets the user requirements.

The design control is regulated by reviews and best practice means that work on the next phase does not proceed until the current tollgate has been passed. The purpose of reviews is to ensure that quality is built into the product from the very beginning. This saves time and money and the end result is better quality devices. When the design is complete, there are three more steps to check if the device is ready: (i) design verification compares the device with the specification; (ii) it provides the assurance that what was built was what was intended; and (iii) validation which checks that the user requirements were satisfied.

The formal definitions from the FDA website are:

820.30(f) Design verification

- Each manufacturer shall establish and maintain procedures for verifying the device design;

- Design verification shall confirm that the design output meets the design input requirements;
- The results of the design verification, including identification of the design, method(s), the date, and the individual(s) performing the verification, shall be documented in the Design History File.

The definition for validation:
§ 820.30(g) Design validation

- Each manufacturer shall establish and maintain procedures for validating the device design;
- Design validation shall be performed under defined operating conditions on initial production units, lots, or batches, or their equivalents;
- Design validation shall ensure that devices conform to defined user needs and intended uses and shall include testing of production units under actual or simulated use conditions.

The user requirements are the basis of the validation plan. For each user requirement, there should be a test performed which shows that the device as made meets that requirement.

One area of substantial effort, which is almost always underestimated, is the effort in Validation and Verification. These activities can constitute 70% of the entire R&D budget and time for the project. The fact that it is such a large part of the effort means that early mistakes in the process (not having design control or missing a user requirement) will result in substantial amounts of rework and re-validation.

Many neurobionic devices incorporate a lead to carry signals, for example from a stimulator to an electrode array. In many devices, the lead is a separate component and has a specific set of user requirements including diameter, length, number of conductors, flexibility, method of anchoring and method for implantation. The lead will be implanted in a biological environment and therefore must be made from biocompatible materials and sterilised before use. It has to be moved from the factory to the hospital; both the packaging and the lead itself need to survive this process. For a simple item like a lead, the list of user and design requirements grows quickly.

To illustrate the process, some example user requirements specifications and headings for verification and validation are shown in Table 10.1. This is certainly not an exhaustive list and there are many other detailed requirements that would flow from a thorough consideration. In the example, we have categorized the requirements according to the performance dimensions discussed previously. The lead must deliver the appropriate therapy and this will depend on the target tissue and its location and the intended stimulus paradigm. The electrical characteristics of the lead, resistance and contact surface area, are factors which need to be considered. The lead resistance needs to be sufficiently low so that the resistive drop across the wires is acceptable. The contact area needs to be sufficiently large so the charge does not exceed the safe charge limit. This is a critical safety issue and is dealt with in Chapter 3. The lead

Table 10.1 The progression from performance categories to user requirements specifications verification and validation using a neuromodulation lead as an example of this process.

Dimension	User Requirement	Specification	Verification	Validation
Performance	The lead shall deliver therapy appropriate stimulation	The lead shall have 8 channels The resistance shall be <100 Ohms/conductor	Design Verified Resistance is measured	Saline and animal testing
Safety	The lead should have a mean time failure >10 years Stimulation shall be safe and non-damaging Shall not cause adverse tissue reaction Device shall be sterile	The lead implanted in position X will survive 10 000 bends of 90 degrees The charge shall not exceed the safe charge limits Lead shall be made from biocompatible materials Device shall be able to be sterilised	Accelerated age flex testing Safe Charge Protocol Biocompatibility study	Chronic device trial Chronic animal stimulation study Target organ biocompatibility Sterilisation protocol
Total Cost of Ownership	The lead shall cost <1000$	Cost of components Costs of assembly	Manufacturing plan	Cost of production cost of implantation maintenance and use
Ergonomics	The lead shall take no longer than 10 min to implant	The lead and surgical tools are designed for ease of use	User acceptance testing	Timing of implantation
Habilitation	The surgeon shall be able to implant after the observation of one surgery	Training material	Training effectiveness	Assessment of the time required for clinicians to learn to use the device

must be made of biocompatible materials and a means to demonstrate biocompatibility needs to be defined. There is a user requirement that the hypothetical lead is safe for the recipient and we have chosen a mean time to failure of 10 years.

In order to specify the lead, an estimate of how much the device is likely to be stretched or bent is required. This depends on the location of the lead and how much movement it is likely to experience. A cardiac pacemaking lead will undergo a large movement with each heartbeat, but a cochlear implant lead moves very little during its lifetime. Establishing the specification requires a rationale for the amount and extent of movement that the device will experience and then the specification can be described in terms of how many flexes and at what angle the device should be able to withstand without failure. This is verified in a protocol where samples of the design would be flex-tested until destruction or a sufficiently wide safety margin has been reached. The components to be tested need to be representative of the components which are intended for implantation and ideally should be samples taken from production batches where the production processes have been validated. The device is also required to be sterile and this process also requires validation. Some sterilization techniques affect the properties of polymers and if this is the case the verification needs to be performed on sterile samples.

For a complete system there may be hundreds of user requirements, all of which must be traced through specification to design verification and validation. The size of this task is often dramatically underestimated. The consequences of doing a poor job can mean the difference between success and failure of the project.

There are numerous examples of medical devices which, despite the application of design control, have been released and result in significant adverse events. Medical device manufactures are required to report to the FDA when they learn that any of their devices may have caused or contributed to a death or serious injury. The definition of what constitutes a serious injury can be ambiguous and depends on the nature of the device. For some devices re-operation to replace the device if it fails can be considered an adverse event, as the exposure to anaesthetic and the wound from surgery can be considered a serious injury; however, replacement of a device as a result of a depleted battery would not be considered an adverse event. Device malfunction must also be reported and if recurrence of the malfunction could cause or contribute to a death or serious injury.

The FDA maintains a publically accessible database, MAUDE, where all device-related reports are available to view. As part of the user requirements, gathering and specification phase of a review of the MAUDE database for adverse events provides an extremely valuable input to the design control process. The FDA, in its guidance document "Do it by design", provides an example of the setting of an oxygen flow control knob at a setting between two levels that resulted in no flow and a hypoxic infant. The control was smooth but flow occurred only on the discrete integer settings.

The MAUDE database provides useful data but often it is difficult to extract overall incidences of issues relating to medical devices. However, the literature provides meta analysis of combined data collected from publications, manufactures and the FDA. For instance, Levy (2011) has provided an analysis of neurologic injury resulting from surgical implantation of paddle leads in the spinal cord. Any project which is directed at developing new leads for new applications is served well by a thorough and detailed review of the literature describing the problems with existing devices.

10.3 PRECLINICAL DEVELOPMENT

The preclinical development phase moves the idea from the bench into animals with the aim of obtaining sufficient data to support use of the device. During this phase the fundamentals of biocompatibility, efficacy in animals and validation of the approach are addressed. Chapter 6 in this volume deals with biocompatibility and ensuring safety of devices. The choice of materials for the fabrication of the device can have an enormous impact on the timescales for regulatory approval. Use of new materials for long-term implants requires extensive biocompatibility testing. In almost all neurobionic applications, the device is constructed with the use of another party's product (e.g. ingredient, subassembly or accessory) or someone else's facility is used in the manufacture of the device. In order that a sound scientific evaluation may be made of the regulatory device submission, the review of data and other information related to the other party's product will be required. The other party, while willing to allow FDA's confidential review of this information, may not want the regulatory applicant to have direct access to the information. To help preserve the trade secrets of these suppliers and at the same time facilitate the sound scientific evaluation of medical devices, the FDA established the device master file system. The existence of an FDA master file does not remove the responsibility for biocompatibility testing. Target organ biocompatibility is still required and a rationale as to why the processing of the material to form the finished device does not affect the biocompatibility. Very few start-up ventures have sufficient funds to consider obtaining approval for a new material and so choose to fabricate devices with FDA master files.

The animal and clinical studies in this phase are directed at establishing the feasibility of the therapy with devices which function as they may be expected to in their final form. These test items do not need to be completed products but rather prototypes that have the functionality which is being tested but may not have the same form as the final devices for human chronic trial. The choice of animal model needs to be carefully considered and justified. Data that is collected and may support the therapy may not be convincing if the animal model is not relevant.

10.4 CLINICAL TRIALS AND APPROVAL TO SELL

If all has gone well to this stage, the preclinical work will have been successful and a product will have been developed with good design control. The next phase is to seek regulatory approval for the conduct of a clinical trial. This process depends on the country and the nature of the device. In the USA, the FDA require manufacturers to obtain pre-market approval (PMA) or Premarket Notification 510(k) to import and sell medical devices. The FDA provides a great deal of guidance for application and approval of devices. The regulations are complex and change from time to time; it is beyond the scope of this chapter to explain the process in full detail.

The first step in the process is to apply to the FDA for an Investigational Device Exemption (IDE) 21CFR Part 812; this allows the investigational device to be used in a clinical study in order to collect safety and effectiveness data required to support a PMA or 510 k application to FDA. The choice of 510 k versus PMA is based on the risk of the device to the patient and the ability to show that the new device is substantially equivalent to a predicate device. The majority of neurobionics devices are active implantable devices and will therefore fall into the Class III device category requiring approval through the PMA process (Figure 10.8).

Clinical studies with devices of significant risk (i.e. Class III devices) must be approved by the FDA and by an Institutional Review Board (IRB) before the study can begin. Studies with devices of non-significant risk must be approved by the IRB only before the study can begin. The IDE application provides the information to the FDA that allows them to assess the risk to patients in the performance of the study and the scientific basis for the study.

The study itself is defined by a protocol, but the application needs to contain a great deal of additional information. The device description includes how it works, how it is used and how it is tested. The ethics of conduct of clinical trials need to be considered carefully by all concerned (Chapter 11). The IDE application includes the informed consent and any documents relating to the relationships between the company and the institutions conducting the study. Assembling an IDE application is a simple task if approached in the correct way, and can be quite onerous if not.

For individuals new to the regulatory process, the level and standard of documentation required to support applications to the FDA can seem to be completely

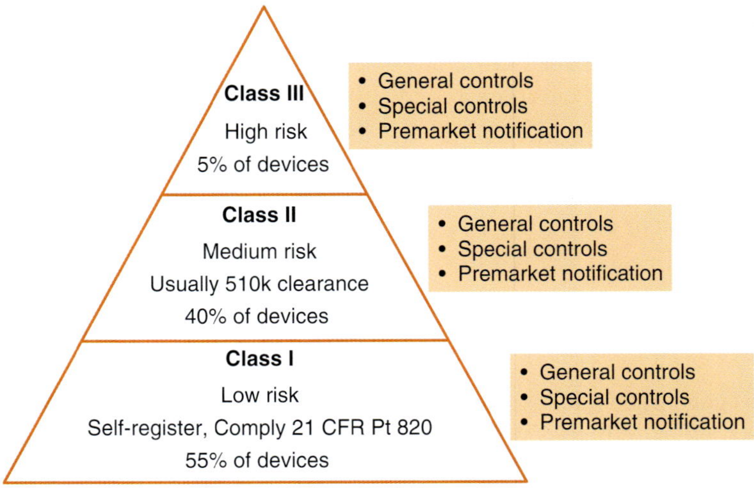

Figure 10.8 FDA classification of medical devices. Neurobionic devices are typically considered Class III devices and must pass through the FDA's PMA process in order to get into clinical trial.

overwhelming. Some documentation even appears to be written in a different language as it can be full of references to standards and acronyms. This is one key area where the effort in commercialization is consistently underestimated.

Although the IDE application requires a detailed description of the device and the protocol, it is a shorter document to prepare than a full PMA. The PMA application provides the FDA with the evidence collected in the trial and any additional testing done by the applicant, including the pre-clinical results, to show that the device and procedure are safe and efficacious. This can be a substantial amount of documentation and includes all the analysis of the trial results.

The only efficient way to assemble documentation in support of an application to the FDA (or other regulatory process) is to ensure that they are written properly in the first place. The percentage of original PMA applications which receive a Major Deficiency letter has dramatically increased from 40% 10 years ago to over 70% today. A major or minor deficiency letter details concerns and additional information that is required by the FDA during the review (21 CFR 814.44). A Major Deficiency letter will hold up the review cycle and can cause major delays while companies respond to the request.

A great submission is no guarantee of approval; the product or therapy has to meet its end point for success. "This is a colossal disappointment" was the only comment from Medtronic's director of endovascular services on the failure of their US clinical trial for renal denervation. This announcement wiped 2.4% (more than $1B) off the company's share price. A clinical trial failure for one of the large companies is manageable, but for a small company it can be catastrophic. The controlled clinical trial is only one part of the approval process.

The gold standard in clinical trial design is the randomised cross-over study. This is a comparison between the therapy and either the best alternative therapy or non-treatment or placebo. A population is selected and controlled with inclusion criteria and then separated into groups, one which receives the therapy and the other which will receive the control and so there are two arms to the study. At some point, the groups are swapped and the efficacy is recorded periodically during the study. Clinical trial design is complex and is its own speciality. In many neurobionic applications it is almost impossible to blind the subject to the treatment arm of the trial. If the subject can feel the effect of the device, for example they can hear with a cochlear implant but cannot with non-treatment or control, then this can bias the results.

The statistical power of the trial must be scientifically-based and this will set the number of participants required. The cost of running a trial can be very large and it may take several years to complete and so a small firm cannot contemplate beginning a trial until there is a high level of confidence in the outcome of the study. Smaller preclinical studies are the way to do this where the trial design is tested; the survey instruments and the tools to measure the performance are also tested. The clinical trial should be conducted when the outcome is almost certain and it is designed to provide definitive proof of efficacy. Approval times and market access for medical devices vary across jurisdictions as illustrated in Figure 10.9.

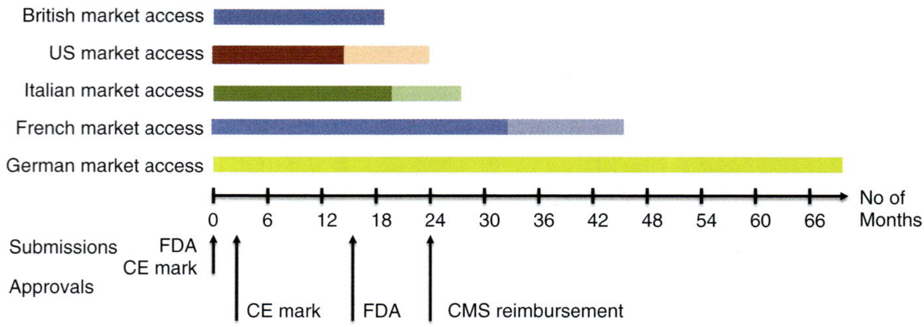

Figure 10.9 Approval times and market access for medical devices in a number of jurisdictions. The figure was redrawn from data in Basu and Hassenplug (2012). CMS refers to the Centers for Medicare and Medicaid Services, which is a US government health insurance program that covers 100 million people.

A start-up venture rarely gets more than one chance to run a pivotal clinical trial and failure to reach endpoint is normally catastrophic. Northstar Neuroscience, a small medical device company, folded in 2008 after it failed to meet endpoint goals for its Everest trial of a novel cortical brain stimulation product for stroke recovery. The early reports of the effectiveness of therapy were remarkable and the rationale for the intervention was soundly based. Northstar's product delivered electrical stimulation to the motor cortex from electrodes placed outside the dura to accelerate stroke recovery. Northstar was founded in 1999 and had raised \$193 M before it was dissolved and returned a couple of dollars per share of unused capital and sold its assets to St Jude Medical for \$2 M.

Northstar had developed a product and had extremely encouraging pre-trial data (Levy *et al.* 2008) and followed the commercialization path in Figure 10.1, but failed completely at the last step. The commentary on why makes interesting reading. A *post hoc* analysis of the data from the trial, demonstrated that stimulation of the intended brain target was obtained in only 14% of the 151 implanted subjects in the trial (Plow *et al.* 2009). The failure to stimulate the intended target was correlated with poor electrode placement and complications as a result of the changes in anatomy due to the stroke (brain atrophy and stroke volume). In those subjects where stimulation of the cortical target was confirmed, the study was highly successful, with 70% of treated subjects reaching success criteria (compared with 30% of controls) and exceeding the studies endpoint by 100%.

The pressure to market and commercialize has been identified as the cause of the trial failure. The net result is a frustrated clinical community who has lost access to a potentially highly promising therapy for their patients and loss of value for the shareholders of the company. James Cavuoto points out that the Northstar failure has had consequences for the entire industry. It is harder to support other potentially valuable innovations in cortical stimulation and any potential investor in similar projects is going to be much harder to convince.

The lesson from the Northstar experience is that the science must come first when designing and conducting a clinical trial. The principles of identifying and controlling the variables that affect the outcome of the trial must have priority over the commercialization and business objectives. In the end there is no business if the trial fails.

10.5 BUILDING A BUSINESS NOT A PRODUCT

Funding all the activity from preclinical to clinical trials in a changing economic climate can be extremely challenging. Venture capital has historically been the traditional source of funding for new projects; however, since the global financial crisis, venture funding has become much harder to obtain while the costs associated with device development and approval have grown. In a recent example, NeuroPace obtained FDA approval after 16 years of effort and a total investment of $200 M. Post-FDA approval, the company must generate sales and build the infrastructure to support and grow and this requires even more investment.

Investment in the medical device and equipment industry in the USA fell to $2.14 billion in 2013, down more than 40% from 2007 and the sharpest drop among the top five industry's which receive venture funding. The shift away from medical devices together with increasing costs has forced entrepreneurs to look for alternative sources of funding. Creative ways of using research funding, looking for special interest grants and high net worth individuals and private equity funds are more common sources of finance in the early stages.

Start-ups that survive early stages with minimal financing are able to negotiate better deals. SetPoint Medical, developer of a nerve-stimulation device to treat rheumatoid arthritis, was turned down by dozens of venture firms over three years before it presented promising clinical data from a small study in 2013. In August 2013, SetPoint made a deal to receive $27 million in financing from a group of investors including Boston Scientific, Covidien PLC and GlaxoSmithKline PLC. Even with top-tier companies investing money on unproven technology, no firm has been given exclusive acquisition rights.

Figure 10.10 illustrates typical funding sources, expected returns and amounts over the commercialization pipeline. There has been a definite shift towards having to achieve much more for less in the early phases of development. The cheapest resources to deploy on the commercialization journey are those that you do not have to pay for. Leveraging resources in other organizations, including academic institutions, is perhaps the only way to lower the risk and spread the workload. The nature of the collaborations and how they are structured depends on the development stage of the project and achieving co-operation and results across a network of collaborators requires careful management. Achieving a truly functional, value adding collaborative network requires that the IP ownership is clear from the outset. There is no clearer IP position than to own the patents.

The initial development of the idea and clinical proof of concept are cheap and quick by comparison to the rest of the commercialization journey, and one of the barriers to achieving effective collaboration is over-valuation of the contribution.

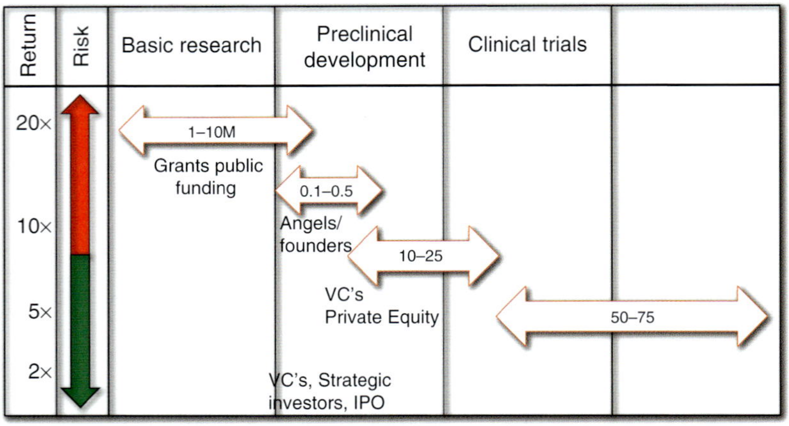

Figure 10.10 Funding sources, expected returns and amounts.

Academic collaborators rarely understand the level of scientific and intellectual rigor required to take an idea to product and then to market and often price a valuable contribution well above a new enterprise's capacity to pay.

Grants and public funding are the predominate sources of funds during basic research; get more done in this phase with less is the key to future success. At some point during preclinical development, pure commercial activities will need to be funded. Establishment of companies getting agreements in place for IP, simply creating a commercial vehicle which can receive future funding rounds, requires some investment and this is normally left to angels and founders to fund.

The drop in funding available from VCs has led to an increase in the number of private equity deals that have been done to support new companies. Some of the gap in financing as a result of the decline in the VC funding has been filled with corporate investment moving to earlier stages in the new company lifecycle, although the amount of money invested by the corporates has remained static. The major companies, Medtronic, St Jude and Boston Scientific, have a strong interest in ensuring that there is a vibrant community of small developing companies. The cost of bringing a device to a commercialization endpoint for a start-up is a fraction of the cost for one of the majors, and small companies can take more risk as they are not protecting a huge revenue stream or a loyal customer base. The large corporates who grow through acquisition need to encourage thriving medical device company nurseries in order to create targets for later acquisitions, and at the same time they do not want to be exposed to the risks associated with new ventures. Early involvement of strategic investors in a business can cap the upside value for the company, but in lean investment times this trade off can be the difference between financing the next stage of development or not. An example is the St Jude Spinal modulation deal consisting of \$40 M dollar in equity injection in exchange for exclusive rights to distribute the product and an option to buy the company for \$300 M after commercialization milestones are reached.

GlaxoSmithKline, one the world's largest pharmaceuticals companies, announced their intention to seed funds for the development of "Electroceuticals". They are effectively catalysing an effort to discover and provide proof of principle studies for new neuromodulation targets. The research funding is backed by the formation of Action Potential investments, which is a venture arm, aimed at providing funding for new start-up companies.

Professional and strategic investors will conduct due diligence on a new product or venture. They will ask:

- Does the product work?
- Is there a market for it?
- Will someone pay for it?
- Is the device going to be competitive?
- IP – Landscape and freedom to operate?
- Can the team actually do the work?

How well these questions are answered will determine the difference between successfully attracting funding or not. The last question is one of the most important. The capacity of the team and their experience and understanding of the commercialization pathway is the only assurance that an investor has that timing costs and effort are not underestimated.

10.6 CONCLUSIONS

The chapter began with observation that the time and effort required to reach commercial success with a medical device venture is almost always underestimated. Understanding what is required and ensuring that it is done as efficiently as possible is the only path to success. In a nut shell:

1. Work on the right idea in the first place and ensure that targets for success are well defined
2. Patent early
3. Engage with industry (before you begin)
4. Find first-class clinical partners to work with as early as possible
5. Ensure quality system development occurs early (during basic research) and is aligned with regulatory requirements
6. Verification and Validation is most of the work and expense in the development cycle
7. Explore every option for funding before seeking professional funding
8. Generate early clinical results
9. Good Science is of paramount importance in clinical trial design
10. Aim to achieve more for less with effective collaboration.

The time from completion of the product to where it is reliably generating revenues can take years. The competitive landscape can change completely in that time and so it is essential not to build a product which is needed in today's environment, but one which is needed at that future time to ensure that the need has not disappeared.

REFERENCES

Basu, S. and Hassenplug, J.C. (2012) Patient access to medical devices – a comparison of US and European review processes. *New Eng. J. Med.*, **367**(6): 485–488.

Granstrand, O. (2000) *The Economics and Management of Intellectual Property: Towards Intellectual Capitalism.* Edward Elgar Publishing.

Kleiner, G. (2006) Subthalamic nucleus deep brain stimulation: summary and meta-analysis of outcomes. *Movement Disorders*, **21**(S14): S290–S304.

Levy, R. *et al.* (2011) Incidence and avoidance of neurologic complications with paddle type spinal cord stimulation leads. *Neuromodulation: Technol. Neural Interface*, **14**(5): 412–422.

Levy, R. *et al.* (2008) Cortical stimulation for the rehabilitation of patients with hemiparetic stroke: a multicenter feasibility study of safety and efficacy. *J. Neurosurg.*, **108**(4): 707–714.

Meads, C. and Hyde, C. (2003) What is the cost of blindness? *Br. J. Ophthalmol.*, **87**(10): 201–204.

Mortimer, J.T., Shealy, C.N. and Wheeler C. (1970) Experimental nondestructive electrical stimulation of the brain and spinal cord. *J. Neurosurg.*, **32**(5): 553–559.

Pantev, C. (2006) Dynamics of auditory plasticity after cochlear implantation: a longitudinal study. *Cerebral Cortex*, **16**(1): 31–36.

Plow, E.B. *et al.* (2009) Invasive cortical stimulation to promote recovery of function after stroke. *Stroke: a Journal of Cerebral Circulation*, **40**(5): 1926–1931.

Pollo, C. (2014) Directional deep brain stimulation: an intraoperative double-blind pilot study. *Brain: A Journal of Neurology*, **137**(7): 2015–2026.

Rousseax, M. (2004) Effects of subthalamic nucleus stimulation on parkinsonian dysarthria and speech intelligibility. *J. Neurol.*, **251**(3): 327–334.

Sassi, F. (2006) Calculating QALYs, Comparing QALY and DALY calculations. *Health Policy and Planning*, **21**(5): 402–408.

Shultz, M.D. (2012) Sensor-driven position-adaptive spinal cord stimulation for chronic pain. *Pain Physician*, **15**(1): 1–12.

Shealy, C.N., Mortimer J.T. and Reswick, J.B. (1967) Electrical inhibition of pain by stimulation of the dorsal columns. *Anesth. Analg.*, **46**: 489–491.

Temel, Y. (2006) Behavioural changes after bilateral subthalamic stimulation in advanced Parkinson's disease. *Parkinsonism & Related Disorders*, **12**(5): 265–72.

Wyatt, R.J. (1996) Cost utility of the multichannel cochlear implant in 258 profoundly deaf individuals. *The Laryngoscope*, **106**(7): 816–821.

WEBLIOGRAPHY

http://www.accessdata.fda.gov/scripts/cdrh/cfdocs/cfcfr/CFRSearch.cfm?CFRPart=820&showFR=1&subpartNode=21:8.0.1.1.12.4 – "CFR– Code of Federal Regulations Title 21," http://www.accessdata.fda.gov/scripts/cdrh/cfdocs/cfmaude/search.cfm – "MAUDE – Manufacturer and User Facility Device Experience,"

http://www.aleva-neuro.com/index.html – "Aleva Neurotherapeutics SA I HOME,"

http://www.emergogroup.com/blog/2012/12/fda-report-claims-improved-510k-pma-review-times – "New FDA Data on 510k and PMA Medical Device Reviews,"

http://www.fda.gov/MedicalDevices/DeviceRegulationandGuidance/GuidanceDocuments/ucm080735.htm – Center for Devices and Radiological Health, "Guidance Documents (Medical Devices and Radiation-Emitting Products) – Use of International Standard ISO-10993, 'Biological Evaluation of Medical Devices Part 1: Evaluation and Testing'

http://www.fda.gov/medicaldevices/deviceregulationandguidance/guidancedocuments/ucm070627.htm – Center for Devices and Radiological Health, "Guidance Documents (Medical Devices and Radiation-Emitting Products) – Design Control Guidance For Medical Device Manufacturers,"

http://www.fda.gov/MedicalDevices/Safety/ReportaProbLem/default.htm – Center for Devices and Radiological Health, "Medical Device Reporting (MDR),"

http://www.fda.gov/medicaldevices/deviceregulationandguidance/guidancedocuments/ucm094957.htm – Center for Devices and Radiological Health, "Guidance Documents (Medical Devices and Radiation-Emitting Products) – Do It By Design – An Introduction to Human Factors in Medical Devices,"

http://www.fda.gov/medicaldevices/deviceregulationandguidance/guidancedocuments/ucm341918.htm – Center for Biologics Evaluation and Research Center for Devices and Radiological Health, "Guidance Documents (Medical Devices and Radiation-Emitting Products) – Types of Communication During the Review of Medical Device Submissions – Guidance for Industry and Food and Drug Administration Staff,"

http://www.neurotechreports.com/pages/publishersletterJan09.html – "Northstar Neuroscience EVEREST Trial Stroke Motor. Rehabilitation Primary Endpoint, Cortical Plasticity, Mechanisms Deep Brain Stimulation (DBS),"

http://online.wsj.com/news/articles/SB10001424052702304470504579164122110690860 – "Funding Dries Up for Medical Startups – WSJ.com,"

http://www.sapiensneuro.com – "Sapiens Steering Brain Stimulation,"

http://www.sf-36.org/tools/sf36.shtml#LIT – "The SF Community – SF-36® Health Survey Update,".

http://www.smithsonianmag.com/science-nature/how-human-echolocation-allows-people-to-see-without-using-their-eyes-1916013/-- "How Human Echolocation Allows People to See Without Using Their Eyes," *Smithsonian*

http://www.startribune.com/business/244397211.html – "Pacemaker-like Device Offers Hope to Some Epilepsy Patients."

11

ETHICAL CONSIDERATIONS IN THE DEVELOPMENT OF NEURAL PROSTHESES

FRANK J. LANE AND KRISTIAN P. NITSCH
Illinois Institute of Technology Rehabilitation Psychology, Chicago, IL, USA
MARCIA SCHERER
University of Rochester Medical Center, Rochester, NY, USA

11.1 INTRODUCTION

Neural prostheses, encompassing cochlear implants, vision implants, deep brain stimulation (DBSs), and the like, are in different stages of development. The cochlear implant for example, has been developed over the past 30 years and has been implanted therapeutically in approximately 330,000 deaf individuals worldwide. In contrast, vision prostheses are regarded as largely experimental; the first retinal implant commercially available, developed by Second Sight, only recently received a device exemption from the Food and Drug Administration (FDA) in 2013.

Because neural prostheses are in different stages of development and commercial availability, the authors of this chapter decided to present the discussion in two main sections. The first focuses on the importance of including people with disabilities in the development of technology, and the second focuses on the ethical treatment of individuals with disabilities in clinical research trials. Each section independently warrants discussion and both are equally important. The first section outlines information that investigators should understand regarding disability culture and the disability rights movement as it pertains to the inclusion of individuals with

Neurobionics: The Biomedical Engineering of Neural Prostheses, First Edition.
Edited by Robert K. Shepherd.

disabilities in prosthesis development. The second part of the discussion outlines the basic ethical principles of biomedical research, the informed consent process, research that addresses the information important to present the prospective participant in a research trial, and the possible challenges of understanding complex information associated with neural implants.

11.2 INDIVIDUALS WITH DISABILITIES AND TECHNOLOGY DEVELOPMENT

Any discussion regarding the development of neural prostheses would be incomplete without addressing the importance of understanding individual and environmental factors that influence their use, and how these factors interact with the societal and cultural landscape in which they operate. The following section will provide a succinct overview of the context in which most neural prostheses are used and by whom; the role and purpose of these devices in the lives of individuals with disabilities; and the importance of integrating direct feedback from potential device users during the development stages. A proposed mechanism for integrating consumer feedback into the device development process, and finally a historical example of divergent reactions within specific disability populations to the introduction of implantable devices will then be reviewed.

11.2.1 Assistive technology in the context of disability

An essential step in the process of developing neural prostheses should be to establish an understanding of the context in which the device will be used, and who it will be used by. Because the primary consumers of neural prostheses, which are fundamentally assistive technology devices (ATDs), are persons with disabilities, considering these factors within a disability context is imperative. Divergent beliefs regarding how disability should be conceptualized have given rise to a number of classification systems driven by medical, social and functional perspectives (Masala and Petretto 2008). However, the global presence and integrative nature of the International Classification of Functioning, Disability and Health (ICF) from the World Health Organization, establishes the ICF as a strong candidate model for conceptualizing disability (WHO 2013). The ICF offers a universal nomenclature and conceptual basis for the measurement of disability, and emphasizes the integration of biological, psychological and social factors in understanding the onset and subsequent impact of disability (WHO 2013).

11.2.2 International classification of functioning, disability and health

The ICF is organized into two main divisions, including: (i) "Functioning and Disability"; and (ii) "Contextual Factors". Each of these domains contains two additional divisions. "Functioning and Disability" which is further divided into: (a) "Body Functions and Body Structures", which are understood as the physiological functions and structures of the body; and (b) "Activities and Participation", which

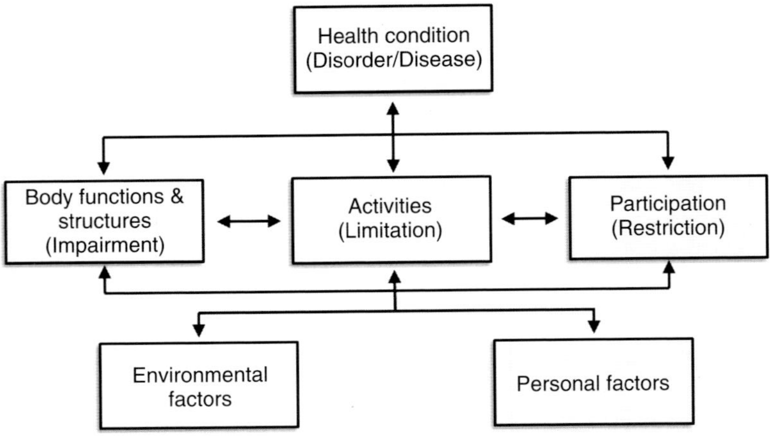

Figure 11.1 ICF model and interaction of components.

represent the execution of a task by an individual, and involvement in life activities, respectively. According to the ICF, the term "impairment" refers to "problems in body functions and structures such as significant deviation or loss", whereas "activity limitations" exist when an individual experiences difficulties in executing specific activities. Finally, "Participation Restrictions" describe difficulties or problems an individual might experience in involvement in life circumstances (WHO 2013). Importantly, an individual's overall level of functioning in varying areas is thought to result from interactions between that individual, their unique health condition, and additional extraneous contextual factors (Figure 11.1). These interactions are dynamic, unpredictable and bidirectional. That is, an individual's personal factors can influence their choice of environment just as easily as the environment can impact personal factors, while both simultaneously affect the individual's experience of a health condition (WHO 2013). By considering these domains, technology developers can take steps to ensure the utility of neural prostheses, while addressing concerns and potential device shortcomings before an end product is delivered to the consumer.

The ICF describes "Functioning" and "Disability" as dimensional constructs that represent an individual's interaction with their environment, rather than endogenous characteristics of that individual (WHO 2013). This conceptualization is a dramatic divergence from early medical models, which describe disability within the individual due to diagnosed impairments related to injury, disease or developmental processes (White *et al.* 2010). As defined in the ICF, "Functioning" refers to positive aspects of the interaction between an individual and various components of that individual's context. Conversely, "Disability" encompasses the "Impairments", "Activity Limitations", "Participation Restrictions", and other negative aspects of the interaction between an individual and that individual's "Contextual Factors" (WHO 2013). These "Contextual Factors" are categorized as: (a) "Environmental Factors", which make up the physical, social and attitudinal environments in which people live and conduct their lives; and (b) "Personal Factors", which have not yet been outlined in the ICF.

Neural prostheses, irrespective of their technological complexity, are ATD; that is, they are pieces of technology specifically designed to restore or supplement functioning in individuals with disabilities. Throughout the remainder of this chapter, the terms "neural prostheses" and "assistive technology devices" will be used interchangeably. The ICF considers assistive technology and ATDs to be an integral component of the "Environmental Factors" domain (WHO 2013).

Considerable conceptual overlap exists between the ICF's "Products and Technology" classification, and the definitions of assistive technology previously put forth by documents such as the US Technology-Related Assistance of Individuals with Disability Act of 1988, and the International Organization for Standardization (ISO 9999). This convergence suggests a consensus of what constitutes ATDs, and who the primary consumers of these technologies are. In the ICF, "Products and Technology" refers to "any product, instrument, equipment or technology adapted or specially designed for improving the functioning of a disabled person, which may lead to increased participation in activities" (WHO 2013). This definition allows for broad interpretation of what is considered an ATD, ranging from "low-tech" devices to "high-tech" devices such as neural implants. However, the defining characteristic of this definition is the focus placed on the functional improvement and increased opportunity for regular participation in desired activities and societal integration. It is this perspective, that individuals living with disabilities should have equal opportunity to engage in activities commonly accessible to individuals without disability, which gave rise to the Independent Living (IL) model.

11.2.3 "Nothing About Us, Without Us"

The progression towards independent living was born from the disability rights movement in the USA throughout the 1950s and 1960s, which was propelled by the creation of membership organizations such as the National Federation for the Blind and Americans Disabled for Accessible Transportation (Meyer *et al.* 1991; Pelka 1997). These organizations catered to the specific needs and priorities of its membership, and provided a unified voice to advocate on the behalf of specific disability populations. Prior to this time, disability was thought of as terrible hardship that simply needed to be endured (Barnes and Mercer 2005). Disability activism spearheaded by individuals with disabilities and disability groups acted as a catalyst for the move from the dependence seen in antiquated medical models towards independence, autonomy and consumer-centered service provision.

Perhaps one of the most widely-cited examples of individuals with disabilities advocating for social justice and civil rights is Ed Roberts (White *et al.* 2010). Roberts, who had developed polio in the 1950s, wanted to advance his education at the University of California, Berkeley, but was denied admittance due to his physical disability. However, Roberts made arrangements for personal aides to assist him in preparing for classes and took up residence in the university hospital. With the help of other students who wanted to secure civil rights and entrance to the university, Roberts was able to start a disability rights advocacy group known as the "Rolling Quads" (Oxford and McDonald 1999).

The Rolling Quads primary objective was to advocate on the behalf of students with disabilities to remove barriers that precluded their involvement and access to the resources provided by the university. Roberts went on to obtain federal funding from the Rehabilitation Administrations, which he used to establish the inaugural Center for Independent Living (CIL) (Shapiro 1994). The Oakland CIL stressed the importance of personal independence, consumer choice and autonomy; and its success served as the catalyst for the establishment of CILs across the USA (Willig-Levy 1988). Personal choice and consumer empowerment continue to be the fundamental tenants of CIL service delivery. However, the IL model is more than a social justice movement; it also represents a philosophy embodied by people with disabilities that places the focus of rehabilitative services on the individual, making it a consumer-centered practice (Deegan 1992).

Within medical models of disability, clinicians are considered experts who normally possess unilateral decision-making power regarding treatment planning and rehabilitative intervention. This approach precludes individuals with disabilities from being active participants in their own health care, and establishes an authority role for the clinician (White *et al.* 2010). Conversely, conceptualizations of disability predicated on the IL framework view disability as resultant from problems endogenous to the environment, and assert that many of the issues facing individuals with disabilities stem from dependence on professionals and family when making decisions (DeJong 1979). However, given the frequent long-term nature of disability, individuals become experts on their own body, facilitating opportunity for direct involvement in making informed health care choices. Beyond health care, the IL framework asserts that individuals with disabilities should have the right and equal opportunity to makes choices related to housing, employment and community engagement among many other life domains (Frieden 1980). This gained expertise allows people with disabilities to advocate on their own behalf and provide integral information to ATD developers based on insight generally unavailable to individuals that do not share similar disability experiences.

The IL model removes individuals with disabilities from the role of passive recipient, and establishes them as active, knowledgeable and empowered consumers who are fully engaged in managing their own lives. Involving prospective consumers of ATDs in the development process is consistent with the IL approach to disability through creating opportunities for direct engagement in meaningful decision-making. Because individuals with disabilities are experts on their own lives, they are uniquely qualified to provide information and recommendations throughout the development process of ATDs. By incorporating this feedback, technology developers can ensure that the most germane concerns and preferences expressed by prospective end-users are considered prior to the completion of a final product. Moreover, this process also protects against erroneous or inaccurate assumptions made by technology developers regarding the desired functioning of a device, the level of risk prospective users would be willing to accept when using the device, and what characteristics are likely to result in device abandonment or poor consumer satisfaction. This approach to device development has recently been implemented by researchers exploring the user-defined preferences and device characteristics for an intra-cortical visual prosthesis through the use of focus groups (Lane *et al.* 2011, 2012). One framework that has been developed to capture consumer feedback regarding ATDs, The Matching Person and Technology (MPT) model, may serve as a theoretical starting point for ways to consider and integrate vital consumer input during device development.

11.2.4 Matching Person and Technology: applications to neural prosthesis development

The MPT model proposes a comprehensive evaluation of the contextual, cultural, technological and personal factors that impact the perceived success, functional gain, and ultimately the use of many technologies designed to assist individuals with disabilities (Scherer 2004). This model has close conceptual ties to the ICF, and addresses the many facets identified as potentially influential on the use of ATDs. While this model has historically been used to match individual consumers to presently available technology modalities, a study implementing the principle tenants of the MPT model suggest that it may have potential application during the development process of new neural prostheses as well (Kilgore *et al.* 2001).

The MPT model posits that consumers of assistive technologies have highly diverse and unique expectations which are shaped by individual needs, abilities, limitations, preferences and past experiences with technology (Scherer and Cushman 2001). The MPT model aims to evaluate and account for these idiosyncrasies in technology use through the implementation of a comprehensive and collaborative assessment process that allows the consumer to provide direct feedback and information regarding their specific needs and wants (Scherer and Cushman, 2001). This mechanism for incorporating consumer input into the ATD selection process allows specific technological needs and preferences to be addressed, maximizing the utility and success of the ATD in daily use.

The MPT model emphasizes evaluation of three distinct but interrelated components which have been found to influence the use of ATDs (Figure 11.2). These components include:

1. the characteristics of the person who will be using the technology;
2. the technology itself; and
3. the environment in which the consumer will be using the technology (Scherer *et al.* 2005).

Each of these domains, independently and in combination, contributes negative or positive influences on the consumer's satisfaction with the technology in use. For optimal utilization and satisfaction with a specific technology, each domain should be evaluated to ensure that the technology is successfully fulfilling the specific needs and wants of the consumer (Scherer *et al.* 2005). While designing advanced implantable neural prostheses on an individual basis may represent a daunting and potentially unrealistic enterprise, the underlying tenant of integrating the opinions of prospective

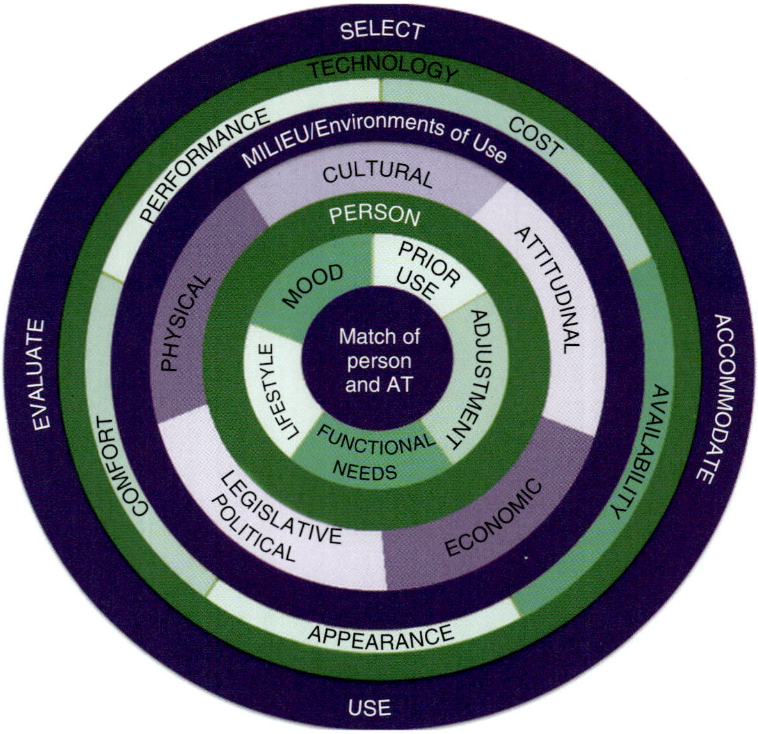

Figure 11.2 The MPT model emphasizes evaluation of three distinct but interrelated components which have been found to influence the use of ATDs (See Figure 11.2).

device consumers remains integral. Doing so during the development process provides a means for addressing concerns and recommendations which may be prevalent within the specific disability population the device is intended for.

The development of bionics devices that follows a "one size fits all" mentality, without including recommendations and considerations from stakeholders and individual users, may be more likely to be abandoned or result in decreased user satisfaction. When a technology does not fulfill the need of a consumer, or does so at the cost of other areas of functioning, the device is often abandoned and goes unused. In fact, approximately one-third or all commercial assistive technologies are abandoned by their users within one year of being received (Scherer and Glueckauf 2005). Perhaps more than any other form of ATD, the development of implantable neural prostheses, which are intended to become at least semi-permanent fixtures in the human body, should be informed by the recommendations of the to-be implanted stakeholders. Unlike a long-cane or wheelchair, which can be more readily replaced or adapted to fit the needs of the user, the inherent nature of implantable prostheses often precludes easy upgrade or removal. Furthermore, involving people with disabilities in the development process serves as a mechanism for understanding how individuals that the device is being designed for are likely to react to and adopt the device into daily use. Historically, ATDs intended to restore function in individuals with disabilities have not been unanimously welcomed.

11.2.5 Disability culture: the cochlear implant

While it may be assumed that ATD developers are inherently well-intentioned when producing technologies designed to restore or supplement functioning in individuals with disabilities, history has proven that ATDs have not been universally accepted and adopted into disability cultures. In fact, the cochlear implant, which is a "high-tech" ATD, faced considerable resistance from the Deaf culture community when first introduced, and continues to be a topic of debate and contention among many individuals who considered themselves part of the Deaf community. According to the United Nations Conventions on the Rights of Persons with Disabilities, individuals who consider themselves members of Deaf culture "shall be entitled, on equal basis with others, to recognition and support of their specific cultural and linguistic identity, including sign languages and deaf culture". Rather than viewing deafness as a disability, constituents of Deaf culture assert that being deaf is a variation in human experience, and any technology intended to "cure" deafness suggests that deaf experience is inferior to the experience of hearing individuals (Nunes 2001). This example illustrates the necessity of collaboration between researchers and the individuals intended to use neural prostheses; inviting deaf people, including those who identify as having been raised in Deaf culture and post-lingual deaf individuals who are part of the hearing community, to discuss the progression of the cochlear implant during the developmental phase may have provided insight into the resistance the designers faced, and how to work towards solutions that address the concerns expressed.

11.3 ETHICAL PRINCIPLES OF BIOMEDICAL RESEARCH

Neural prostheses are unique because they are implanted into the nervous system; a number of these devices can alter our sense (of) and perception (of) the world. Retinal implants designed to restore visual percepts and cochlear implants designed to restore hearing are among the more common neural prostheses (Chapter 7). Neural prostheses are not new. The modern cochlear implant, developed in several centers around the world, including in 1978 by Graeme Clark in Melbourne, Australia, recently celebrated its 30th anniversary. The development of vision prostheses implanted in the visual cortex of human participants began in the late 1960s (Brindley and Lewin 1968) (Chapter 1), with the majority of devices being implanted between the 1970s (Seelig and Dobelle 2001) and 2005 (Lane *et al.* 2013). Advances in retinal prostheses have shown the most promise with one company, Second Sight, receiving FDA approval in 2013 (Marcus and Koch, 2014), while a second company, Retina Implant AG, received CE mark of approval in 2013. Other examples of neural implants include DBS for treatment of individuals with intractable neuropsychiatric diagnoses, Parkinson's disease and epilepsy (Chapter 8); and neural implants to restore sensation for individuals with spinal cord injury. While some of these technologies are in various stages of development, and are preparing for human implantation, others, such as DBSs for Parkinson's disease, represent mature technologies that have been granted FDA approval.

The ethical implications of neural implants have been debated in the literature with each type of implant having a different focus. As noted above, cochlear implants have received widespread criticism from the Deaf culture community as evidence of an attitude towards deaf people that deafness is a disease that should be cured. Individuals concerned about DBS implants have focused on concerns about the potential change in cognition and emotion that, in turn, may alter an individual's personality.

Ethicists and disability advocates have asked why an individual would consent to participate in experimentation that is high in risk and low in direct benefits to the participant. This has given rise to a discourse about what motivates an individual to participate in research trials and whether protected classes such as children, elderly and people with disabilities are capable of consenting to such procedures.

What can be said is that neural implants are complex devices that require a sophisticated understanding of the unique profile of benefits, risks and alternatives for each device before an individual consents to receive one. Moreover, the decision to have a neural prosthesis implanted is both impacted by and has an impact on the society and culture in which an individual lives, works and learns.

11.3.1 Principles of biomedical ethics

Beauchamp and Childress (2001) have long been viewed as the grandparents of modern bioethics. They have written about four basic principles in bioethics: autonomy, non-maleficence, beneficence and justice. These ethical principles, outlined in Table 11.1 below, will also be addressed within the context of the discussion that follows. The authors of this chapter also believe the principle most relevant to this discussion is the principle of autonomy (Beauchamp and Childress 2001). The application of these principles to healthcare will be discussed, as well as their application to clinical research trials. As a prospective participant in a clinical trial or experimental procedure, an individual's autonomy can be thought of as being exercised through the informed consent process (Beauchamp and Childress 2001), but there are also aspects of clinical research trials that are different from clinical care.

11.3.1.1 Autonomy According to Beauchamp and Childress (2001), autonomy also includes but is not limited to protecting confidential information, obtaining informed consent from research participants, and helping individuals who are in need of assistance with decision-making (Beauchamp and Childress, 2001, p. 65). These guidelines should be used by investigators to respect and honor the autonomy of the research participant.

In order to ensure the participant has made an autonomous decision, their decision to participate must be *voluntary* and free of "control by others" and contain

Table 11.1 The four basic principles in bioethics

Ethical Principle	Brief Definition	Example
Autonomy	Right to make a decision free from control by others	Inadequate understanding that prevents informed choice
Non-maleficence	Do not harm	An adverse effect on an individual's interests
Beneficence	Help others and further their interests	Provide healthcare to individuals who are in need
Justice	Distribution of benefits and burdens to society	An individual with a need for an organ transplant receives one

sufficient *information* (Beauchamp and Childress (2001) in a manner that the participant can understand. This has particular relevance to neural implant trials, because the information can be very complex.

In addition to freedom from the control of others and sufficient information, the individual must also be competent. For example, whether an individual with a severe injury to the frontal lobe, which has lost executive functioning, is incapable of making an autonomous decision. The individual's *decision-making capacity*, or *competence*, is compromised (Beauchamp and Childress 2001). In the example above, the prospective participant is not capable of making an autonomous decision. In this example, the health care professional or research investigator may consider behaving in a paternalistic manner. Such a decision may include involving a parent, spouse or other type of guardian. Typically, paternalism is considered if it is believed that the benefit to the individual is outweighed by his or her capacity to make an autonomous decision. Paternalism will be discussed in greater detail in Section 11.3.1.4 below.

A prospective research participant makes the decision to enter a clinical research trial. It needs to be evaluated whether the participant has made an autonomous decision, if he or she has made the decision *voluntarily* (free of control), has been *informed* in a language they can understand, and comprehension has been evaluated and determined to occur, and if they possess the *capacity* to make decisions (i.e. he or she is *competent*) (Beauchamp and Childress 2001). It is important to note that while obtaining informed consent in a research context, where expectations for personal benefit from the experimental procedure can be easily misunderstood, it is essential to address the expectation of possible benefits from the implantation of the device versus the potential risks. It must also be made clear that the investigators are not necessarily there for the benefit the research participants. This is also a common form of misunderstanding among research participants.

11.3.1.2 Non-maleficence Non-maleficence means above all, "do no harm." Healthcare professionals, especially, are expected to, and by way of codes of ethics, are responsible for not causing harm to their patients, clients, participants, etc. Non-maleficence says, "One ought not to inflict evil or harm" (Beauchamp and Childress 2001, p. 116). Here, harm is defined as 'an adverse effect on an individual's interests". Any harmful act is considered wrong. However, there are some circumstances where the harm may be considered justifiable. The concept of justifiable can be complex. A common example of justifiable harm includes one individual harming another in self-defense. We must also consider the potential for harm in clinical research trials, particularly those involving neural implants. Examples of this will be discussed in Section 11.3.1.4 on weighing beneficence and non-maleficence.

11.3.1.3 Beneficence Beneficence means that healthcare professionals have a duty to help others and to further their interests (Beauchamp and Childress 2001). Unlike the caution to avoid harm as with the principle of non-maleficence, beneficence urges or encourages healthcare professionals, as well as research investigators, to act in favor of or in support of the individual.

Beauchamp and Childress (2001) make a distinction between two types of beneficence: specific and general. *Specific beneficence* refers to obligations. The key in the concept of specific beneficence is whether the healthcare professional and individual patient or research participant has a special relationship. Some examples include the special relationship between a psychologist and client, a physician and patient, lawyer and client, etc. In these circumstances, the reliance on the professional is based in large part on trust. For this reason, when we are engaged in a special relationship, we are obligated to practice the skills in which each professional is trained in a manner that benefits the individuals with whom we work. Healthcare professionals have a specific obligation to benefit their patients, because the expectation of the intervention is that it will be therapeutic in nature. Healthcare professionals and research investigators have a special relationship with their patient or research participant. However, while healthcare professionals are obligated to utilize interventions that are therapeutic in nature and benefit the patient, the same may not be true for a research investigator.

General beneficence is directed beyond the special relationships covered under specific beneficence and directs the attention of the professional to all individuals. General beneficence speaks to the belief that healthcare professionals are not obligated to help others when the professional does not have a special relationship.

11.3.1.4 Weighing Beneficence and Non-maleficence in Research How would a healthcare professional handle a situation where s/he is an investigator in a research study and where there is no likelihood of a therapeutic personal benefit to an individual participant? There is a potential for harm to the individual participant, but the benefit of the research project is to society. This apparent dilemma occurs in research trials or clinical trials and is reviewed by an institutional review board (IRB), whose job it is to weigh the risks to individual participants against the benefits to society. In addition to professional review, the investigator must also inform the prospective participant as to the risks and benefits of the trial.

The distinction between therapeutic and research trials can be confusing to participants and the role that healthcare professionals play in each can also be confusing. There is a high potential for misunderstanding and a lack of comprehension. In this section, it is important to mention that achieving the delicate balance between beneficence and non-maleficence is not easy and must be carefully managed. An individual has the right to decide to participate in a clinical research trial if there is a potential for risk, even if the benefit is to society and not to the individual. An individual's right to decide to participate in research that involves risk can be thought of as respecting the autonomy of the individual making the decision. Our role as healthcare professionals is to ensure that the prospective participant is sufficiently informed, and understands and comprehends the information before making a decision. The matter will be discussed in greater depth in the following sections on *informed consent*, *disappointment* and *therapeutic misconception*.

Paternalism is generally considered when it is believed that the benefit to the individual is greater than the individual's ability to make an autonomous decision. Beauchamp and Childress (2001) define paternalism as "the intentional overriding of one person's known preferences ... by another person, where the person who overrides justifies the action by the goal of benefiting or avoiding harm to the person whose preferences ... are overridden" (p. 178). Paternalism may be considered if the following are present: (i) "a patient is at risk of a serious, preventable harm; (ii) the paternalistic action will probably prevent the harm; (iii) the projected benefits to the patient of the paternalistic action outweigh its risks to the patient; and (iv) the least autonomy-restrictive alternative that will secure the benefits and reduce the risks is adopted" (Beauchamp and Childress, 2001, p. 186).

Examples of paternalism that are relevant to this chapter would be a parent making a decision to have a bionic ear implanted in their deaf child. Another example might be the appointment of an advocate or guardian for an individual who is considering vision prosthesis, but because s/he sustained a traumatic brain injury the cognitive impairment has created problems of competency.

11.3.1.5 Justice The *principle of justice* speaks to the distribution of burden and benefit to individuals and society in general (Beauchamp and Childress 2001). A common example that is discussed is the distribution of healthcare benefits to society. A common question that is asked pertains to issues related to access to healthcare. Specifically, is it just that there is disparity in access to healthcare because of wealth. For example, you could have two individuals with similar health conditions. One individual is from a high socioeconomic status and the other is from a lower socioeconomic status. The individual from the lower socioeconomic status receives less healthcare benefits than the one from a higher status. Is this just? Another common example pertains to the implantation of organs. For example, you have two patients, both in need of a liver, of which there is only one available. Who receives it? What if one of the potential recipients was an alcoholic and has cirrhosis? Would it make a difference if the alcoholic had been sober for the past two years? If an individual contributes to his or her condition, does that make the two patients different in terms of their need for the liver?

Aristotle is credited with the theory of formal justice (Beauchamp and Childress 2001). Within formal justice theory, individuals who are equal should receive an equal amount of resources. In contrast, individuals who are unequal should not receive an equal amount of resources, but the resources they do receive should be proportional to the degree of inequality. The example above regarding the liver transplant shows that this theory can be extremely difficult to apply in real life. How unequal is the potential recipient with the cirrhosis of the liver compared to the potential recipient who did not contribute to their liver failure? If we believe the individual with the cirrhosis of the liver is unequal to the other prospective participant, what is the proportional benefit to the unequal individual? Answers to these questions depend on the theory of justice being applied and is commonly debated by ethicists. However, these debates must also be a part of our conversation regarding neural implants. For example, what if you have two prospective participants for a clinical trial involving a vision implant? One prospective participant lost his/her sight due to diabetic retinopathy. The second prospective participant lost his/her sight due to an automobile accident and this individual was driving while under the influence of alcohol. There is only one space left in the clinical trial, but both are eligible candidates. Who gets the last place in the trial?

11.3.2 Informed consent in clinical research trials

Informed consent is a process and the informed consent document is a document that contains multiple components. The informed consent process is when the intervention, the potential benefits, possible risks, alternative options and anticipated outcomes are described to the prospective participant and a discussion involving question and answers to ensure comprehension is included (Hyde and Power 2000). Intervention here is the intervention in the study, which may be therapeutic to the individual or not.

11.3.3 Information and informed consent

Investigators are responsible for ensuring the prospective participant in the research study is aware of the potential risks and benefits of the research and what alternatives exist for that participant. The term materiality refers to the extent to which investigators place a sufficient amount of emphasis on each of these areas. Consequently, the evaluation of materiality can arise in matters of malpractice. What makes information material can also vary, based on the specific statutes in a particular state. For this reason, the authors of this chapter do not claim to establish the materiality of any information contained herein from a legal perspective. Rather, the authors use the principle of autonomy to argue that the investigator's obligation is to disclose all known information about a specific intervention to the research participant. According to Beauchamp and Childress (2001), legal, philosophical and medical literatures suggest the following as the necessary components of informed consent: (i) competence to act; (ii) thorough disclosure; (iii) comprehension of the disclosure; (iv) voluntariness; and (v) consent.

Competency refers to an individual's ability to make a particular decision. Competency is also decision-specific; using the example in Section 11.3.1.4, an individual with a traumatic brain injury with associated cognitive deficits may not be competent to make a decision about participating in a clinical trial where the potential risks are significant, but be capable of making a decision regarding where he or she lives.

Chell (2004) believe that a prospective participant in a research study is determined to be competent if: (i) he or she has an understanding of the situation and the consequences of the decision; and (ii) the decision is based upon rational reasons.

In situations where the comprehending complex information in a clinical trial is of concern, it is also important to find the balance between autonomy and non-maleficence and make sure that the goal of avoiding harm does not infringe on the participant's right to autonomy. For example, an IRB, in reviewing a protocol involving a vision prosthesis, believe the technical complexity of the implant and its potential risk might prevent an individual from fully comprehending the risks. The concern over whether it is possible for a prospective participant to be adequately informed to provide consent (autonomy) might result in an IRB denying the protocol or to require procedures for ensuring understanding that are onerous and convey a lack of confidence in the prospective participant's ability to make independent decisions. In this type of situation, the investigators may decide to have an "advocate" assist the participant in weighing the potential risks and benefits to ensure the participant is not unknowingly exposed to risk (non-maleficence), while at the same time maintaining participant autonomy.

11.3.4 The process of obtaining informed consent

Information about clinical trials involving the implantation of biomedical devices is complex and not always easy to understand or to comprehend. It has been argued that the informed consent process is more to safeguard the researchers and sponsors (Bhutta 2004) and not inform the participant. Moreover, the consent process of signing informed consent documentation is largely to ensure there is written documentation that information has been shared rather than ensuring the potential participant understands and comprehends the information. Because the goal is not to truly inform the participant, the information contained in the consent document is often written in a language that is too complex to understand

Flory and Emanuel (2004) published a review of research with the goal of improving the participant's ability to comprehend the material contained in the informed consent documents. The interventions were the use of multimedia, enhanced consent forms, extended discussion with project personnel, and testing participants following the informed consent process and providing feedback. When investigators spent more time discussing the proposed research, risks and benefits with prospective participants, they showed an improved comprehension of information. However, Flory and Emanuel (2004) noted that further research is needed in this area due to the poor quality of some of the studies reviewed.

Paris *et al.* (2009) conducted a study to improve the comprehension of the information presented by reducing the complexity of the information. A phase III biomedical clinical trial with 159 participants were assigned to one of three groups, including: (i) a "Working Group"; (ii) a "Flesch Reading Improvement Group"; or (iii) a control group. The Working Group intervention consisted of ethicists and participant representatives rewording the consent document to improve understanding. The alternate treatment groups focused on improving the Flesch reading index of the document. The investigators found no improvement in comprehension of participants in either intervention group compared to individuals in the control group.

11.3.5 Decision-making

The process of decision-making by prospective participants in clinical trials is not well researched and, in fact, an area where we know very little. It does, however, appear to be highly individualized with social and cultural components influencing it.

For some individuals, the process of weighing the risks and benefits of the research study and arriving at a decision is very personal. Individuals might consult with family members, significant others, health or allied health professionals, or seek spiritual counseling. For some, while consultation is important, they ultimately reserve the final decision for themselves (Lane *et al.* 2011, 2012, in press). Conversely, for other individuals contemplating participation in prosthesis research, family members, significant others and health professionals play a much larger and central role in the final decision.

11.3.6 Influence of culture and country

The informed consent process has been argued to focus on the Western view of autonomy in terms of independent decision-making. In non-Western societies, it is common to find that family members and community leaders are involved in the decision-making process (Marshall *et al.* 2006). In a study of decision-making and motivation to participate in biomedical research in southwest Nigeria, women were twice as likely to obtain "permission" from another individual before participating in a clinical trial.

11.3.7 What information is material?

The question of what information is necessary to be included in the informed consent process can vary, depending on whether we are looking through a legal or an ethical lens. From the perspective of the authors, an ethical approach is in the best interests of the individual and includes providing the prospective participant with all information that is available.

11.3.7.1 Motivation Motivation to participate in a clinical trial for research purposes can vary significantly. This section will address altruism, restoration of vision-therapeutic benefit, and adventure as primary motivators:

1. *Altruism*: According to Beth Seelig and her colleagues, the true psychological benefit of participating in a clinical trial is altruism (Seelig and Dobelle 2001; Seelig and Rosof 2001). There are many different types of altruism, including pathological altruism, and for this reason, a clinician trained in its assessment

must evaluate it. Several studies of prospective participants of cortical vision implant projects have reported altruism as a motivating factor (Lane *et al.* 2011; 2012) and some retrospective studies of participants in clinical trials who report the same (Lane, in press). In the studies conducted by Lane and his colleagues, the participants reported having developed a concern for individuals who are blind (family and otherwise) that came about in middle age and they wished to assist researchers to advance the science of vision.

2. *Restoration of Vision*: The most salient motivator for individuals to participate in a clinical trial is the restoration of sensation or perception and the functional ability that results. While cochlear implants have developed to the point that they are a mature technology, vision implants resulting in restoration of visual percepts are an emerging technology (Chapter 7). Lane and colleagues conducted a series of focus groups to determine how far an individual's vision would have to be restored to participate in a vision prosthesis research trial. Participants who were blind reported that any restoration of vision would be beneficial. For example, with bare light perception an individual would be able to detect objects, thereby improving orientation and mobility (Lane *et al.* 2011, 2012).
3. *Adventure*: Some individuals who participate in clinical research trials report being motivated by the adventure of participating in pioneering and exciting research. Altman (1986) suggested that a sense of adventure can be used to explain why physicians have historically experimented on themselves. A few studies have reported that adventure is a motivating factor for some individuals participating in spinal cord implants (Kilgore *et al.* 2001) and vision implants (Lane *et al.* 2011, 2012). This adventurous motivation to pursue neural implants has also been documented among optic nerve implant recipients living in Belgium, which suggests that this phenomenon may be evident across cultures, at least in one other Western country (Lane, in press).

11.3.7.2 Risks A number of health-related risks exist for any type of surgical procedure where a neural implant is involved. It is not our intent to address those risks outside our area of expertise. We will address those risks that may not be as obvious:

1. *Experience of Loss*: A total of 21 (known) individuals were implanted with different versions of a vision prosthesis developed by William Dobelle between 1980 and 2005. When the investigator died and the project ended, the devices stopped working over time, because they required repair and there was no one associated with the project to facilitate the repairs. The individuals from the Dobelle study who reported visual percepts in the form of phosphenes and were able to use them functionally also reported experiencing feelings of loss similar to when they initially lost their sight (Dagnelie *et al.* 2013; Lane *et al.* 2013). Some might argue that the restoration of a few percepts is insufficient to elicit a process of readjustment to vision loss because, stated simply, visual percepts are not vision. However, individuals who are blind report that dots of light in their visual field, no matter how rudimentary to a sighted person, is vision if

it improves independence by way of restored function and safety (Lane *et al.* 2011, 2012).

Feelings of "re-loss" resulting from the loss of visual percepts due to device failure or removal occurred among several of the Dobelle study participants for a period of least five years; however, to date, we are not aware of any previous discussions related to "re-loss" in the literature relative to vision implants. The significance of this finding cannot be over-emphasized. Researchers have a clear responsibility to make sure that participants in vision implant studies are aware of the potential for experiencing loss should they lose any regained visual percepts. Moreover, the investigators have a responsibility to ensure the maintenance of implanted devices is accounted for beyond the life of the study.

2. *Disappointment*: Participant expectations are best discussed in the following section on benefit. However, there is a relationship between participant expectations and disappointment, particularly when the expectations are unrealistic. There are a number of reasons why a participant can have expectations that are unrealistic based on the technology. Information about a device can often be complex and understanding the research processes can be challenging. In addition, factors such as adjustment to blindness can also play a role in disappointment.

An example of how participant expectations can be managed to avoid disappointment was conducted at the Universite Catholique De Louvain located in Brussels, Belgium. This experimental study investigated the feasibility of an optic nerve implant using a cuff electrode on the optic nerve. Participants in the study were carefully screened and a clinical psychologist, functioning in the capacity of a patient advocate, was available to ensure complete understanding of the device and research procedures. Prospective participants were evaluated for motivation for participating, adjustment to blindness and the expectations of therapeutic benefit (Lane *et al*, in press).

The process of informing a potential participant in a research study who has lost their vision as a result of a traumatic event (e.g. automobile accident) must be handled carefully. Some participants reported that after losing your sight, "you want it back so badly, you will do anything" (Dagnelie *et al.* 2013; Lane *et al.* 2013). Lane *et al.* (2013) found in a preliminary analysis of interviews of a vision prosthesis study that individuals who were well adjusted and had realistic expectations were not disappointed, even though the outcome for them personally was not what they hoped for. In contrast, individuals who had not adjusted well to blindness, often had unrealistic expectations, and the outcome, while possibly successful from the perspective of a researcher, resulted in disappointment. Participants with unrealistic expectations coupled with poor adjustment to blindness often experienced the feelings of anger, resentment, self-doubt, depression and anxiety (Lane *et al.* 2013).

It is the job of the investigation team to properly screen prospective participants and evaluate for appropriate expectations and adjustment to blindness. If there is no likelihood of personal therapeutic benefit, participant expectations must reflect this reality. However, investigators must be careful not to remove participant hope in the process (Lane *et al.* in press). While a participant may not expect the likelihood of personal benefit, they may still *hope* for a personal benefit. This is an important distinction because to kill an individual's *hope* is to remove their investment in the belief that someday technology will likely restore vision (Lane *et al.* in press).

3. *Therapeutic misconception*: The idea of therapeutic misconception (TM) was first presented by Appelbaum *et al.* (1982). They found that individuals entering a research project in a psychiatric setting expected there to be personal therapeutic benefit to them. In other words, the participants did not understand that unlike a therapeutic interaction, where the physician's goal is to benefit the individual, the goal of a physician in clinical research is to collect valid data to inform the research. Moreover, there are various aspects of research studies that may not be in the individual's best interest. The investigators also found that despite being comprehensively informed about the methodological aspects of the study, and expressing a clear understanding of study information, several of the participants' thinking seemed distorted; they maintained a belief that their experience would be therapeutic.

Lidz (2006) examined new research on therapeutic misconception conducted within the past few years and determined that of the studies they reviewed, 62% of the participants had evidence of a therapeutic misconception. In order for the participants to have a TM, they had to make a statement that indicated:

1. a belief that the treatment would be individualized to the participant; and/or
2. unreasonable assessment of benefit, based on a failure to grasp the potential consequences of the study.

Kim *et al.* (2013) found different results. Twenty-seven participants in a clinical trial using gene therapy for the treatment of Parkinson's disease were tested for comprehension of the likelihood that they would receive the experimental treatment versus the control. They found that 83% of the participants were able to correctly respond to a semi-structured interview to elicit understanding of trial randomization.

11.3.7.3 Benefits The benefits associated with clinical trials changes depending on the nature of the trial. In the informed consent process for a research trial, the investigator must delineate the potential benefits for the participant. In the event that the research presents no direct benefit to the participant, the investigators must determine whether the benefit to society is such that is outweighs the individual risk to the participant:

1. *Participant expectations*: The largest body of knowledge in participant expectations is on parental expectations of cochlear implants for their children. Participant expectations are important to understand. Kampfe *et al.* (1993) stressed the importance of educating the parents of a deaf child before having the cochlear implant procedure because of high expectations. An empirical study conducted

by Hyde *et al.* (2010) surveyed 247 Australian parents whose child received a cochlear implant and discovered that the majority of parents had high expectations of the cochlear implant and their expectations were met following implantation. There were a small number of parents who were disappointed in the implant. A qualitative analysis of 27 of the families found that of those parents who were dissatisfied with the implant, the focus was on the child's social development and participation with peers.

A survey of 35 mothers of deaf children, by Zaidman-Zait and Most (2005), found that the mother's expectations of the cochlear implant outcomes were high. Mothers also expressed a high degree of understanding with regard to the demanding rehabilitation process after implantation. Maternal satisfaction was associated with their child's social and communication skills. Maternal stress was associated with communication difficulties and lower satisfaction with professionals associated with the project.

In a study consisting of 11 in-depth qualitative interviews with recipients of the Dobelle vision implant in the visual cortex (Dagnelie *et al.* 2013; Lane *et al.* 2013), participant expectations often reflected what the participants reported were told to them during the informed consent process. Some individuals reported they were told not to expect any therapeutic benefit from the procedure. The individuals in this category were well adjusted to the outcome, whether positive or negative. There were, however, individuals who reported being told they would be able to drive an automobile and see a sunset. The individuals reporting receiving this information during the consent process experienced a profound negative reaction.

11.3.7.4 Social and cultural factors The discussion of social and cultural implications of neural implants is the most robust around the implantation of cochlear implants in children. The practice of hearing parents consenting on behalf of their deaf children is of concern to ethicists. However, this section will be devoted to the information about Deaf culture that must be considered in the informed consent process. As mentioned in the section on informed consent, the discussion about a therapeutic intervention must also include a description of the alternative which, in this case, would be the option for the child to lead a deaf life.

The objection of cochlear implants by the Deaf culture community is deeply seated in the challenge to the "medical" community's belief that deafness is a disease that needs to be "cured." Such a belief is symbolic of an attitude about deafness (Hyde and Power 2006). From the perspective of the Deaf culture community, they are merely "different" and do not believe they should be robbed of their right to their own history and culture. Moreover, a deaf child can live a viable and satisfactory deaf life without a cochlear implant. In order for parents to understand this, they must be presented with information that includes the social, linguistic and cultural characteristics of being deaf (Hyde and Power 2006) and that parents of deaf children must receive this information to be truly informed (Hyde and Power 2000).

11.3.8 Restoration versus enhancement and mental change

Most professionals in the general field of rehabilitation (rehabilitation engineers, physical therapists, occupational therapists, rehabilitation psychologists and rehabilitation counselors) would argue that the process of restoring function to an individual with a disability is not only humane but also laudable. The concept of restoration, however, may have a glass ceiling. What happens if engineers were to develop a prosthetic limb, for example, that is better than a normal limb? The idea may have seemed unrealistic ten years ago, but a recent article published in *Scientific American* asked if the prosthetic legs worn by Oscar Pistorius should disqualify him from the Olympics because he had a competitive advantage over the other runners (Eveleth 2012).

What if the individual with a disability were to hear better with a cochlear implant, or see better with a vision implant? The likelihood that an implant will enhance function becomes greater as implant technology improves. Discussion on ethical issues surrounding enhancement and mental change in the literature has focused on enhancement of memory and change to mental functioning, particularly personality change.

11.3.8.1 Enhancement There is general agreement that technology aimed at restoring function to some perceived "normal" level is acceptable, but that technology designed to enhance function above such a level is regarded as ethically questionable (Hansson 2005; Wolpe 2002). For example, implanting a 'brain chip" that is designed to improve the mental functioning of a patient with Parkinson's disease is regarded as restorative and without ethical concern (Northoff 1996). However, if a brain chip were to be implanted in an individual with normally functioning memory to enhance the individual's ability to store and recall information, such a device would be regarded as ethically problematic (Hansson 2005; Wolpe 2002).

Hansson (2005) argues that enhancement is ethically problematic for medical professionals, because the goal of medicine is to treat disease or malfunction and to restore the individual to their initial level of functioning. To what extent then is it advisable to improve on an individual's memory above his or her normal capacity? What if enhancing an individual's memory would result in an individual's ability, for example, to recall the post-traumatic amnesia that followed head trauma associated with an automobile accident? What would be the psychological effects of an individual's recall of post-traumatic amnesia? If there is a potential for harm, then how can such an endeavor be ethical for a healthcare professional charged with not doing harm?

11.3.8.2 Mental change Implants designed to change mental functioning are often treated as suspect, based on the premise that human consciousness and personality exist in the brain and directly determine the physical, mental, emotional and social characteristics of the individual (Hansson 2005; Northoff 1996; Wolpe 2002). Some scientists who are concerned about ethical implications of brain implants have questioned whether the introduction of any foreign material such as an electronic chip could change or alter the basic mental functioning of the individual (Hansson 2005). They argue that to transplant an individual's entire brain would be unethical because the person, as we knew him or her, would cease to exist (Northoff 1996; Wolpe 2002). Although this example is an extreme to prove their point, the question "How much technology can be introduced into a person's brain before his or her consciousness is altered" remains unanswered and a subject for debate.

Studies documenting the use and benefit of DBS for individuals with neuropsychiatric disorders are increasing in the literature. The use of DBS as a treatment for individuals with Obsessive Compulsive Disorder and Major Depressive Disorder, particularly when they are resistant to psychotherapy and pharmacological interventions, has surged according to Synofzik and Schlaepfer (2011). One of the chief ethical concerns raised with DBS is the potential for alteration of the individual's personality (Synofzik and Schlaepfer 2011). Neural motor prostheses and brain-computer interfaces have similarly been criticized as also having the potential to alter one's personality (Clausen, 2008; Haselager *et al.* 2009).

Although visual implant technology may not result in a mental change in an individual's personality or cognitive functioning, it is possible that restoration of perception could have psychological consequences. An example of this is when individuals with hearing loss show irritability and frustration once auditory perception is improved. Individuals who are blind may experience a similar reaction if visual perception is restored (Kurson 2007).

11.4 CONCLUSIONS

If there is one phrase that best describes the history and culture of the IL and disability rights movement it is "nothing about us, without us." It speaks of a history of exclusion and oppression and the importance of including people with disabilities in any and all conversations regarding the development of assistive technology. By welcoming and acknowledging the unique contributions which can only be made available by including individuals with disabilities in the development process of neural implants, technology researchers can perpetuate the ideals held by the IL and disability rights movement. Moreover, collaborating with individuals with disabilities during the development phase, technology developers may gain access to invaluable information regarding what is important to consider when designing a device to be used by specific disability populations. By obtaining this information prior to dedicating time and resources into the device development process, researchers can be more certain that the device characteristics and potential functional benefits offered by the device are congruent with the needs and preferences of the end-users.

The World Health Organization ICF model best describes the belief that disability is a life experience that transcends culture and country and is environmentally based. Whether an individual with a disability is able to participate in an activity is largely due to an individual's attitude about disability and whether the technology exists to assist them to navigate the environment. The MPT model was designed to assess the contextual factors and match technology to the individual based on personal and environmental factors. The MPT model can be applied to the development of neural implants to underscore the importance of involving people with disabilities in technology development.

The importance of individuals receiving thorough information that is relevant to the clinical trial must be underscored. When neural prostheses are being tested in clinical research trials, the ability of the individual with a disability to function as an autonomous person is based largely on the information s/he receives in the informed consent process. Individuals with disabilities may be largely motivated to participate in a clinical trial because of the belief that they will personally benefit from the intervention and it will result in a restoration of function. Participant expectations should be managed carefully so there are no misconceptions. While prospective participants may benefit personally from an experimental intervention, they should be clear what the potential risks and benefits are for the particular trial they are considering and that generally the goal of experimental research is not personal benefit.

Individuals should understand that it is easy to develop a misconception about the goal of a research trial. Information shared with prospective participants about trials involving neural prostheses is complex. The number of individuals who appear to misunderstand the information imparted by investigators from available literature is alarming. Because of the experimental nature of many neural implants, great care must be taken when informing prospective participants and ensuring their understanding and comprehension of the information.

Participants should understand potential psychological and social risks to participation. The example where participants in a clinical trial experienced restored visual percepts, which were lost when the project ended, can elicit an adjustment to loss reaction similar to when they acquired their disability. Participants should also be clear on the alternatives to the intervention, particularly if the potential impact may be on social and cultural interaction.

Investigators must exercise due diligence to ensure that participant expectations are realistic. Research suggests that outcomes that match expectations, no matter how high, result in satisfaction, particularly with therapeutic trials. However, those individuals that report outcomes not matching their expectations in experimental trials reported profound psychological distress that included anxiety, depression, anger and resentment. Evaluations of prospective participants must include a clinical psychologist or other mental health professional, trained at the graduate level, who is skilled in such assessment.

REFERENCES

Appelbaum, P.S., Roth, L.H. and Lidz, C. (1982) The therapeutic misconception: informed consent in psychiatric research. *Int. J. Law Psych.*, **5**: 319–329.

Altman, L.K. (1986) *Who goes first?* Random House, New York.

Barnes, C. and Mercer, G. (2005) Disability, work, and welfare: challenging the social exclusion of disabled people. *Work. Employ. Soc.*, **19**: 527–545.

Beauchamp, T.L. and Childress, J.F. (2001) *Principles of Biomedical Ethics*, 5th Edition. Oxford University Press, Oxford.

Bhutta, Z.A. (2004) Beyond informed consent. *Bull. WHO*, **82**: 771–778.

Brindley G.S. and Lewin, W.S. (1968) The sensations produced by electrical stimulation of the visual cortex. *J. Physiology* (London), **196**: 479–493.

Chell, B. (2004) Competency: What it is, what it isn't, and why it matters, in *Healthcare Ethics: Critical Issues for the 21st Century*, Monagle, J.F. and Thomas, D.C. (Eds), Jones & Bartlett, Sudbury MA.

Clausen, J. (2008) Moving minds: ethical aspects of neural motor prostheses. *Biotech. J.*, **3**: 1493–1501.

Dagnelie, G., Jones, B., Greuschat, D. and Lane, F. (2013) *Prosthetic Vision: Where we are now and where we are going*. Panel presentation for the Low Vision Section Symposium of the Annual meeting of the American Academy of Optometry (AAO), Seattle WA, October 25.

Deegan, P.E. (1992) The Independent Living Movement and people with psychiatric disabilities: taking back control over our lives. *Psychosoc. Rehab. J.*, **15**(3): 3–19.

DeJong, G. (1979) Independent living: from social movement to analytic paradigm. *Arch. Phys. Med. Rehab.*, **60**: 435–446.

Eveleth, R. (2012, July) Should Oscar Pistorius's prosthetic legs disqualify him from the Olympics? Scientific American. Retrieved electronically http://www.scientificamerican.com/article/scientists-debate-oscar-pistorius-prosthetic-legs-disqualify-him-olympics/

Flory, J. and Emanuel, E. (2004) Interventions for improve research participants' understanding in informed consent for research: a systematic review. *JAMA*, **292**: 1593–1601.

Frieden, L. (1980) Independent living program models. *Rehab. Lit.*, **41**: 169–173.

Hansson, S.O. (2005). Implant ethics. *J. Med. Ethics*, **31**: 519–525.

Haselager, P., Vlek, R., Hill, J. and Nijboer, F. (2009) A note on ethical aspects of BCI. *Neural Networks*, **22**: 1352–1357.

Hyde, M. and Power, G. (2006) Some ethical dimensions of cochlear implantation for deaf children and their families. *J. Deaf Stud. Deaf Educ.*, **11**: 102–111.

Hyde, M. and Power, G. (2000) Informed parental consent for cochlear implantation of young deaf children: social and other consideration in the use of the bionic ear. *Austral. J. Soc. Iss.*, **35**: 117–127.

Hyde, M., Punch, R. and Komesaroff, L. (2010) A comparison of the anticipated benefits and received outcomes of pediatric cochlear implantation: parental perspectives. *Am. Ann. Deaf*, **155**: 322–338.

Kampfe, C. M., Harrison, M., Oettinger, T., Ludington, J., McDonald-Bell, C. and Pillsbury, H.C. (1993) Parental expectations as a factor in evaluating children for the multichannel cochlear implant. *Am. Ann. Deaf*, **138**: 297–303.

Kilgore, K.L., Scherer, M., Bobblitt, R., Dettloff, J., Dombrowski, D.M. *et al.* (2001) Neuroprostheses consumers' forum: consumer priorities for research directions. *J. Rehab. R&D*, **38**(6): 655–660.

Kim, S., de Vries, R., Wilson, R., Parnami, S. Frank, S. *et al.* (2013) Research participants' irrational expectations: common or commonly mismeasured? *IRB Ethics Hum. Res.*, **35**: 1–9.

Kurson, R. (2007) *Crashing Through: A true story of risk, adventure, and the man who dared to see*. Random House, Inc., New York.

Lane, F.J., Huyck, M.H, Troyk, P.R. and Schug, K. (2012) Responses of potential users to the intracortical visual prosthesis: final themes from the analysis of focus group data. *Disab. Rehab.: Assist. Tech.*, **7**: 304–313.

Lane, F.J, Huyck, M.H. and Troyk, P.R. (2011) Planning for the first human intracortical visual prosthesis by using pilot data from focus groups of potential users. *Disab. Rehab.: Assist. Tech.*, **6L** 139–147.

Lane, F.J., Nitsch, K.P., Huyck, M.H. and Troyk, P.R. (2013, May) *Participants' experience in a clinical trial for vision restoration: Motivation to participate, visual perception and functional use, and experience of loss following termination.* Poster presented at the annual conference of the Association for Research in Vision and Opthamology (ARVO), Seattle WA, May 8.

Lane, F.J., Nitsch, K.P., Huyck, M.H., Troyk, P.R. and Schug, K. (2014) Perspectives of optic nerve prostheses: an ethnographic perspective of an optic nerve vision prosthesis project. *Disability and Rehabilitation*, 1–9. DOI: 10.3109/17483107.2014.961178.

Lidz, C.W. (2006) The therapeutic misconception and our models of competency and informed consent. *Behav. Sci. Law*, **24**: 535–546.

Marcus, G. and Koch, C. (2014, October) The future of brain implants. *Wall Street Journal.* Retrieved from: http://online.wsj.com/news/articles/SB10001424052702304914904579435592981780528

Marshall, P.A., Adebamowo, C.A., Adeyemo, A.A., Ogundrian, T.O., Vekich, M. *et al.* (2006) Voluntary participation and informed consent to international genetic research. *Am. J. Public Health*, **96**: 1989–1995.

Masala, C. and Petretto, D.R. (2008) From disablement to enablement: conceptual models of disability in the 20th century. *Disab. Rehab.*, **30**(17): 1233–1244.

Meyer, L.H., Peck, C.A. and Brown, L. (1991) *Critical Issues in the Lives of People with Severe Disabilities.* Brookes, Baltimore MD.

Northoff, G. (1996) Do brain tissue transplants alter personal identity? Inadequacies of some standard arguments. *J. Med. Ethics*, **22**: 174–180.

Nunes R. (2001) Ethical dimension of pediatric cochlear implantation. *Theor. Med. Bioethics*, **22**: 337–349.

Oxford, M. and McDonald, G. (1999) *Two paths – One journey: Ending oppression … one accessible step at a time.* Kansas Department of Health and Environment, Topeka.

Paris, A., Brandt, C., Cornu, C., Maison, P., Thalamas, C. and Cracowski, J. (2009) Informed consent document improvement does not increase patients' comprehension in biomedical research. *Br. J. Clin. Pharmacol.*, **69**: 231–237.

Pelka, F. (1997) *The ABC-CLIO companion to the disability rights movement.* CA: ABC-CLIO, Santa Barbara.

Scherer, M.J. (1999) Outcomes of assistive technology use on quality of life. *Disab. Rehab.*, **18**(9): 439–448.

Scherer, M.J. (2004) *Matching person and assistive technology process and accompanying assessment instruments*, Revised Edition. Institute for Matching Person and Technology, Webster NY.

Scherer, M.J. and Cushman, L.A. (2001) Measuring subjective quality of life following spinal cord injury: a validation study of the assistive technology device predisposition assessment. *Disab. Rehab.*, **23**(9): 387–393.

Scherer, M.J. and Glueckauf, R. (2005) Assessing the benefits of assistive technologies for activities and participation. *Rehab. Psychol.*, **50**(2): 132–141.

Scherer, M.J., Sax C., Vanbiervliet, A., Cushman, L.A. and Scherer, J.V. (2005) Predictors of assistive technology use: the importance of personal and psychosocial factors. *Disab. Rehab.*, **27**(21): 1321–1331.

Seelig, B.J. and Dobelle, W.H. (2001) Altruism and the volunteer: psychological benefits from participating as a research subject. *ASAIO J.*, **47**(1): 3–5.

Seelig, B.J. and Rosof, L.S. (2001) Normal and pathological altruism. *J. Am. Psychoanal. Assn*, **49**: 933–959.

Shapiro, J. (1994) *No Pity*. Times Books, New York.

Synofzik, M. and Schlaepfer, T.E. (2011) Electrodes in the brain-ethical criteria for research and treatment with deep brain stimulation for neuropsychiatric disorders. *Brain Stim.*, **4**: 7–16.

White, G.W., Simpson, J.L., Gonda, C., Ravesloot, C. and Coble, Z. (2010) Moving from independence to interdependence: a conceptual model for better understanding community participation of centers for independent living consumers. *J. Disab. Pol. Stud.*, **20**(4): 233–240.

Willig-Levy, C. (1988) *A people's history of the independent living movement*. Research and Training Center on Independent Living, Lawrence KS.

World Health Organization (2013, October) How to use the ICF: a practical manual for using the International Classification of Functioning, Disability and Health (ICF). Exposure draft for comment. WHO, Geneva.

Wolpe, P.R. (2002) Treatment, enhancement, and the ethics of neurotherapeutics. *Brain and Cogn.*, **20**: 387–395.

Zaidman-Zait, A. and Most, T. (2005) Cochlear implants in children with hearing loss: maternal expectations and impact on the family. *The Volta Rev.*, **105**: 129–150.

APPENDIX: EXAMPLES OF COMPANIES DEVELOPING AND/OR MARKETING BIONIC DEVICES*

Company Name	Description	Website
Advanced Bionics Corp.	Manufacturer of cochlear implant devices.	www.advancedbionics.com
Advastim	Provide hardware and software technology platforms for neurostimulator device developers.	http://advastim.com
Aleva Neurotherapeutics	Developing neurostimulation technologies to improve therapies for neurological disease, in particular, a microDBS technology for Parkinson's disease.	www.aleva-neuro.com

Neurobionics: The Biomedical Engineering of Neural Prostheses, First Edition.
Edited by Robert K. Shepherd.

Company Name	Description	Website
Alfred Mann Foundation	Developed a microstimulator for neuromuscular disorders. Also developing an implantable MyoElectric Sensor for use with electromyographic (EMG)-controlled prosthetic devices.	www.aemf.org
Autonomic Technologies Inc.	A medical device company developing a Neurostimulation system for the treatment of severe headache.	www.ati-spg.com/
BioControl Medical	Developed implantable medical devices to treat disorders of the autonomic nervous system. First device developed to treat the heart via direct vagus nerve stimulation.	www.biocontrol-medical.com
Bioinduction	Manufacture implantable neuromodulation devices, including Picostim, a unique skull-mounted DBS platform.	http://bioinduction.com/
Bioness Inc.	Implantable pain management device.	http://www.bioness.com/Home.php
Bionic Enterprises	Developing DBS technology for Parkinson's disease and epilepsy detection devices.	http://www.bionicenterprises.com.au/
Bionic Vision Technologies	Developing a retinal prosthesis.	www.bionicvis.com
Blackrock Microsystems LLC	Blackrock provides electrode arrays and enabling tools for neuroprosthetics research and clinical applications.	www.blackrockmicro.com
Boston Scientific Neuromodulation BSX	Developing microelectronic implantable technologies for managing chronic neuropathic pain and neurological diseases.	www.bsci.com

Company Name	Description	Website
Cardionomic	Developing an acute neuromodulation therapy for the treatment of Acute Decompensated Heart Failure.	http://cardionomicinc.com/
CerebralRx	Developing FitNeS - an implantable neurostimulation system for treatment-resistant depression.	www.biocontrol-medical.com
Cirtec Medical Systems	Contract manufacturing company for implantable devices in the cardiovascular, neurological, gastrointestinal, orthopaedic and diagnostic instrumentation sectors.	http://www.cirtecmed.com
Clinatec	Developing implantable neuroprosthetic devices for the treatment of neurodegenerative diseases such as Parkinson's disease.	http://www.clinatec.fr/en/clinatec_/
Cochlear Corp. COH	Manufacturer of cochlear implant devices.	www.cochlear.com
CorTec	Developing CorTec Brain Interchange, an implantable technology that measures and stimulates brain activity. For the treatment of epilepsy, brain tumours and paralytic diseases.	http://cortec-neuro.com/en/
CVRsx Inc.	Developed an implantable technology, called Barostim neo, for the treatment of high blood pressure and heart failure.	www.cvrx.com

Company Name	Description	Website
Cyberonics, Inc.	Developed Vagus Nerve Stimulation (VNS) Therapy system, for the treatment of refractory epilepsy and treatment-resistant depression.	www.cyberonics.com
Endostim	Developing neural stimulation therapy for the treatment of severe gastroesophageal reflux disease.	http://www.endostim.com/
EnteroMedics, Inc.	Developed a neurometabolic therapy, an implantable, weight loss treatment for obesity and obesity-related risk factors.	http://www.enteromedics.com/
Envoy Medical Corporation	Developed a fully implantable hearing aid for patients with moderate to severe sensorineural hearing loss.	http://esteemhearing.com/
Finetech Medical Ltd.	A specialist manufacturer of implanted functional neuro-stimulators for the treatment of spinal cord injury, stroke and multiple sclerosis.	www.finetech-medical.co.uk
Functional Neuromodulation Ltd.	Developing the use of Deep Brain Stimulation in Alzheimer's disease.	http://www.functionalneuromodulation.com/
Greatbatch Medical	Developing Algostim: a spinal cord stimulation system for the treatment of chronic pain. Manufacturer of implantable batteries and electrode arrays.	http://www.greatbatchmedical.com/
ImThera Medical Inc.	Developed an implantable neurostimulator for the treatment of obstructive sleep apnoea.	http://imtheramedical.com/

Company Name	Description	Website
Inspire Medical Systems, Inc.	Developed a fully implanted system for the treatment of obstructive sleep apnoea.	www.inspiresleep.com
IntraPace Inc.	Developed the abiliti® system: an implanted system for the treatment of obesity.	www.intrapace.com
Mainstay Medical, Inc.	Manufacture ReActiv8, an implanted device for the treatment of chronic lower back pain.	http://www.mainstay-medical.com/
MED-EL GmbH	Manufacturer of cochlear implant devices.	www.medel.com
Medtronic Inc.	Manufacture implantable neurostimulation devices for the treatment of a variety of neurological disorders including motor disorders and chronic pain.	www.medtronic.com
Metacure	Developed an implantable medical device that reduces blood glucose levels for the treatment of Type 2 diabetes.	www.metacure.com
MicroTransponder, Inc.	Developing implantable devices to treat tinnitus, stroke and chronic pain.	www.microtransponder.com
MXM Laboratories	Develops Active Implantable Medical Devices -AIMD-dedicated to neurostimulation.	www.mxm.eu/
Neural Signals Inc.	Developing brain-computer interfaces for locked-in patients.	http://www.neuralsignals.com/
Neurodan A/S	Manufacturer of ActiGait®: a foot lifter stimulator implant.	www.neurodan.com
NeuroPace, Inc.	Manufactures implantable devices for the treatment of neurological disorders such as epilepsy.	www.neuropace.com

Company Name	Description	Website
Neuros Medical Inc.	Developing therapy for chronic pain using an implantable device that provides an electrical nerve block.	www.neurosmedical.com
NeuroSigma, Inc.	Developing an implantable subcutaneous trigeminal nerve stimulation system for the treatment of drug resistant epilepsy.	www.neurosigma.com
Nevro Corp	Developed a neuromodulation platform for the treatment of chronic pain.	http://www.nevro.com/ous/
Nidek Co., Ltd.	Developing a retinal prosthesis.	http://www.nidek-intl.com/visual_prosthesis/index.html
Nurotron	Develop and manufacture a multi-channel cochlear implant.	http://en.nurotron.com/
Oticon Medical/ Neurelec S.A.S.	Manufacture a multichannel cochlear implant system	www.oticonmedical.com
Pixium Vision	Developing both epi-retinal and sub-retinal prostheses for the treatment of retinitis pigmentosa and aged-related macular degeneration.	http://www.pixium-vision.com/en
Respicardia, Inc.	Developed an implantable stimulation device for the treatment of central sleep apnoea.	http://www.respicardia.com/
Retina AG	Developed a sub-retinal prosthesis for the treatment of retinitis pigmentosa and macular degeneration.	http://www.retina-implant.de/en/about/default.aspx
Ripple LLC	Developing a number of implantable recording and stimulating neuroprostheses for the treatment of rare neurological disorders.	www.rppl.com
SPR Theraputics	Percutaneous peripheral nerve stimulation system for pain management	www.sprtherapeutics.com/

Company Name	Description	Website
Saluda Medical	Developing Evoke™, a closed-loop neuromodulation for the treatment of chronic pain and other difficult-to-treat conditions.	http://www.saludamedical.com/
Second Sight Medical Products	Developed the Argus® II Retinal Prosthesis System, for use in patients with severe to profound retinitis pigmentosa.	www.2-sight.com
Set Point Medical Corp	Developing implantable neuromodulation devices for the treatment of inflammatory diseases such as Crohn's Disease and rheumatoid arthritis.	www.setpointmedical.com
Spinal Modulation Inc	Developed a neuromodulation therapy for pain management.	http://www.spinalmodulation.com
SPR Therapeutics, LLC	Developing an implantable device for interventional pain management.	http://www.sprtherapeutics.com/
St Jude Medical Neuromodulation SJM	Developed implantable neuromodulation devices for the treatment of chronic pain, and neurological disorders such as Parkinson's disease.	www.sjm.com
Stimwave Technologies, Inc	Manufacturer of a spinal cord stimulator system for pain management	http://stimwave.com/index.php
Synapse Biomedical Ltd.	Developed NeuRx DPS®, an implantable device designed to help spinal cord injury patients decrease their dependency on mechanical ventilation.	www.synapsebiomedical.com

Note this list contains organisations with an active website (as of August 2015) that are developing or have commercialised implantable neural prostheses suitable for long-term clinical application.

INDEX

Neurobionics: The Biomedical Engineering of Neural Prostheses, First Edition.
Edited by Robert K. Shepherd.